HIGH-SPEED CIRCUITS
FOR LIGHTWAVE
COMMUNICATIONS

SELECTED TOPICS IN ELECTRONICS AND SYSTEMS

Editor-in-Chief: **P. K. Tien**

Selected Topics in Electronics and Systems – Vol. 13

HIGH-SPEED CIRCUITS FOR LIGHTWAVE COMMUNICATIONS

Editor

Keh-Chung Wang

Rockwell Semiconductor Systems,
California, USA

World Scientific
Singapore • New Jersey • London • Hong Kong

Published by

World Scientific Publishing Co. Pte. Ltd.

P O Box 128, Farrer Road, Singapore 912805

USA office: Suite 1B, 1060 Main Street, River Edge, NJ 07661

UK office: 57 Shelton Street, Covent Garden, London WC2H 9HE

British Library Cataloguing-in-Publication Data
A catalogue record for this book is available from the British Library.

HIGH-SPEED CIRCUITS FOR LIGHTWAVE COMMUNICATIONS

ISBN 981-02-3536-4

Printed in Singapore by Uto-Print

INTRODUCTION

Economic growth, technology and competition in the global market have been the topics of intensive research and hotly debated for several decades. It started from the neoclassical theory of Solow[1] as early as 1957. As better data and methodologies for economic growth became available in more recent years, the theory was modified and new conclusions were drawn by economists, for example, Jorgenson, Gollop and Fraumeni,[2] Durlauf,[3] and Boskin and Lau.[4] Although a quantified model is difficult to establish, some consensus does emerge, "economic growth requires investments in hardware, software and people" (Jorgenson[5]) and, "the investment prospers only when the cost of capital is low" (Hatsopoulos[6]). As for global competition, the general recommendations are, "speed of adjustment to changes in market demand, shortening of product cycles, and greater attention to quality improvement and reliability" (Landau[7]).

The message is then clear that we, engineers, need to do more and to learn more. *To learn more means to broaden the knowledge base.* "The development and commercialization of new products and services have always been multidisciplinary efforts. The most successful practitioners in one field are likely to be those who can identify and draw upon complementary or supporting technical advances in other fields".[8] Another issue which we often face today is "survival" and "expansion" in the marketplace. In that case, *to learn more means to acquire a portfolio of flexible skills, making it easier for the engineers to manage the ups-and-downs of the profession.*

With these notions in mind, a new publication format, that is a combination of journal and books was started. The journal is the *International Journal of High Speed Electronics and Systems* which has been in public domain for nine years. It has served well those who regularly read the latest research and developments in the diverse fields of electronics and systems. The books form the series entitled *Selected Topics in Electronics and Systems.* The articles in the books are either selected from or published concurrently with the journal. Each book deals with a specific topic and is edited by a Guest Editor who is well-known in the field of interest. It contains one or two tutorial reviews and a series of articles covering various aspects of the field. The books are, therefore, valuable for graduate students who wish to learn a new field, to the professors who teach the related courses and the professionals who use it as reference material. We wish to thank Dr. K. K. Phua and Professor M. Shur who supports the publication, and Mr. C. Y. Chiam who manages the production of the Journal and the books in this series.

References

1. Robert Solow, *Review of Economics and Statistics*, **39** (1957) 312–20.
2. D. W. Jorgenson, F. Gollop and B. Fraumeni, *Productivity and US Economic Growth*, Harvard University Press, 1987.
3. S. Durlauf, "International differences in economic fluctuations", *Conf. of Economic Growth and the Commercialization of New Technologies*, Center for Economic Policy Research, Stanford University, 1989.
4. M. J. Boskin and L. T. Lau, "Post war economic growth of five countries: A new analysis", Working Paper, Department of Economics, Stanford University, 1990.
5. D. W. Jorgenson, "Investing in productivity growth", *Technology and Economics*, National Academy Press, 1991.
6. G. N. Hatsopoulos, "Technology and the cost of equity capital", *Technology and Economics*, National Academy Press, 1991.
7. Ralph Landau, "How competitiveness can be achieved: Fostering economic growth and productivity", *Technology and Economics*, National Academy Press, 1991.
8. "Prosperity in a global economy", Committee on Technology Policy Options in a Global Economy, National Academy of Engineering, National Academy Press, 1993.

P. K. Tien
Series Editor-in-Chief

FOREWORD

Lightwave communications offer ultrahigh capacity transmission of voice, data, and video. It is the technology of choice for the backbone network of telecom and datacom to meet ever increasing bandwidth demand of global internet traffic. High speed electronic circuits are key components in lightwave communications. They often dictate the bandwidth of transmission. Tremendous resources and efforts have been invested in developing ever faster transistor technologies and integrated circuits for optical communication by a large number of industrial and academic institutes. Performance of Lightwave communication ICs approaching 100 Gb/s has been demonstrated in research labs. Commercial lightwave products using high speed circuits of 10 Gb/s and beyond are readily available.

In this Special Issue, we present 12 invited papers by over 50 experts actively working in this area. These papers describe the latest information on the circuits design, fabrication, measured results, applications, and product development. They cover electronic and opto-electronic circuits for transmission, receiving, and cross-point switching. These circuits were implemented with various state-of-the-art IC technologies, including Si and SiGe BJT, GaAs MESFET, HEMT, HBT, as well as InP HEMT and HBT. These papers were prepared for graduate students, researchers, and engineers who are interested or work in this exciting and challenging field of optical communications.

High speed circuits for lightwave communications consist of circuits for transmitter, receiver, and switch. At the electronic front-end, the transmitter circuits include a multiplexer (mux) and a laser diode driver or a external modulator driver. The mux takes parallel data inputs and serializes them into one output data string at a high data rate. The laser driver takes the serial data string as input and supplies modulation current and bias current to a laser diode. The directly modulated laser diode converts an electrical signal into an optical one. Another method of conversion is to use a laser diode operating in CW mode and an external modulator with its transparency controlled by a voltage input. A modulator driver amplifies the serialized data string from the mux and outputs a large voltage signal to control the modulator. In the receiver front-end, the electronic circuits include a preamplifier, a main amplifier, a clock recovery/data regeneration IC (CDR), and a demultiplexer (demux). A photo-detector converts the optical signal back into electrical current. The preamplifier, mostly a transimpedance amplifier in high data rate applications, converts the photo-current into a voltage signal. The main amplifier, mostly a limiting amplifier or an automatic gain control (AGC) amplifier, enlarges the voltage signal that represents the received data. The CDR recovers the frequency and phase of clock signal from the data, and uses the recovered clock to regenerate the data with a decision circuit. The demux converts the data string into parallel outputs of low data rate. Switches are used in access nodes and inter-connection nodes of optical networks. It facilitates add-drop functions and data routing. Switches can be implemented with mechanical, electronic, or optical approaches. Electronic switches feature easy synchronization, flexibility, low latency, data regeneration, and reliability. They are key elements in lightwave communications.

The papers in this Special Issue are presented in the following order. The 1st paper provides a tutorial review. Papers 2–10 describe the high speed circuits and are organized according to IC technologies that the circuits are implemented with. Paper 11 reviews current status of optoelectronic integrated circuits (OEICs). Paper 12 presents an example of further advance of transistor technologies for future lightwave communication ICs.

In the first paper, *Dr. Ken Pedrotti* presents a tutorial review of technical advances in the telecommunication networks, a description of a typical optical link and electronic circuits involved, a survey of IC technologies, and technical requirements and issues of these circuits. In particular, the requirements and design approaches of laser drivers, preamplifiers, and clock/data regenerators (CDRs) are described in detail.

Prof. Hans-Martin Rein gives an overview of Si and SiGe ICs for 10 to 40 Gb/s optical-fiber TDM (time-division multiplexing) links in the 2nd paper. He showed that all ICs in 10 Gb/s TDM systems can be fabricated in production Si-bipolar technologies. For speed-critical ICs in 20 Gb/s, laboratory Si-BJT technologies are required if the circuit specifications (apart from data rate) remain unchanged. He also showed that all 40 Gb/s TDM ICs can be realized in advanced SiGe technologies. This was demonstrated by relaxing some critical specifications, and by using optical amplifiers and improved opto-electronic components, as well as a modified system architecture to eliminate some of the speed-critical circuits. He even realized 60 Gb/s 2:1 mux and 1:2 demux circuits. All circuits reported in this paper were measured in packages using conventional wire bonding. Accurate modeling of transistors, parasitics of on-chip metal interconnects, Si-substrate coupling, and packaging contributed greatly to the success of these circuits.

Mr. Toru Masuda and co-authors describe a 10-GHz Si-bipolar preamplifier design in detail. The design emphasizes low transimpedance (Tz) fluctuation in frequency response over a large dynamic range of photo-current. Using measured data, they optimized current density of transistors and open loop voltage gain to obtain desired transimpedance and bandwidth. The design also employs a low-loss (high resistance) pad structure with U-grooves to enhance bandwidth. With such a design approach, the authors realized a preamplifier of 10.2 GHz bandwidth, less than 0.5 dB Tz fluctuation, and an input current range of up to 1.6 mA, using a 35-GHz Si bipolar technology

In Paper 4, *Dr. Taiichi Otsuji and co-authors* describe 20–40 Gb/s GaAs MES-FET digital ICs for optical fiber communications. They invented new IC designs of data selector, static and dynamic flip-flops. With these new designs and novel broadband data/clock buffer designs, the authors realized a DC-to-44 Gb/s 2:1 data multiplexer, a DC-to-22 Gb/s static decision IC, and a 20-to-40 Gb/s dynamic IC. There circuits were implemented with the state-of-the-art 0.12 μm gate-length GaAs MESFET technology. They demonstreted record speed performance of GaAs MESFETs. The results show that GaAs FET is a potential candidate for 20–40 Gb/s optical communications.

Dr. Zhihao Lao and co-authors present nine 20–40 Gb/s GaAs-HEMT ICs for optical date receivers in Paper 5. They describe the 0.2-μm HEMT technology, circuit design techniques, circuit schematics, layouts, and measured results. The ICs demonstrated are: a 22-GHz 76 dB-ohm transimpedance amplifier, a 20 Gb/s OEIC front-end optical receiver, a 25 Gb/s automatic-gain-control (AGC) amplifier, a 27.7 GHz 26 dB limiting amplifier, a 20–40 Gb/s clock recovery IC, a 24 mW 20 Gb/s decision circuit, a 20–40 Gb/s parallel decision circuit, a 40 Gb/s 1:4 demultiplexer, and a 30 GHz static frequency divider. With novel circuit architecture, design optimization, and careful layout, the authors demonstrated stable operation for 20–40 Gb/s and low power consumption. New receiver ICs with higher operation speed and higher level of integration are being developed.

Using a manufacturable hybrid digital/microwave AlGaAs/GaAs HBT process, *Mr. Klaus Runge and co-authors* designed 25-GHz to 40 Gb/s circuits for optical communications. The circuits reported in Paper 6 are: 40 Gb/s 4:1 multiplexer, > 30 Gb/s 1:4 demultiplexer, DC-26 GHz variable gain amplifier, DC-25 GHz transimpedance amplifier, 30 Gb/s clock and data regenerator, 40 Gb/s differential-and-rectify IC, and 40 Gb/s delay-and-multiply IC. The authors also described the HBT process, device characteristics, and 40-GHz packaging technology. Comparison of circuit architectures for main amplifier (limiting amplifier vs. AGC amplifier) and clock recovery (filter type vs. phase-locked loop) are presented.

Electronic cross-point switches play key roles in lightwave communications. They are used in the backbone of optical networks and in access nodes. In Paper 7, *Dr. Charles Chang and co-authors* present an outstanding review of technology and performance of electronic cross-point switches with data rates above 1 Gb/s. They review performance criteria, architectures, IC technologies and packaging of electronic switches. Scaling of cross-point switches to a large number of in/out ports are described. As a particular example, they report a 120 Gb/s 12×12 switch IC implemented with AlGaAs/GaAs HBTs. The design, packaging, testing, and performance of this switch are described in detail.

Drs. John Sitch and Robert Surridge describe GaInP/GaAs HBT ICs for the high-speed portion of 10 Gb/s OC-192 fiber communication products. In Paper 10, they present the rationale of using HBT ICs for OC-192 equipment and the philosophy behind HBT IC introduction. They faced the challenge and risk of incorporating a new IC technology in a new OC-192 product. The HBT technology provided very fast logic, precise and repeatable analog circuits, high driving capability, high reliability, and reasonable noise, required by OC-192. The authors presented epitaxial structure, processing sequence, circuit designs, testing, and reliability of the product development. It is a "must read" for people interested in production of HBT ICs.

A lot of progress has been accomplished recently on InP based HFET and HBT IC technologies and integrated circuits. In Paper 9, *Dr. Eiichi Sano and co-authors* review the current status of InP-based lightwave communication ICs in terms of device, circuit, and packaging technologies. Based on a 0.1 μm gate-length InP HFET process, novel IC design, and broadband packaging technologies, all transceiver ICs except the modulator driver have achieved 40 Gb/s operation. A successful demonstration of 40 Gb/s 300-Km optical transmission using ICs implemented with these is described. This showed the feasibility of these ICs in real 40 Gb/s optical networks. The authors also estimate future IC performance based on the relationship between electronic device figure-of-merit and IC speed. To realize 100 Gb/s IC performance, inter- and intra-chip interconnect technologies are being addressed.

In Paper 10, *Dr. Mehran Mokhtari and co-authors* describe InP-HBT ICs for 40 Gb/s Optical Links. They present an overview of HBT technologies, and review InP HBT technologies and monolithic optoelectronic receivers. Using an AlInAs/GaInAs single heterojunction bipolar transistor technology, they developed an IC chip set to demonstrate the potential of InP HBTs in ultrahigh speed lightwave communications. Among the ICs reported here are: a differential buffer, a 35 Ghz 6-dB broadband amplifier, a 40-GHz 16-dB narrowband amplifier, a 40 Gb/s edge detector, a 25 Gb/s 2 × 2 crosspoint switch, a 40 Gb/s 2:1 mux, and a 25 Gb/s 1:2 demux. Speed performance of these circuits is evaluated with on-wafer probing and is mainly limited by measurement equipment. The designs were based on SPICE simulation with proper interconnect modeling. These circuits were operational with a low 3-volt supply voltage.

Optoelectronic integrated circuits (OEIC) receivers are promising for high speed optical communications. This is due to the inherent advantages of integration that minimize the interconnect parasitics of the photodetector and preamplifier, while making use of the intrinsic speed of optical and electronic devices. *Mr. Bob Walden* presents an excellent review of InP-based OEIC receiver front-ends. He described clearly the relationship and trade-offs of key OEIC performance parameters (sensitivity, data rate, etc.) and device parameters (responsivity, noise, etc). Recently published InP-based OEICs and OEIC-arrays were summaried. It is evident that they can be useful in high speed WDM, TDM, and analog transmission systems.

An approach to enhance HBT speed performance is to reduce extrinsic base-collector capacitance. *Dr. Bipul Agarwal and co-authors* introduce a novel concept of transferred substrate HBT and developed fabrication process to realize such HBTs. These HBTs are based on the AlInAs/GaInAs material system on InP substrate. In Paper 12, they describe the concept, the process, transistor DC and RF characteristics, and a gainblock IC. They demonstrated HBTs with 0.6 μm-wide emitter and 0.8 μm-wide collector. RF measurement showed Ft of 134 GHz and Fmax of 520 GHz. The gainblock showed 13 dB gain and 50 GHz 3-dB bandwidth. This technology, when fully developed in the near future, should be capable of supporting 100 GHz ICs for lightwave communicaitons.

I sincerely acknowledge *all authors and reviewers* for their dedicated efforts in making this special issue a great success. I would like to thank *Dr. P. K. Tien* for offering me the opportunity of being a guest editor, and for his encouragement. It has been a learning and enjoyable experience. Many thanks to *Mr. Chiam Choon Yong and his staff* in World Scientific Publishing Company and *Ms. Gretchen Fay-McLaughlin* of Rockwell for helping me put this volume together. I thank my wife, *Jenny*, for her support and patience during the production of this special issue.

 **K. C. Wang** received a B.S. degree in physics from National Taiwan University in 1972, and a Ph.D. degree in physics from California Institute of Technology in 1979. He was a research physicist at University of California, Irvine, where he investigated the physics of neutrinos. He joined Rockwell Science Center in 1985. His colleagues and he pioneered the development of GaAs heterojunction bipolar transistor (HBT) technology. He was the Manager of the High Speed Circuits Department there from 1990–97, and was responsible for the development of optical communication circuits and A to D converters. He is currently the Director of Lightwave Products Department of Rockwell Semiconductor Systems. He has authored or co-authored more than 126 journal and conference publications in electronics and physics, and holds one U.S. patent.

Dr. Wang was a recipient of Rockwell's 1994 Engineer of the Year Awards and 1995 Chairman's Team Awards. He also received a 1996 R&D 100 Award. He served in the technical program committees (TPC) of the IEEE LEOS 1995 Summer Topical Meeting on Lightwave ICs for New Age, the IEEE GaAs IC Symposium 1995–96, and a workshop on TDM optical fiber transmission in OFC'97. He currently serves in the TPC of the IEEE International Microwave Symposium. He was a guest editor of IEEE J. of Solid-State Circuits in 1996. Dr. Wang is a senior member of the IEEE and a member of the American Physical Society.

CONTENTS

International Journal of High Speed Electronics and Systems, Vol. 9, No. 2 (1998) 313–346
© World Scientific Publishing Company

HIGH SPEED CIRCUITS FOR
LIGHTWAVE COMMUNICATIONS

KENNETH PEDROTTI

Rockwell Science Center, 1049 Camino Dos Rios,
Thousand Oaks, CA 91360, USA

Recent increase in the demand for bandwidth in the telecommunications network has stimulated both research and commercial activity in high-speed electronics. This paper summarizes the technical issues that bear on high-speed circuits in this context, as well as the device technologies available for these circuits in both laboratory and commercial processes. Finally a tutorial treatment is given of the main design requirements pertinent to circuits required for lightwave systems.

1. Introduction

This paper begins with a brief review of technical advances in the telecommunications network that affect the demand for high bandwidth electronic circuits. This is followed by a description of a typical optical link and the electronic circuits involved. Section 2 surveys the electronic technologies available and their individual suitability for the needs of circuits for lightwave communication. The remaining sections comprise a survey of the design requirements and basic considerations for the circuits required for lightwave communication. The treatment is intended to be a tutorial, and to provide a context, and serve as a reference for the other papers on this subject that follow.

In the last several years a largely unanticipated explosion in the demand for bandwidth in the telecommunications system has occurred worldwide. This is due to the increasing number of connections carrying voice and data traffic resulting in the continual upgrading of the network backbone to accommodate higher and higher speeds. Historically the network backbone has evolved by the installation of higher speed time-division-multiplexed (TDM) circuits. This first occurred over copper cable, and has continued with optical fiber. The first optical fiber systems were installed as 45 Mb/s point-to-point links and the rate increase has continued to this day to 9.95 GB/s OC-192 SONET or SDH-64 links. The choice of TDM over Wavelength-Division-Multiplexing historically triumphed because it was usually cheaper to get 4× the capacity using TDM for less than 4× the cost that is naturally incurred with WDM approaches. Recently however this long-term trend has paused as for the first time a significant volume of WDM point to point links are being installed.

The increase in WDM links at this time can be traced to several technical challenges for high bit rates TDM, and breakthroughs for WDM approaches. The first is the recent availability of Erbium-Doped-Fiber-Amplifiers (EDFA) which can simultaneously amplify many WDM channels almost as easily as a single TDM channel. This technology combined with dispersion compensating fiber allows very long transmission distances using "optical regeneration" before electronic regeneration of the signal is required. The emergence of high-performance optical multiplexers, demultiplexers and laser wavelength stabilization techniques has also contributed to the current proliferation of WDM systems.

Simultaneously with the emergence of these optical technologies, a number of problems emerged that appeared to limit the practicality of TDM systems at the next higher bit rate, 10 Gb/s. These included pulse dispersion in the fiber, which, for long-reach applications, requires the use of external modulators due to the high frequency-chirp encountered with directly modulated DFB lasers. The $4\times$ increase in modulation rate in a typical upgrade gives rise to a $16\times$ increase in the dispersion penalty. At lower bit rates, prior to optical amplifiers, link spans were mostly limited by attenuation. At higher rates dispersion is more important and links must be compensated on a fiber-by-fiber basis to increase the rate. As the rate increases it also becomes more difficult to compensate for polarization mode dispersion in the optical fiber using optical techniques.

Additionally there is no ready supply of 10 Gb/s electronic parts for use by systems integrators, for the following reasons: the recent increase in bandwidth demand was generally unanticipated; lightwave circuits was one of the few applications driving the development of these devices for digital applications; a perceived small market for these devices, which hindered their commercialization; and the intrinsic difficulty in building higher speed transistors. Additionally the cost of installing fiber is very high; upgrading a link from a 2.5 Gbit/s OC-48 to a 10 Gb/s OC-192 link gives only a $4\times$ increase in capacity, while today WDM point to point links running 40 OC-48/OC-192 channels are available, with 64 channels no doubt soon to be available commercially. The choice of WDM thus gives a carrier a clear upgrade path that can accommodate the recent unanticipated increase in demand.

The forgoing events that have occurred in the optical domain have had a clear impact on demand for high-speed optical circuits. Developers of electronics ignore the developments and problems in the optical arena at their peril. The majority of the high-speed circuit demand is now firmly located at 2.488 Gb/s and is expected to grow as more and more links are converted over to WDM. 10 Gb/s systems are starting to emerge, and as production volumes increase the costs will drop, and performance will increase. At speeds beyond 10 Gb/s, particularly at 40 Gb/s, there is a lively research community, with higher speeds and improved performance reported monthly. However, given the aforementioned difficulties, the historically predictable step from 10 Gb/s to 40 Gb/s is by no means assured in the near term. Several electronic technologies are competing in this arena. Circuit demonstrations are occurring in both III-V compound semiconductors and in silicon using both FET

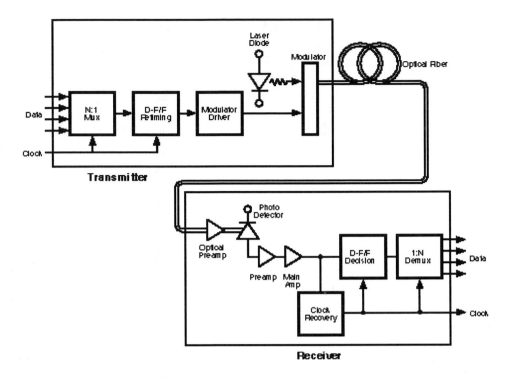

Fig. 1. A block diagram of a typical high-speed lightwave transmission link.

and bipolar transistors. The current small volumes of these parts give no particular cost advantage to one technology over the other, with most of the choice between approaches being decided on technical considerations and relative maturity of the approach.

A simplified block diagram of the high-speed portions of a single wavelength optical link is shown in Fig. 1. This is restricted to those portions of the circuit that contain the high-speed circuitry, typically high-speed multiplexers are used to concentrate a number of lower speed signals for transmission. Once the high-speed bit stream is produced it is often retimed, to remove any phase jitter, and then applied to a modulator (or laser) driver to boost the voltage and current to the level required by the opto-electronic device. The driver usually includes a number of compensating feedback loops to control for temperature, aging and data transition density effects in the opto-electronic device. The optical signal is then modulated by this bit stream and transmitted through the optical fiber where the pulses suffer from attenuation and dispersion. The job of the receiver is to create a faithful replica of the input digital bit streams at its output. Figure 1 shows a receiver consisting of an Erbium Doped Fiber Amplifier (EDFA) which directly boosts the optical signal before it is applied to the photodetector. A low noise preamp followed by either an Automatic Gain Control Amplifier (AGC) or a limiting amplifier then amplifies the signal from the photodetector. A filter is often used to suppress noise and reduce Inter-Symbol Interference (ISI) either between the pre and post amplifiers or after the AGC amplifier. The now constant amplitude bit stream is applied to both the clock recovery and decision circuit. The clock recovery circuit extracts the underlying clock from the data stream. This clock signal is then applied to the decision circuit whose function is to make an optimal decision, in the face of the degradations that the signal has endured, as to whether the bit should be a 0 or a 1. Finally the reconstructed signal and its clock can be used to demultiplex the bit stream to a lower rate for further decoding and processing by more conventional electronics.

The following sections give an overview of the various high-speed technologies available for the construction of these circuits, followed by a description of the design concerns involved with the various circuits themselves.

2. High Speed Electronic Circuit Technologies

There exist a wide variety of high-speed transistors from which the designer may choose to implement the circuits required for lightwave communication. More in-depth reviews of these devices with some circuit design considerations are given in Refs. 1–3. These can be divided into two broad categories, bipolar transistors and FETs. Within the bipolar category we will cover Silicon Homojunction Bipolar Junction Transistors (BJTs), III-V Heterojunction Bipolar Transistors (HBTs) and Silicon-Germanium HBTs. The candidate FET technologies include GaAs MESFET and Heterojunction FETs (HFETs) including Pseudomorphic HFETs (often referred to as PHEMTs). For each device family the general characteristics of relevance to lightwave circuits will be described. In general all the above technologies can prove useful for digital links at 2.488 Gb/s and higher, although for some specific circuits the different technologies have definite strengths and weaknesses.

Two figures of merit are commonly used to characterize transistor speed. The frequency at which the device current gain has decreased to unity, known as f_t, and the frequency of unity power gain $F_{\max}$. $F_{\max}$ is given in terms of the base resistance, R_b (gate resistance for FETs) and the base-collector capacitance, C_{bc}, (gate-drain capacitance for FETs) as:

$$F_{\max} = \sqrt{\frac{f_t}{8\pi R_b C_{bc}}} .$$ (1)

A more exact expression for FETs, Eq. (3), takes into account their output conductance as well.

In general the relative importance of $F_{\max}$ increases in the determination of the circuit switching speed or attainable gain which depends on the voltage swing for logic circuits or voltage gain for amplifiers. More detailed consideration of circuit figures of merit is discussed in the section on decision circuits.

2.1. *Silicon bipolar transistors*

Silicon bipolar transistors have progressed significantly in recent years due principally to the scaling of the devices to smaller dimensions. Lateral scaling reduces
the base spreading resistance under the emitter, decreases the junction capacitances
which improves the speed-power product, increases the packing density which
reduces wiring capacitance leading to faster circuits, and permits small emitters
to be run at higher current densities because of their lower thermal resistance.
Technological improvements have allowed the reduction of process related parasitics such as the extrinsic base-collector capacitance, extrinsic base resistance, and
contact resistances resulting in impressive overall performance. F_t and $F_{\max}$ for
Si bipolar devices are available in commercial processes in the 25–30 GHz range.
For example the trench isolated double polysilicon GST-2 process of Maxim[4] F_t of
27 GHz and $F_{\max}$ of 28 GHz is available. Similarly the Siemens[5] B6HF process
has F_t and $F_{\max}$ of 27 GHz and 34 GHz respectively. F_ts of over 100 GHz have
been reported and combined results of F_t of 82 GHz and $F_{\max}$ of 92 GHz have been
reported by Hitachi.[6]

To further increase speed the collector structure is thinned; this reduces drive
capability, which becomes limited by the collector breakdown voltage. This makes
circuits requiring relatively high voltage drive such as laser and modulator drivers
particularly challenging for silicon BJT. The low $1/f$ noise of silicon transistors
makes them well suited for clock and data recovery and low phase noise VCOs in
particular. The good Vbe matching, typically ~ 1 mV, and excellent process uniformity and repeatability make them attractive in general for analog design. Low
surface recombination velocity allows high current gain. The lack of an insulating
substrate results in lossy transmission lines and increased wiring parasitics. Complementary PNP devices are commonly available, and their use as active loads in
operational amplifiers allows high levels of integration to be achieved in clock and
data recovery loop filters and in the feedback loops of laser drivers.

2.2. *GaAs MESFET*

The high f_t and maturity of GaAs MESFETs has resulted in extensive application
to lightwave circuits at 2.488 Gb/s. The speed advantage of this technology stems
from the superior electron transport properties of GaAs and the availability of semi-
insulating substrates. Unlike silicon devices which rely on an oxide insulated gate,
the gate of GaAs MESFETs is formed from a Schottky diode which is reverse biased
to pinch off and control the current in the channel. The intrinsic f_t of a FET is
given by

$$f_t = \frac{g_m}{2\pi \left(C_{gs} + C_{dg}\right)}, \tag{2}$$

where C_{gs} and C_{dg} are the gate-source and drain-gate capacitances respectively.
The transconductance, g_m, increases as the gate length is made shorter. The gate
capacitances both decrease as the gate length decreases. This results in a desire for

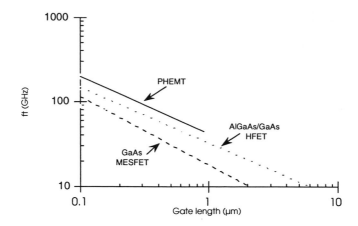

Fig. 2. Approximate graph of the dependence of F_t on gate length for MESFETs and HFETs.

short gate lengths for MESFETs. The dependence of f_t on gate length is shown in Fig. 2.

The smallest gate length that can be produced with acceptable yield is usually used. The power gain for FETs is characterized by f_{max} given by:

$$f_{\mathrm{max}} = \frac{f_t}{2\left(G_0\left(R_g + R_s\right) + 2\pi f_t C_{dg} R_g\right)^{1/2}}, \tag{3}$$

where G_0 is the output conductance, and R_s and R_g are the parasitic source and gate resistances. The use of "t" gates reduces R_g. The same fine lithographic dimensions that give rise to small gates and high f_t are used to create a small gate-source spacing reducing R_s. These factors combine to yield $f_{\mathrm{max}} \gg f_t$ in most high speed processes, resulting in switching times governed mostly by f_t, frequently the only figure of merit, besides g_m and breakdown voltage, quoted for FET processes.

These transistors, while possessing high speed, suffer from threshold variation, gate leakage, back gating, and rely on small lithographic dimensions, resulting in greater process variation, for their speed. These properties pose significant challenges for the designer. MESFETs, however, are widely used in commercially available lightwave circuits, among which are those of Oki, with a .2 μm gate, and a 65 GHz f_t; Triquint QEDA2 process with $l_g = 0.6$ μm and $f_t = 21$ GHz, Vitesse with their H-GaAsIV process which has $l_g = 0.45$ μm and $f_t = 44$ GHz, and NEL with $f_t = 50$ GHz in a 0.15 μm process. The availability of large substrates, a simple process, and volume manufacturing experience have contributed greatly to the current success of this technology.

2.3. *III-V heterojunction field effect transistors*

The addition of a wide bandgap material beneath the gate results in high speed and lower noise FETs. These structures are known by a variety of names including

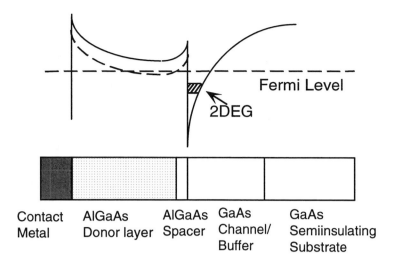

Fig. 3. Band diagram of one type of HFET.

HEMTs, MODFETs, SDHT, TEGFET and the term *I* will use here, Heterojunction FETs (HFETs), a term which includes these structures and others as well. The cross section of a typical HFET is shown in Fig. 3.

The wider bandgap material is doped and some of the carriers transfer into the narrow bandgap channel layer. The confinement of the conducting channel to a narrow ~ 100 Å layer results in higher transconductance. The separation of the ionized dopant atoms from the channel carriers, which are confined to a narrow layer, also improves the channel mobility and hence device speed and channel noise. The confinement is actually more important than the mobility enhancement as is demonstrated by the excellent performance obtained with doped-channel HFETs.[7] The majority of work on these devices has been done with n-channel devices in the AlGaAs/GaAs material system. A wide variety of structures have been investigated under the rubric of HFETs; for a more complete survey the interested reader is referred to Ref. 1.

The high f_t and low noise, which are related, make these devices well suited for preamplifiers. These still suffer from many of the same problems mentioned above for the MESFET, although second order effects such as backgating can be less pronounced. The high $1/f$ noise of these devices can cause problems with VCO stability and poor device matching, and uniformity can cause problems for clock and data recovery circuits.

The use of a narrow gap strained layer in pseudomorphic heterojunction FETs (PHEMTs) further improves their performance. The use of a thin strained layer of InGaAs for the channel results in improved transport properties and a greater conduction band discontinuity. The best results in the HFET family have been reported for this type of device. One notable recent example reports a PHEMT device using an InGaAs channel with a 0.1 μm gate with an F_t of 200 GHz and $F_{\max}$ of 400 GHz. So far however there have been few reports of HFETs used to make integrated circuits for lightwave applications besides those in preamplifiers.

2.4. *Heterojunction bipolar Transistors*

The use of a wider bandgap material in the emitter or, in the case of Si-Ge HBTs, a narrower gap material in the base, allows improvement in the performance of a bipolar transistor.[8,9] Figure 4 shows the doping profile and band diagrams for typical Si bipolar and AlGaAs/GaAs bipolar devices. For the compositionally graded heterointerfaces shown below, the maximum $\beta(\beta_{\max})$, which corresponds to the ratio of the current due to electrons injected from the emitter into the base (I_n) to the hole current due to the holes injected from the base into the emitter (I_p), is given in Eq. (4) in terms of the mean speeds (v_{nb}, v_{pe}) of the carriers and their concentrations (N_e, P_b).

$$\frac{I_n}{I_p} = \beta_{\max} = \frac{N_e}{P_b} \frac{v_{nb}}{v_{pe}} \exp(\Delta\varepsilon_g/kT). \tag{4}$$

$\Delta\varepsilon_g$ corresponds to the difference in the energy gap between the emitter and base in the case shown. For $\Delta\varepsilon_g > kT$ the base doping can be increased and the emitter doping decreased relative to a homojunction transistor ($\Delta\varepsilon_g = 0$) as shown in the

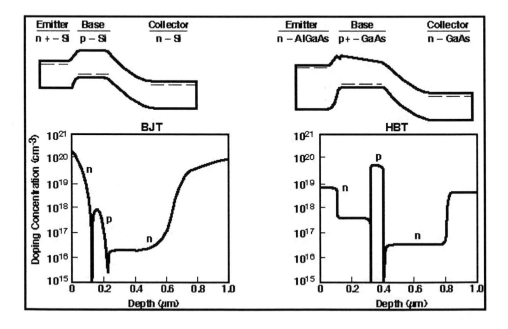

Fig. 4. Bandstructure and typical doping levels of Si BJT and AlGaAs/GaAs HBT.

doping profiles in Fig. 4. The lower emitter doping decreases the base to emitter capacitance (C_{be}) increasing f_t. The higher base doping allows for a lower base resistance which increases $f_{\max}$ (cf. Eq. (1)). High base doping also shields the base emitter junction from the collector voltage leading to a high early voltage.

These are inherently vertical devices and do not rely on small lithographic structures for their speed but rather on the well controlled and highly accurate epitaxial technologies which have demonstrated monolayer resolution.

2.4.1. *GaAs based HBTs*

AlGaAs/GaAs HBTs are the most well investigated and advanced members of the HBT family. The ability of GaAs and AlGaAs to be grown lattice-matched over the full compositional range by a variety of processes allows a number of strategies involving "band-gap" engineering to be applied. Among these is the use of a compositionally graded base layer. This can be used to add a permanent drift field in the conduction band of the base, which aids diffusion in the base transport of the injected electrons reducing the base transit time. Launchers that inject carriers from the emitter through the base ballistically are also possible, as well as a host of collector structures that take advantage of the velocity overshoot in GaAs/AlGaAs to decrease the collector transit time.

GaAs based HBTs have demonstrated high f_t and $f_{\max}$, good threshold matching and uniformity, $1/f$ noise performance intermediate between Si bipolar and FETs, and high breakdown voltage. The good matching and adequate $1/f$ noise performance make them well suited for clock and data recovery applications. The availability of semi-insulating substrates makes them well suited for the larger multiplexer, demultiplexer, and switching circuits as well. The high breakdown voltages make them well suited for power applications such as laser and modulator drivers, both conspicuous problems for Si bipolar transistors. One limitation relative to silicon is that they are often not biased to their region of peak f_t because of thermal and current density constraints. For fair comparison their f_t at the maximum usable bias is the figure to use; for device comparison this is the value that is often quoted. The intrinsic speed advantages of this material system result in much higher speeds relative to silicon when transistors are run at comparable current densities. Large lithographic dimensions can also be used 1–2 μm which still allow the performance of Si BJT at 0.6 μm to be exceeded. The increasing production volumes of these devices are making mature processes available to designers. Examples of commercial HBT processes are those of Rockwell with f_t and $f_{\max}$ of ~ 40 GHz, and TRW with f_t and $f_{\max}$ of 23 GHz and 55 GHz. Laboratory processes have demonstrated very high values of f_t and $f_{\max} \sim 100$ GHz such as in the process developed at Rockwell and applied directly toward 40 Gb/s lightwave circuits, or the recent values of f_t of 134 and $f_{\max}$ of 400 GHz realized in an advanced transferred substrate collector-up structure.[10]

2.4.2. *InP based HBTs*

InP based HBTs have for a given characteristic lithographic and layer dimensions higher f_t and f_{max} than GaAs HBTs. In these HBTs the emitter is typically formed from the wide bandgap materials InAlAs, GaInP or InP and the base from InGaAs. Compared to GaAs based HBTs they exhibit a number of advantages including lower power consumption, higher speed and larger gain. The lower power is the result of a lower built-in potential in the base emitter junction stemming from the lower bandgap of InGaAs and the generally small conduction band offset in that heterojunction, resulting in turn-on voltages of ~ 0.7-0.8 V compared with the ~ 1.3 to 1.4 V for GaAs/AlGaAs HBTs. The intrinsic speed advantage for this system is due to the superior carrier transport properties stemming from lower carrier effective masses, and greater Γ-L valley energy band separation. Low surface recombination velocity and fewer traps result in higher current gain and lower $1/f$ noise. They share the good threshold uniformity of the bipolar family, and their lower turn-on voltage results in lower power consumption at a given speed, making them well suited for larger circuits such as clock and data recovery circuits, multiplexers and demultiplexers.

The most important drawback of this technology is that the single heterojunction HBTs (SHBTs) suffer from low collector to emitter breakdown voltage, which limits their suitability for laser and modulator drivers. The use of InAlAs or more commonly InP in the collector can increase the collector breakdown voltage. For example Chau *et al.*[11] report devices with InP collectors and a superlattice grading layer to the base, that have f_t of 69 GHz f_{max} of 166 GHz and a collector breakdown voltage with the emitter open of 29 V.

Their $1/f$ noise performance and noise figures are comparable to that of GaAs HBT. An important advantage is that photodetectors can be monolithically integrated and are sensitive to both the 1.3 and 1.55 μm optical fiber transmission bands. Some practical disadvantages at present are the lack of large substrates, the complexity of the crystal growth required relative to other approaches and the relative immaturity of this technology. Currently there are no commercially available parts or foundry processes in this transistor family.

2.4.3. *Si-Ge based HBTs*

Some of the advantages that are available for HBTs in the AlGaAs/GaAs system have been realized by the use of strained Si-Ge base layers. Two approaches have been pursued. One is to use a rather high level of Ge in the base, allowing the increase in the base doping as for GaAs based HBTs. The other, pursued by IBM, is to use a lower Ge fraction but to grade its concentration in the base to lower the base transit time. This latter approach is preferred if maximum compatibility is desired with existing processes. Typically the use of a Ge in the base gives a 20% to 30% speed improvement over a comparable Si BJT. One of the main reasons for this besides the aforementioned advantages of HBTs is the superior transport properties afforded by the strained base layer. Strain further decreases the bandgap of the base layer while simultaneously decreasing the electron effective mass.

Using 0–8% Ge grading transistors with $f_t = 75$ GHz have been realized,[12] raising the Ge grading to 0–25% yielded devices with f_t of 117 GHz; f_{max} of ~ 60 GHz was obtained for both.[13]

The advantage of speed improvements, and compatibility with the advanced processing technology available to the Si material system are powerful incentives for development of this technology. The problems that the designer faces are much the same as for Si BJT among which are the conducting substrate. These devices can be operated at bias levels corresponding to their peak F_t.

3. Transmitters/Modulator Drivers

Whereas LEDs often provide a cost-effective solution at low speeds (< 1 Gb/s) and short distances (< 10 Km), higher speeds and distances require the use of either a Distributed Feedback Laser (DFB) or an optical modulator. Advances in both of these devices have mirrored the steady increase in fiber capacity.

With the advent of optical amplifiers it is now cost effective to provide very long fiber links without electrical regeneration for terrestrial as well as submarine links. The elimination of fiber loss as the limiting factor in link length has left optical pulse distortion as the major issue. The principal distortion encountered in TDM links is pulse dispersion. The inherent difference of the propagation speed of light of different wavelengths, along with the unavoidable bandwidth occupied by modulated data, results in the different Fourier components of the pulse arriving at the receiver at different times. The resulting transfer of energy between the time slots occupied by adjacent bits is known as intersymbol interference. Typically, however, the bandwidth occupied by the modulated signal is larger than one would expect from these fundamental considerations. Additional phase and frequency shifts accompany the switching of a semiconductor laser. These shifts, commonly known as chirp, result in bandwidth spread and rapid deterioration of the pulse shape, limiting the reach of directly modulated laser transmitters to approximately 200 km at 2.5 Gb/s in a conventional single mode fiber when operating at the low loss window around 1550 nm. To exceed these limits, various dispersion compensation schemes have been developed; or one can use external modulators, which can be made to exhibit only a small amount of chirp. In general the effect of dispersion becomes worse as the square of the bit rate, meaning that the deleterious effects of pulse distortion are 16 times worse at the OC-192 rate as compared with OC-48. This means that at higher bit rates it might not be possible to increase the capacity of a link merely by changing the terminal equipment.

Additionally, due to the noise and sensitivity properties of receivers, a high extinction ratio is desired from the optical transmitter. To maintain constant/reliable link performance, the effects of temperature, wavelength drift and aging on the optical device must be controlled. The design of the driver circuits for transmitters is largely centered on the need to compensate for the deficiencies of the optical devices.

3.1. *Laser drivers*

The light output from a semiconductor laser above threshold is roughly proportional to the current through it. For this reason a current drive, or at least a driver with an output impedance significantly greater than that of the laser (3–5 Ω typically) is desired. DFB lasers, while having undergone a remarkable evolution, still present a number of problems to the designer of optical transmitters. A more in-depth treatment of lasers in general is given in Refs. 14, 15 and some useful papers that concern the design and characteristics lasers for high-speed modulation are given in Refs. 16 and 17.

The chirping of DFB lasers is dependent on the laser design, bias and the rate at which the device is switched; for this reason, the current pulse applied to the laser should have a rise time no faster than necessary. A trade-off can exist here between Inter-Symbol Interference (ISI) induced by laser chirp and dispersion in the fiber and ISI induced by response "tails" extending into adjacent time slots caused by insufficient bandwidth.

Lasers have a non-linear light-current (L-I) relationship as depicted in Fig. 5. The threshold current at which the laser turns on, and the slope of the L-I curve, change with temperature and aging. Additionally, the bias level in the "off" state affects the rapidity with which the device can be modulated. If the laser is turned

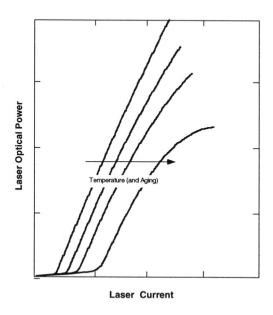

Fig. 5. A typical light versus current curve for a laser diode illustrating qualitatively the effects of increasing temperature and aging.

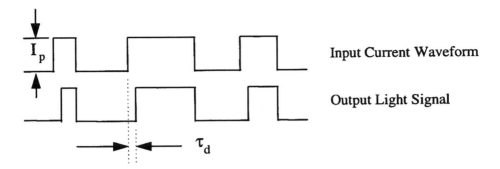

Fig. 6. Idealized laser input and output wave forms illustrating the effect of laser turn-on delay.

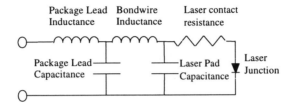

Fig. 7. Simplified electrical model for packaged laser diode.

from completely "off" to "on", it will exhibit a random delay known as "turn-on delay". This is illustrated in Fig. 6.

For an applied current pulse of amplitude I_p the turn-on delay is given by:

$$\tau_d = \tau_{th} \ln \left(\frac{I_p}{I_p - I_{th}} \right) . \tag{5}$$

With a bias current applied this becomes:

$$\tau_d = \tau_{th} \ln \left(\frac{I_p}{I_p + I_b - I_{th}} \right) , \tag{6}$$

where τ_{th} is the delay at threshold (2 ns typically) and I_{th} is the threshold current. Thus to reduce the turn-on delay one should use a low threshold laser and make I_p large. The use of a high bias current will however result in a lower extinction ratio.

Figure 7 shows a schematic of a simple model of a packaged laser. The light output for circuit design purposes can be considered to be proportional to the current through the diode junction. For this reason it is preferred to drive the laser using a current source. The high output impedance, good current drive capability and speed of bipolar transistors have made them the preferred device for driving lasers.

Two commonly used methods for high-speed laser drives are shown below. Figure 8 shows the laser diode driven by a transmission line. Typically 50 Ω is used, but to conserve power, impedances as low as 25 Ω are sometimes seen. One advantage

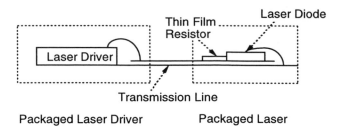

Fig. 8. One approach for driving a laser diode using a transmission line and matching resistor.

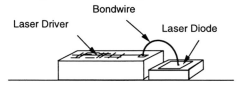

Fig. 9. Alternate laser driver approach using a single wire bond.

of using a transmission line at high speeds is that the laser driver output and laser diode input ports can each be separately engineered for a good broadband 50 Ω match. If a back match at the laser driver and/or a series resistance at the laser ($\sim$ 45 Ω) is used, significant power is then wasted in these matching components.

An alternate approach, Fig. 9, often used below 1 GHz, is to bond the laser driver output directly to the laser diode, with no attempt at impedance matching. The parasitic inductance of the bond wire and the diode capacitance can significantly degrade the rise time of the current pulse and cause ringing. Another challenge that is then encountered is thermal, with the need to have a low inductance ground connection opposed by the desire for thermal isolation of the laser and driver so that their temperatures can be separately controlled.

Typically the design of laser driver circuits incorporates the use of various feedback loops to compensate for the effects of variation of the number of 1's and 0's in the data stream (mark density), temperature and aging. One simple approach is shown in Fig. 10. Here the laser is shown directly connected to the collector of one transistor of a differential pair, with the bias current supplied by the transistor in parallel with the drive transistor. A photodiode is used to monitor the power from the back facet of the laser. Feedback is then used to adjust this bias to maintain a constant output power. This will maintain the output power levels of the 1's and 0's constant as long as the slope of the L-I curve and the relative density of 1's and 0's remain sufficiently constant.

The average laser power will be linearly related to the mark density. If this changes excessively, a circuit such as that shown in Fig. 11 can be used. Here the data is applied to the driver circuit as well as to a circuit whose output is proportional to the mark density. This is used as the reference to the bias control

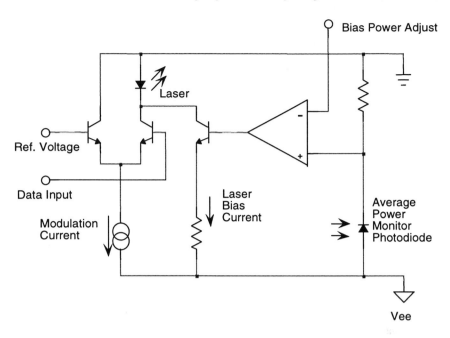

Fig. 10. Laser driver output stage with feedback control of the average output power.

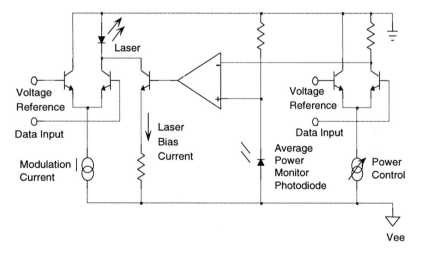

Fig. 11. Laser driver with both constant average power output compensated for the data mark density.

circuit above. Now the amount of bias applied is less if there are more 1's in the data stream and greater if there are more zeros.

At high modulation rates it is often desirable to have a significant bias applied to the laser in the 0 state to increase the speed of the laser and reduce the amount of ringing due to laser self-resonance. The penalty in terms of extinction

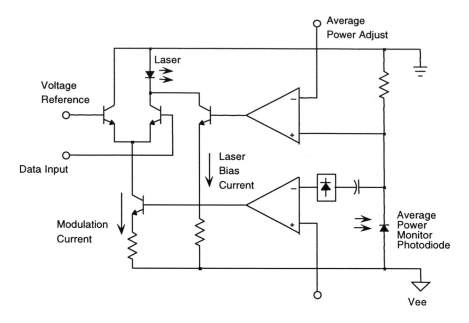

Fig. 12. Laser driver output stage and feedback loops that maintain constant peak and average powers.

ratio is more than compensated by the reduction in chirp and the transmission penalties posed by chromatic dispersion. Under these conditions nonlinearities in the L-I characteristic can lead to variations in the power levels associated with 0's and 1's. One approach to correct for this is shown in Fig. 12. Here the average power is used to control the laser bias, and the peak power is used to adjust the peak modulation current. This will keep the extinction ratio constant and maintain the high and low output powers constant. The disadvantage of this approach is the need for a high-speed monitor photodiode. Mark density compensation as above can also be applied to this design if necessary.

3.2. *Modulator drivers*

Two main types of optical modulators are often used. Electroabsorption modulators rely on the Franz–Keldysh effect or the Quantum Confined Stark effect to produce a variable absorption in a reverse biased semiconductor diode. The similarities of materials and structures make these attractive candidates for direct integration with the laser. Interferometric types typically use Mach–Zehnder interferometers fabricated in lithium niobate. Here the optical signal is split and it traverses a portion of waveguide that also functions as a phase modulator. When a 180° phase shift has occurred on the two paths, the signals combine destructively.

Electroabsorption modulators are attractive at high modulation rates because the wavelength chirp is much reduced relative to lasers and the drive voltage requirements are typically lower than that required for interferometric types. This latter consideration is important given the common trade-off between the maximum voltage swing and the speed encountered in the available transistor technologies.

Interferometric modulators require higher voltage drive but can be driven so that they produce no chirp in the modulated signal. Because of the material incompatibilities the laser, modulator driver and modulator are fabricated on separate substrates and must be interconnected by transmission lines.

Both of these modulator types are controlled by the applied voltage and present a predominantly capacitive load to the driver circuit. A low output impedance driver is thus desired to insure that the low-pass characteristic of the resulting pole does not unduly limit the bandwidth.

Some Mach–Zehnder interferometers are made so that the drive voltage is supplied by a transmission line structure serving as the modulator electrodes. These traveling wave structures can exhibit very high bandwidths and more constant broadband input impedance.

For more in-depth treatments of laser and modulator drivers consult the works by Price and Pedrotti[18] or Shumate.[19]

4. Receivers/Preamplifiers

The receiver of a lightwave system usually consists of a preamplifier, a post amplifier, a clock recovery circuit, a decision circuit, and a demultiplexer. In the following section we will present the preamplifier. The function of the preamplifier circuit is to take the weak signal from the photodetector and to amplify it sufficiently that a decision can be made as to whether each bit is a one or a zero. The usual design goals are for the device to be high bandwidth, low noise, have a large dynamic range, and a high transimpedance. The mutually exclusive nature of these requirements results in design compromises based largely on considerations related to the proposed system environment of the preamplifier. For more information on receiver design consult Refs. 20–25.

Figure 13 shows the three principle types of receivers that are used, along with their common designation and their usual features. The low impedance type can be realized using a back terminated photodetector connected using transmission line to a 50 Ω amplifier. This type is useful in laboratory situations where sensitivity is not of great concern. Its chief virtue is that it can be easily made using commercially available components. The low impedance at the input, however, contributes a large amount of current noise, resulting in low sensitivity.

Greater sensitivity can be obtained using the high impedance design. Here a larger termination resistor is used to reduce the current noise. This also reduces the bandwidth considerably. To correct this, an equalizer is used to compensate the input pole. The equalizer must be designed to carefully match the rolloff of the input pole. This approach is sometimes used but tends to have a limited dynamic range and is difficult to integrate due to the need for component matching.

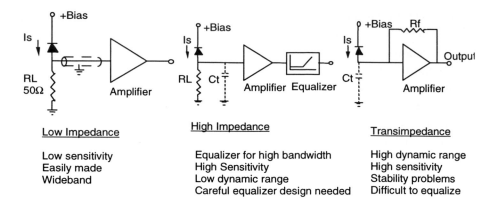

Low Impedance

Low sensitivity
Easily made
Wideband

High Impedance

Equalizer for high bandwidth
High Sensitivity
Low dynamic range
Careful equalizer design needed

Transimpedance

High dynamic range
High sensitivity
Stability problems
Difficult to equalize

Fig. 13. Three approaches used to realize optical receiver preamplifiers.

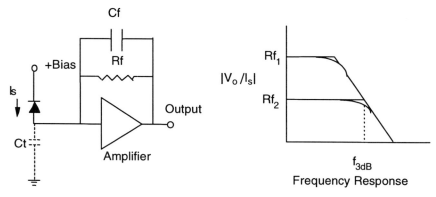

Fig. 14. Simplified schematic of a transimpedance amplifier stage constructed around an amplifier block with voltage gain of A. The frequency response curve shows the effect of the change of the value of the feedback resistor.

The transimpedance amplifier is the most frequently used because of its combination of high dynamic range, sensitivity equal to that of the high impedance approach and high bandwidth.

An idealized version of a transimpedance amplifier is shown in Fig. 14. The transfer function of this amplifier in the single pole approximation is given in Eq. (7).

$$\frac{V_0(f)}{I_s(f)} = \frac{-R_f}{1 + j2\pi f R_f \left(C_f + \frac{C_T}{A}\right)} . \tag{7}$$

From Eq. (8) for the 3 dB point of this function, it can be seen that as the feedback resistor is decreased, the bandwidth increases.

$$f_{3dB} = \frac{A}{2\pi R_f (C_T + A C_f)} . \tag{8}$$

The current noise is controlled by the value of R_f as will be explained further below. The bandwidth however is controlled by the input impedance of the amplifier and the parasitic capacitance of the photodetector and the amplifier. The open loop gain of the amplifier, A, decreases the input impedance of the closed loop amplifier, thus increasing the speed with little noise penalty. Because the parasitic capacitance in the feedback network, C_f, is multiplied by the open loop gain, care must be exercised to ensure that this remains small.

The detailed dependence of the noise and sensitivity depends on the type of transistor technology used. The sensitivity of the preamplifier is more important in receivers that do not use an optical preamplifier prior to the photodetector. For ultimate sensitivity optical amplification is used.

4.1. *Bipolar preamplifiers*

For bipolar amplifier sensitivity we will present the results of Muoi.[23] These results are the same whether the high impedance or transimpedance approach is used. The sensitivity is calculated by computing the contribution of each of the noise generating resistances to an equivalent input current noise. The ratio of the optical signal current to this equivalent input current noise gives the signal to noise ratio. From this the error rate of the decision process, given by the probability that noise will cause the decision circuit to misidentify a 1 as a 0 or vice versa, can be calculated.

The equivalent input current noise of a bipolar preamplifier is given by:

$$\langle i^2 \rangle_a = \frac{4kT}{R_L} I_2 B + 2qI_b I_2 B + \frac{2qI_c}{g_m^2}(2\pi C_T)^2 I_3 B^3 + 4kT r_{bb'}[2\pi(C_d + C_s)]^2 I_3 B^3 \quad (9)$$

Figure 15 shows a common high frequency model for a bipolar transistor. This is shown to define the various elements that are used in the sensitivity calculation.

C_t the total input capacitance is defined as:

$$C_t = C_d + C_s + C_{b'e} + C_{b'c}, \quad (10)$$

where

$$C_{b'e} = C_{je} + \frac{\tau_f I_c}{V_t}. \quad (11)$$

Other quantities are

$$
\begin{aligned}
B &= \text{Bit rate} \\
I_b &= \text{base bias current} \\
I_c &= \text{collector bias current} \\
g_m &= \text{transconductance} = I_c/V_T \\
k &= \text{Boltzman's constant} \\
\beta &= \text{transistor current gain} = I_c/I_b \\
T &= \text{Temperature in degrees Kelvin} \\
\tau_f &= \text{forward transit time}
\end{aligned}
$$

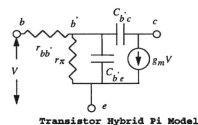

Transistor Hybrid Pi Model

Fig. 15. Bipolar transistor high frequency model showing equivalent circuit elements used in the preamplifier sensitivity calculation.

The terms I_2, I_3 are known as Personick integrals and are tabulated in Ref. 20. These account for the effects of integrating the noise spectra over the amplifier bandwidth.

The origin of the various terms in Eq. 9 can be identified with various physical attributes of the circuit. The first term is identified with thermal noise in the load or feedback resistor. This is the usual Johnson noise that one would expect. This is usually the dominant noise source in the preamplifier. The second term arises from shot noise in the base circuit of the transistor. The third term arises from shot noise in the collector circuit; the derivation of an equivalent current noise generator at the input results in the division of its contribution by the transistor gain and multiplication by the bandwidth. Similarly the base resistance noise contribution is given by the fourth term.

The bias dependence of the base and collector shot noise results in an optimum bias point for the transistor, which occurs when these two terms are equal. Practically for high bandwidth applications the current bias is dictated by the speed required of the transistor, which tends to increase with current bias. The curve of input noise current versus transistor bias is, however, generally quite flat, so this is not a serious defect.

A figure of merit for bipolar amplifiers can be derived and is given by:

$$\text{Bipolar Figure of Merit} = \frac{2f_t}{C_0 + \pi f_t r_{bb'}(C_d + C_s)^2} \approx \frac{2f_t}{C_0}, \tag{12}$$

where

$$C_0 = C_d + C_s + C_{b'c} + C_{je}. \tag{13}$$

From this one can see that to improve the preamplifier sensitivity, low detector (C_d) and stray (C_s) capacitances and high F_t transistors are desired.

The receiver sensitivity is defined as the optical power (usually expressed in dBm, power ratioed to 1 mW) required to achieve a given Bit Error Rate (BER). Usually the performance at a BER of 10^{-9}, one error in a billion bits, is used for comparison, even though in actual links lower error rates are usually required.

The sensitivity of the receiver can be expressed as:

Sensitivity = Average detected optical power for a given bit error rate

$$\eta \overline{P} = \left(\frac{h\nu}{q}\right) Q \langle i^2 \rangle^{1/2}, \tag{14}$$

where

$h\nu$ = Optical Photon energy

and

for pin detectors

$$\langle i^2 \rangle = \langle i^2 \rangle_{\text{amplifier}} + 2q I_d I_2 B \tag{15}$$

which includes the mean square amplifier noise calculated above and the shot noise due to the detector leakage current. Q, known as Personick's Q factor, is closely related to the signal-to-noise ratio and is well approximated implicitly by the following expression for the probability of error.

$$P(E) = \frac{1}{\sqrt{2\pi}} \frac{e^{-(Q^2/2)}}{Q}. \tag{16}$$

For a BER of 10^{-9} $Q = 6$.

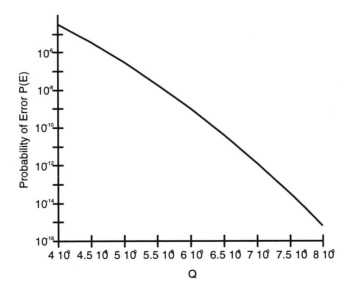

Fig. 16. Probability of error versus Q.

4.2. *FET preamplifiers*

A similar situation holds for FET based preamplifiers. In this case the mean square equivalent input noise current is given by:

$$\langle i^2 \rangle_a = \frac{4kT}{R_L} I_2 B + 2_q I_{\text{gate}} I_2 B + \frac{4kT\Gamma}{g_m} (2\pi C_T)^2 f_c I_f B^2$$

$$+ \frac{4kT\Gamma}{g_m} (2\pi C_T)^2 I_3 B^3 , \tag{17}$$

where the variables are defined as above except for:

$$
\begin{aligned}
I_{\text{gate}} &= \text{FET leakage current} \\
C_T &= \text{total input capacitance including FET, detector and stray} \\
&\quad \text{capacitances} \\
f_c &= \text{FET } 1/f \text{ corner noise frequency} \\
\Gamma &= \text{FET channel noise factor.}
\end{aligned}
$$

The noise now can be seen to be due in the first term to Johnson noise in the feedback or load resistor, and gate leakage shot noise in the second. The third term is due to $1/f$ or flicker noise in the FET and the last, usually the dominant term, to FET channel noise. The g_m of the FET is proportional to its width as is its input capacitance. For this reason the FET-related terms at first decrease as the device width increases but then as the capacitance begins to dominate they increase. The lowest noise point occurs when the detector and stray capacitance just equal the input capacitance of the FET.

The figure of merit for an FET based preamp is:

$$\text{FET Figure of Merit} = \frac{f_T}{\Gamma(C_d + C_s)} \tag{18}$$

which, as in the bipolar case is proportional to the frequency of unity current gain of the device divided by the input capacitance, but now weighted by the channel noise factor.

In general the lowest noise preamplifiers have been produced in HFET technologies.

4.3. *Receivers using optical amplifiers*

As with any amplifier chain the most important noise source lies in the first gain stage. If a lower noise amplifier is used in the beginning of the chain then greater sensitivity is possible. The use of Avalanche photodetectors (APDs) practically results in ~ 6 dB of sensitivity increase with Erbium doped fiber amplifiers giving sensitivity increases of 20 dB relative to a PIN detector and an electrical amplifier. The excellent dynamic range and low noise characteristics of EDFAs[26] allow the design considerations for electrical preamplifiers to be dominated by other than the bit rate/sensitivity trade-off considered above.

5. Clock Recovery/Data Regeneration[27]

In optical communication systems the received signal, which has been attenuated and degraded by noise and imperfections in the modulation and transmission process, must be rxeconstructed before further processing. To do this the signal is typically amplified, filtered and then sampled at the time in the bit slot at which there is the lowest probability of error. This is typically done using a D-flip-flop which is clocked using a clock signal that is derived from the input bit stream. It is important that the clock stay centered at the optimum decision point over time and temperature for the life of the system.

The spectrum of an NRZ signal shown in Fig. 17 has no frequency component at the underlying clock frequency. To produce a frequency component at the clock frequency, a nonlinear process such as squaring or differentiating is usually used.

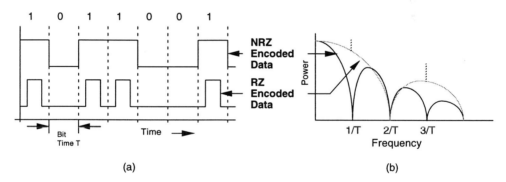

(a)   (b)

Fig. 17. (a) shows the same bit sequence encoded using both Non-return to zero (NRZ) and return to zero (RZ) encoding.

Two main methods are used to extract the clock from NRZ signals. One is to apply the signal to a very high Q filter. The other approach is to use a phase locked loop to lock an oscillator onto the data stream. To regenerate the data the output of this oscillator or filter is used as the clock signal to latch the data in the decision circuit.

A number of good references investigate various aspects of the clock recovery process. Among them is the tutorial paper by Franks[28] which discusses clock recovery not only for PAM but for other modulation schemes as well. The IEEE press has recently published a compilation of many of the important references on clock recovery.[29] Early work includes the papers by Lange[30] and Duttweiler.[31] Good general references include those by Wolaver,[32] Best,[33] and Gardner.[34]

5.1. *Filter based clock recovery*

For the filter approach very high Q filters are required to reduce phase noise in the clock. A filter will pass phase variations up to its rolloff frequency and then attenuate and remove jitter at frequencies above this. The resulting sine wave signal is then applied to a bandpass filter to remove unwanted harmonics and then to a limiting amplifier to remove amplitude noise caused by variations in the data pattern. This is shown in Fig. 18. The output is then routed to the decision circuit using a delay carefully matched to that of the data. Alternatively, a variable phase delay circuit can be used to establish and maintain the proper phase relation between the data transitions and the filter output.

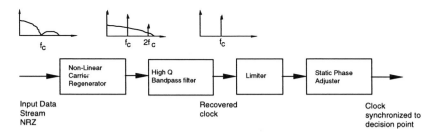

Fig. 18. Clock recovery using a high-Q filter.

For example, for OC-48 systems the clock is centered at 2.488 GHz, but the phase transfer function must have a bandwidth of at least 2 MHz. A minimum filter Q of at least 600 is thus required. For a OC-192 receiver, a Q in excess of 1200 is needed. Typically at lower bit rates up to a maximum of a couple gigahertz Surface Acoustic Wave (SAW) filters are used. For higher frequency applications, filters based on dielectric resonators or waveguide filters are employed.

The bandpass filter centered at the clock frequency serves to determine the jitter transfer and tolerance of the phase transfer function. For symmetrical bandpass filter characteristics, the phase transfer function can be simply derived from the filter pass band by translating the filter center frequency to DC. So a filter with a 1 MHz 3 dB bandwidth centered at 2.488 GHz will have a phase transfer function with bandwidth of 500 MHz measured from DC to the 3 dB point.

Figure 19 shows an example of a filter and its magnitude and phase transfer functions. If the center frequency of the filter shifts relative to the clock component, the phase of the recovered clock signal will shift relative to the data. This results in a departure from the ideal sampling time, possibly resulting in increased error rates if the shift is great enough. Such shifts in the filter characteristics are caused by both temperature and aging. Production induced phase differences are compensated using the static phase shift adjustment. Careful matching of the temperature coefficients of the electrical circuits and filter can be used to ensure that the circuit performance will be maintained over temperature.

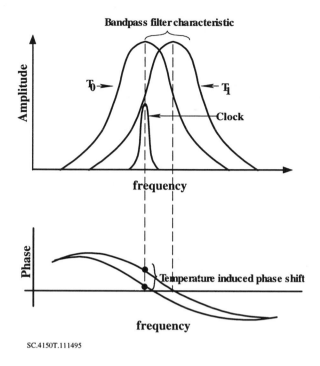

SC.4150T.111495

Fig. 19. Amplitude and phase characteristics for a typical filter transfer function showing the effect of the temperature shift between T_0 and T_1 on the recovered clock phase.

5.2. *Phase locked loop clock recovery*

Phase locked loops can sometimes offer a number of advantages over filter based approaches, depending on the details of the application. Schemes relying on high-Q passive filters dissipate a large amount of power to drive the signal from the nonlinear PRZ generator into the low impedance of the filter; stray coupling of signal components operating at the full data rate require careful packaging; unknown group delay of the filter requires off chip hand tuning; drifts in the filter delay due to temperature must be designed into the device and drifts due to aging are not compensated. Monolithic integration of filter based regenerators has not been demonstrated.

PLL-based clock recovery circuits can overcome all of the above problems. One advantage is that a PLL will track out frequency shifts of both itself and the incoming signal. Another is that PLLs can be monolithically integrated. This can lead to significant reductions in size, power, and system cost. One disadvantage PLLs possess is that the desired loop bandwidths are often smaller than the loop tuning range; for this reason, lock acquisition can be slow even with the addition of frequency acquisition aids. For SONET systems with a service restoration requirement of 50 ms, 10 to 20 ms can be allotted to reacquisition of the clock, so this is not a serious problem. For burst mode systems, it can be a serious consideration.

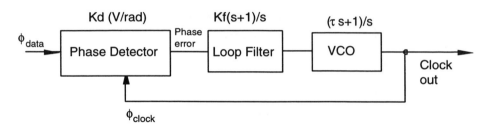

Fig. 20. Block diagram of a second order PLL.

A basic PLL consists of a phase detector, loop-filter and VCO connected as shown in Fig. 20. Of these only the phase detector and VCO must operate at high speed. For a first order PLL the characteristics of the phase transfer function are much the same as for a filter based system. This includes the existence of a static phase offset in the case where the input data clock is detuned from the free-running VCO, much as when the data is detuned from the filter transmission peak in a filter-based system. The static phase error may be eliminated by inclusion of an integrator in the loop filter. This results in a two-pole response for the PLL phase transfer function. The closed loop phase transfer function for the loop shown in Fig. 20 is given by:

$$H(s) = \frac{K_d K_o (1 + \tau s)}{s^2 + K_d K_o \tau s + K_d K_o}, \qquad (19)$$

where K_d is the phase detector gain in volts per radian, τ is the time constant for the loop zero, and K_o is the VCO gain in radians per volt. A second order transfer function will inevitably exhibit some peaking, which is undesirable from consideration of jitter peaking in a chain of regenerators. For this reason the loop is often designed to be over-damped, which tends to further impair its acquisition range. This disadvantage is mitigated by the frequency tracking of the PLL-based approach, one of its main advantages.

The main design trade-offs with PLLs concern those among jitter, acquisition time and bandwidth. To minimize output phase jitter due to external noise the loop bandwidth should be made a narrow as possible. To minimize output jitter due to internal oscillator noise, and to obtain the best tracking and acquisition properties the loop bandwidth should be made as wide as possible. Jitter can be produced both coherently by the low frequency components of the random data, and incoherently from electrical noise. Both sources can induce bit errors at the decision circuit and increase with wider loop bandwidths. The design of a clock recovery circuit consists of trading off these performance criteria subject to the additional requirements imposed by the need to acquire the signal in a timely manner over the expected range of environmental conditions and while tracking the low frequency phase variations in the input data stream.

This simple approach while buildable might not be optimum. Typically this approach suffers from a limited capture range, long pull-in times and the requirement of good temperature stability of the components. To aid signal acquisition, a frequency detector loop is commonly added around the PLL to make a Frequency Locked Loop (FLL)-PLL combination. This allows signal acquisition over a much broader frequency range. This is particularly desirable if significant drifts in the VCO center frequency are expected due to temperature or aging.

5.2.1. *Phase detectors*

The phase detector can in many cases become quite elaborate, Oberst[35] provides an overview of many of the approaches used to realize phase detectors.

For high speed applications two simple phase detector choices that are often used are the four quadrant Gilbert Cell multiplier or a digital EXOR circuit. These are really the same, differing only in whether or not the circuit is operated in the linear regime. These provide a "saw tooth" like characteristic in the case of the EXOR, or a sinusoidal characteristic in the case of a multiplier, that is repetitive in phase. The important design criterion here is to maintain a low phase offset voltage. A large offset results in higher pattern jitter. Because the phase detector must function at twice the data rate, digital phase detectors[36] such as are often used at lower bit rates are not often used for the highest speed applications. Phase detectors with a sawtooth type of characteristic also possess a broader capture range than those with a sinusoidal output easing clock acquisition.

One notable digital phase detector is that developed by Hogge.[37] One of the virtues of this circuit, besides its simple digital implementation, is that when the phase detector output is zero, the clock and data are automatically centered at the first flip-flop. Its output then consists of the correctly retimed data so no decision circuit or additional phase adjustment circuitry is required.

One common deficiency that all the above phase detectors possess is an output voltage that depends on the transition density and consequent deviation of their output voltage from that in phase alignment. This becomes important when considering the jitter tolerance of the circuit. When tracking the incoming phase excursions of the data the phase detector produces voltages that force the VCO to track the incoming phase. Dependence of the phase error signal on the low frequency transition density variations in the data produce an additional phase jitter in the clock.

5.2.2. *VCOs*

The VCO is the next most challenging and critical component for the loop performance. The desirable characteristics of a VCO are a large electrical tuning range, phase stability, linearity of frequency versus control voltage, large gain factor, and wide modulation bandwidth. The requirement of phase stability is in opposition to the other four requirements. A number of choices exist for the VCO. These, in order of increasing phase stability (higher Q) are:

RC multivibrators: They have a wide tuning range and are tolerant to large process variations in circuit fabrication. They are also compact, simple, easily designed and compatible with standard digital processes. The low Q of these oscillators[38] and the fact that they are memoryless make them particularly vulnerable to self-injection locking.

Ring oscillators: These circuits are formed by the series connection of a number of inverters, with the output of the chain coupled back to the input. The chain will oscillate at a frequency corresponding to the inverse of the propagation delay of the chain. Tuning can be accomplished by changing the delay of each element or by summing the output of a number of the stages prior to feeding it back to the input. The phase noise of these circuits is high, corresponding to a relatively low,[39] but still useful Q in the range of 1–1.4. This approach can be widely tunable, and can be designed to naturally generate the quadrature clock components needed for most frequency detectors.

LC oscillators possess moderate to low Q, good tuning range, and good stability. Additionally they are capable of monolithic construction at high frequency. Monolithic inductors present the greatest problem with regard to Q. For monolithic designs care must be taken to ensure that the tuning range of the VCO is wide enough to include the expected range of process variation and frequency shift due to temperature. These oscillators are most often tuned using varactors. One advantage of L-C type VCO is that their phase noise characteristics can be well modeled[40] prior to circuit construction, unlike multivibrators or ring oscillators although a recent paper by Razavi[41] provides a generalized definition of Q and insight into the phase stability of these latter types.

As an alternative to the use of lumped Ls and Cs, a *transmission line stub* can be used as the resonant element. Use of an off-chip transmission line leads to increased packaging complexity, hand tuning, and increased power consumption required to drive the transmission line. At sufficiently high frequencies the transmission line resonator becomes sufficiently short that it can be integrated on-chip. Transmission line loss is typically low, and these types of oscillators exhibit moderately high Q and are well suited to clock recovery applications.

Crystal or *Dielectric Resonators* can be used to build very high Q and very stable VCOs with narrow tuning ranges. These are the lowest phase noise choice but require hybrid circuit construction techniques. In clock recovery applications this type of oscillator is most often used to realize an extremely low jitter transfer bandwidth thus eliminating, for example, jitter that has accumulated on a signal after a large number of regenerations within a network.

5.2.3. *Lock acquisition with PLLs*

One of the main problems with the use of PLLs for clock recovery is their limited lock acquisition range. This is typically on the order of the loop bandwidth, which

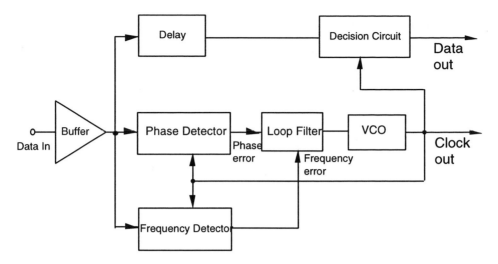

Fig. 21. Generic PLL Clock recovery circuit including a frequency detector to aid lock acquisition.

for clock recovery applications can be quite narrow. For these applications the use of acquisition aids, or some strategy to maintain the VCO free-running frequency to within one loop bandwidth of the data rate is indispensable.

One approach commonly used to acquire lock using PLLs requires the addition of a lock detection circuit and a voltage scanning circuit for VCO control. To acquire the signal, the VCO voltage is scanned until lock is detected. One disadvantage of this approach for clock recovery is that the lock detector can limit the jitter tolerance of the circuit.

A more desirable approach is to add a frequency detector to the loop. This voltage will then slew the VCO within the lock range of the PLL. Once this is achieved, the output from the frequency detector is either disabled or frequently is designed to have a zero voltage output in lock.

The "quadricorrelator" circuit developed by Richman[42] in the early days of color television was proposed for use in clock recovery circuits by Bellisio[43] and successfully implemented by Cordell.[44] The only high-speed portions of this circuit are the two phase detectors, making this an attractive choice for high-data rate applications. Ideally this circuit produces an output that is proportional to $\Delta\omega/8$.

Another approach known as a rotational frequency detector was proposed by Messerschmitt.[45] Here the clock quadrature and in-phase components are used to sample the incoming data transitions. Depending on the relative phase relations between the samples, an output whose average value contains information as to whether the clock rate is too fast or too slow can be generated. This approach is all digital and is not as well-suited for high frequency or low signal-to-noise applications as compared to the quadricorrelator. It is, however, well suited to monolithic integration, and variants have been used in some of the examples to be presented below.

5.3. *Decision circuits*

While not strictly part of the clock recovery circuit, the decision circuit performance is integral to the performance of any regenerator circuit. A typical decision circuit is shown in Fig. 22. Here the data signal is first applied to a comparator or gain stage and then sampled at the rising clock edge to sample the data.

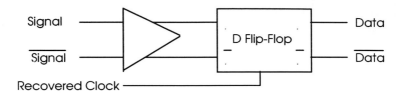

Fig. 22. Block diagram of a typical decision circuit.

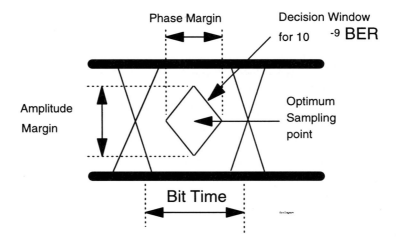

Fig. 23. This shows the data eye and defines the quantities defining the performance of a decision circuit.

The figures of merit for a decision circuit are illustrated in Fig. 23. The phase margin is defined as the maximum phase difference that can be allowed to achieve a given BER. The amplitude margin is the comparable quantity defined for the position of the sampling point with respect to amplitude. The contour that the decision point traces out at constant BER is the decision window.

A typical curve of amplitude margin and phase margin versus bit rate for a decision circuit[46] is shown in Fig. 24.

The phase margin of a part is approximately equal to the bit period minus the sum of the setup (ts) and hold (th) times of the flip-flop. In terms of degrees the phase margin can be expressed as:

$$\Phi_M = 360°(1 - B(\tau_s + \tau_h)),\tag{20}$$

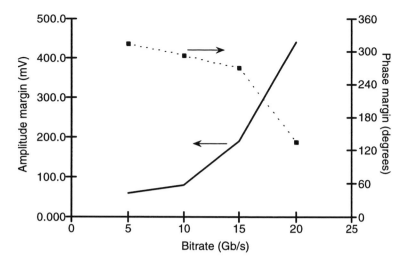

Fig. 24. Amplitude and phase margin for an AlGaAs/GaAs decision circuit as a function of bit rate (Courtesy of Klaus Runge, Rockwell).

where B is the bit rate. This formula can be used to derive a convenient figure of merit for a decision circuit which can be used to compare the results in different technologies and using different circuit designs. From published data of the bit rate and phase margin the sum of the sample and hold times can be inferred. Using those values the bit rate at which a phase margin of $180°$ would be obtained can be calculated. This is close to the maximum frequency that a decision circuit would be likely to use in an actual system; frequently at least $270°$ or better is desired.

One question that is faced by a designer concerns the relationship between device figures of merit and the expected performance that can be obtained from a given process. In a recent paper by Sano *et al.*[47] from NTT figures of merit were derived that correlate well with the switching times of D flip-flops for both FET and Bipolar technologies. Additionally they show that the same figure of merit provides a good indication of the gain-bandwidth product that one can expect from amplifiers. For bipolar technologies the maximum switching speed ($f_{c\,\text{max}}$) is shown to be well approximated by:

$$f_{c\,\text{max}} \approx \alpha \left\{ \frac{1}{f_t} + \sqrt{\frac{2V_{sw}}{0.15} \frac{1}{f_{\text{max}}}} + \left(2 + \frac{V_{sw}}{0.15}\right) \frac{f_t}{4f_{\text{max}}^2} \right\}^{-1}, \tag{21}$$

where V_{sw} is the switching voltage used, and α is an empirical parameter that is fit to experimental data and is found to be ~ 1.38 for HBT D flip-flops.

For FETs

$$f_{c\,\text{max}} \approx \beta f_t, \tag{22}$$

where again β is an empirical parameter found to be ~ 0.24.

Using these expressions, with $V_{sw} = 0.5$ V and the published data where available for the recent high-speed results shown above, $f_{c\,\text{max}}$ versus the calculated bit

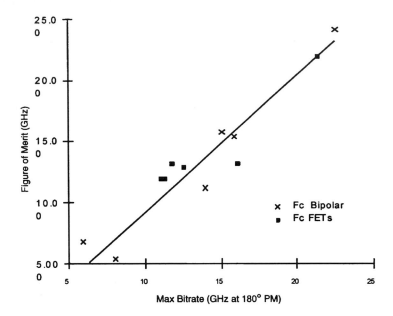

Fig. 25. This shows the good agreement between the figure of merit proposed by Sano *et al.* with the bit rate for 180° phase margin using a D-flip flop as a decision circuit.

rate for 180° phase margin is plotted in Fig. 25. This shows the good correlation of this figure of merit with recent results, giving the device designers guidance as to the space of f_t and $f_{\max}$ values that are desired to achieve a given level of performance and circuit designers guidance as to the suitability of a given process for the realization of their system goals.

6. Conclusion

The rapid increase in demand for bandwidth has been met in the short term by the deployment of WDM systems and the first commercial OC-192 based 10 Gbit systems. To meet future demand there exist a number of viable electronic technologies, all of which can achieve the speeds, and likely the performance required for the next generation or so of lightwave systems. The various technologies have been discussed in the light of their suitability for construction of the circuits comprising optical transmitters and receivers. The salient design concerns for those circuits unique to lightwave systems have been given, as has a means to evaluate the device requirements necessary to support a given lightwave system design.

References

1. S. M. Sze (ed.), *High-Speed Semiconductor Devices*, Wiley, New York, USA, 1990.
2. M. F. Chang (ed.), *Current Trends in Heterojunction Bipolar Transistors*, World Scientific Publishing Co. Pte. Ltd., Singapore, 1996.
3. S. I. Long and S. E. Butner, *Gallium Arsenide Digital Integrated Circuit Design*, McGraw-Hill, New York, USA, 1990.

4. P. McGoldrick, "27-GHz process makes unique parts for RF transceivers", *Electronic Design* July 8, 1996.
5. H. Klose *et al.*, "B6HF: A 0.8 micron 25 GHz/25 ps bipolar technology for 'mobile radio' and ultra fast data link IC-products", *Proc. IEEE BCTM '93*, Minneapolis, Oct. 1993, 125–127.
6. Y. Kiyota *et al.*, "Lamp-heated rapid vapor-phase doping technology for 100-Ghz Si bipolar transistors", *Proc. Bipolar/BiCMOS Circuits and Technol. Mg.*, Minneapolis, 1996, pp. 173–176.
7. H. Inomata, S. Nishi, S. Takahashi, and K. Kaminishi, "Improved transconductance of AlGaAs/GaAs heterostructure FET with Si doped channel", *Jpn. J. Appl. Phys.* **25** (1986) L731.
8. H. Kroemer, "Theory of a wide-gap emitter for transistors", *Proc. IRE* **45** (1957) 1535–1537.
9. H. Kroemer, "Heterostructure bipolar transistors and integrated circuits", *Proc. IEEE* **70** (1982) 13–25.
10. B. Agarwal, R. Pullela, U. Bhattacharya, D. Mensa, Q.-H. Lee, L. Samoska, J. Guthrie, and M. Rodwell, "Ultrahigh $f_{\max}$ AlInAs/GaInAs transferred-substrate heterojunction bipolar transistors for integrated circuit applications", *Int. J. High Speed Electron. Syst.* **9**(2) (1998) 643–670.
11. H.-F. Chau, W. Liu, and E. A. Beam III, "InP-based HBTs and their perspective for microwave applications", *Proc. 7th Int. Conf. Indium Phosphide and Related Materials* (1995).
12. G. L. Patton, H. H. Comfort, B. S. Meyerson, E. F. Crabbe, G. J. Scilla, E. DeFresart, J. M. Stork, J. Y. Sun, D. L Harame, and J. Burghartz, *Electron. Lett.* **28** (1992) 160–162.
13. E. F. Crabbe, B. S. Meyerson, J. M. C. Stork, and D. L. Harame, "Vertical profile optimization of very high frequency epitaxial Si-and SiGe-base bipolar transistors", *IEEE IEDM Tech. Dig.*, Proc. IEDM '93, Washington, D. C., 1993, pp. 83–86.
14. G. P. Agrawal and N. K. Dutta, *Long Wavelength Semiconductor Lasers*, Van Nostrand Reinhold, New York, 1986.
15. J. E. Bowers and M. A. Pollack, "Semiconductor lasers for telecommunications", in *Optical Fiber Telecommunications II*, Academic Press, 1988.
16. J. E. Bowers, "High speed semiconductor laser design and performance", *Solid State Electron.* **30** (1987) 1–11.
17. K. Y. Lau, "Ultra-high speed semiconductor lasers", *IEEE J. Quantum Electron.* **QE-21** (1985).
18. *Ibid.*
19. P. W. Shumate, "Lightwave transmitters", in *Optical Fiber Telecommunications II*, Academic Press, 1988.
20. Personick, "Receiver design for digital fiber optic communication systems, I", *Bell System Tech. J.* **52** (1973).
21. Smith and Personick, "Receiver design for optical fiber communication systems", Ch. 4 in *Semiconductor Devices for Optical Communication*, Springer-Verlag, New York, 1982.
22. T. V. Muoi, "Receiver design for high-speed optical-fiber systems", *IEEE J. Lightwave Tech.* **LT-2** (1984) 243–267.
23. B. Y. Kasper, "Receiver design", in *Optical Fiber Telecommunications II*, Academic Press, 1988.
24. C. D. Hull and R. G. Meyer, "Principles of monolithic wide band feedback amplifier design", *Int. J. High Speed Electron. Syst.* **3** (1992) 53–93.

25. M. H. El-Diwany *et. al.*, "Design of low-noise bipolar transimpedance preamplifiers for optical receivers", *IEE Proc.* **28** (1981) 299.
26. E. Desurvire, *Erbium-Doped Fiber Amplifiers-Principles and Applications*, Wiley, New York, 1994.
27. K. D. Pedrotti, "Clock recovery for high-speed optical communication", in *Optoelectronic Interconnects and Packaging*, eds. Ray T. Chen and Peter S. Guilfoyle, *SPIE Critical Review* **CR62** (1996) 267–296.
28. L. E. Franks, "Carrier and bit synchronization in data communication — A tutorial review", *IEEE Trans. Commun.* **COM-28** (1980) 1107–1121.
29. B. Razavi (ed.), *Monolithic Phase-Locked Loops and Clock Recovery Circuits: Theory and Design*, IEEE Press, Piscataway, N.J., 1996.
30. O. E. D. Lange, "The timing of high-speed regenerative repeaters", *The Bell System Technical J.* **37** (1958) 1455–1486.
31. D. L. Duttweiler, "The jitter performance of phase-locked loops extracting timing from baseband data waveforms", *The Bell System Technical J.* **55** (1976) 37–58.
32. D. H. Wolaver, *Phase-Locked Loop Circuit Design*, Prentice Hall, Engelwood Cliffs, New Jersey, 1991.
33. R. E. Best, *Phase Locked Loops: Theory, Design, and Applications*, McGraw-Hill, New York, 1984.
34. F. M. Gardner, *Phaselock Techniques*, Wiley, New York, 1979.
35. J. F. Oberst, "Generalized phase comparators for improved phase-locked loop acquisition", *IEEE Trans. Communication Technology* **COM-19** (1971) 1142–1147.
36. W. C. Lindsey and C. M. Chie, "A survey of digital phase-locked loops," in., (1981) 410–431.
37. C. R. Hogge, "A self correcting clock recovery circuit", *J. Lightwave Technology* **LT-3** (1985) 1312–1314.
38. B. Razavi, "A study of phase noise in CMOS oscillators", *IEEE J. Solid-State Circuits* **31** (1996) 331–343.
39. *Ibid.*
40. W. P. Robbins, *Phase Noise in Signal Sources*, Peter Peregrinus Ltd., London, UK, 1984.
41. B. Razavi, "A study of phase noise in CMOS oscillators", *IEEE J. Solid-State Circuits* **31** (1996) 331–343.
42. D. Richman, "Color-carrier reference phase synchronization accuracy in NTSC color television", *Proc. IRE* **42** (1954) 106–133.
43. J. A. Bellisio, "A new phase-locked loop timing recovery method for digital regenerators", *IEEE Int. Conf. Commun. Rec.* (1976) 10–17.
44. R. R. Cordell *et al.*, "A 50 MHz phase- and frequency-locked loop", *IEEE J. Solid-State Circuits* **SC-14** (1979) 1003–1010.
45. D. G. Messerschmitt, "Frequency detectors for PLL acquisition in timing and carrier recovery", *IEEE Trans. Commun.* **COM-27** (1979) 1288–1295.
46. K. Runge and J. L. Gimlett, "20 Gbit/s AlGaAs/GaAs HBT decision circuit IC", *Electron. Lett.* **27** (1991) 2376–2378.
47. Sano *et al.*, "Device figure-of-merits for high-speed digital ICs and baseband amplifiers", *IEICE Trans. Electron.* **E78-C** (1995).

International Journal of High Speed Electronics and Systems, Vol. 9, No. 2 (1998) 347–383

Si AND SiGe BIPOLAR ICs FOR 10 TO 40 Gb/s
OPTICAL-FIBER TDM LINKS

H.-M. REIN

AG Halbleiterbauelemente, Ruhr-University Bochum,
D-44780 Bochum, Germany

This paper gives an overview on very-high-speed ICs for optical-fiber systems with restriction to Si-based technologies. As a main aim, the circuit and system designer shall get an impression what operating speeds have already been achieved and, moreover, get a feeling for potential limitations. It is shown that all ICs in 10 Gb/s TDM systems can be fabricated in Si-bipolar production technologies, while for the speed-critical ICs in 20 Gb/s systems, present SiGe laboratory technologies are required if the circuit specifications, apart from the data rate, must remain unchanged. With uncritical circuits like time-division multiplexer (MUX) and demultiplexer (DEMUX), record data rates of 60 Gb/s were achieved with a SiGe laboratory technology, using an adequate mounting and measuring technique. Recent measuring results even showed that all ICs in a 40 Gb/s TDM system (i.e., also the speed-critical ones) can be realized in advanced SiGe technologies. However, compared to ICs in 10 and 20 Gb/s systems, some circuit specifications must be relaxed. This is possible by the use of optical amplifiers and improved opto-electronic components as well as by system modifications, which further make possible the elimination of some of the speed-critical circuits. It should be noted that all the experimental results presented are measured on *mounted* chips, using conventional wire bonding, and that most of the circuits have been used in experimental TDM links.

1. Introduction

There is no doubt that the transmission capacity of long-haul optical-fiber links will be increased further and further. For this to happen, high-speed electrical time-division multiplexing (TDM) systems are required. Based on these TDM systems, the transmission capacity can be further increased by wavelength division multiplexing (WDM). An open question for future systems concerns at which data rates the system designer has to change from TDM to WDM. These "interface data rates" depend on the upper limit for an economical solution of the dispersion problems in optical fibers, on the upper speed limit of the electronic and opto-electronic components, and on the progress in future all-optical networks. Today, in commercial systems the interface data rate is 2.5 Gb/s and 10 Gb/s, respectively. For future long-haul systems, 40 Gb/s are under discussion.[1-3] Further applications for ultra-high-speed TDM systems will be local area networks as well as interconnections in high-speed switching networks[4] and multiprocessor computers.

Therefore, it must be found out which data rate in TDM links can be realized in the near future at reasonable costs. The intention of the author is to give a contribution to this topic from the electronics point of view, with restriction to Si-based ICs. The reader should get an impression of which data rates can be achieved with available Si production technologies as well as with today's SiGe laboratory technologies. The discussion and experimental examples are restricted here to ICs in long-haul transmission links. However, the results may also be useful for the designer of short-distance TDM links, where the demands on circuit performance usually are less stringent at a given data rate, so that power consumption and costs can be reduced.

The paper is organized as follows. In Sec. 2, the state-of-the-art with respect to maximum data rates is summarized for ICs in both Si-based production and laboratory technologies. A few remarks on Si and SiGe bipolar technologies suited for very high operating speed are given in Sec. 3. Section 4 shortly recapitulates the basic circuit concepts and diagrams for digital circuits and broadband amplifiers required in the high-speed part of TDM links. This is also to simplify the succeeding discussions where some of the circuits then can be described by block diagrams only. In Sec. 5, realized circuits and their experimental results are presented with clear distinction between production- and laboratory-technology results. Speed-critical circuits like modulator drivers and amplifiers are stressed, but record data rates of 60 Gb/s for a time-division multiplexer and a demultiplexer are also presented. Most important for future developments are Secs. 6 and 7 where the author answers the question: "Are Si-based technologies fast enough for the complete high-speed electronics in 40 Gb/s long-haul systems, and what measures must be taken to reach this goal?".

2. Si-Based ICs in Long-Haul Transmission Links — State-of-the-Art

The circuits required can be taken from the scheme of an optical-fiber TDM link shown in Fig. 1 for a transmission rate of 10 Gb/s as an example. On the transmitting side, we need a time-division multiplexer (MUX) which combines several channels with lower data rates to a single one. The succeeding power circuit drives the (external) modulator for the laser light.[a] The modulator is either of the electro-absorption type (EAM) or a Mach-Zehnder interferometer. On the receiving side, the small signal current generated by the photodiode is amplified by a low-noise transimpedance amplifier and a succeeding main amplifier. The latter is either of the limiting or automatic-gain-control (AGC) type, in order to guarantee a constant output voltage swing (independent of the input amplitude). For retiming the data stream, a decision circuit is required, which is often realized as a master-slave D-type flipflop (MS-D-FF). The succeeding demultiplexer (DEMUX) splits the transmitted data stream into the original lower-rate channels. For driving the decision circuit and the DEMUX, a clock extraction circuit and a high-speed frequency divider are required.

---

[a]Direct laser modulation is less common at and above 10 Gb/s due to dispersion problems.

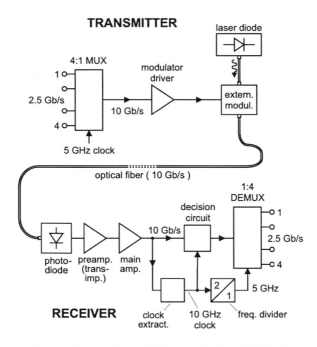

Fig. 1. Scheme of a 10 Gb/s optical-fiber TDM link.

Table 1. High-speed ICs for optical-fiber systems in Si-based bipolar technologies — state-of-the-art. The data rates in the second column were achieved with Si production technologies. Record speeds in a laboratory SiGe technology are given in the last two columns (RUB + Siemens). All the results were obtained by measuring on *mounted* chips.

Circuit	Max. Speed / Si Production Technology		Remarks, Specifications	SiGe Laboratory Technology	
Multiplexer	30 Gb/s (RUB + HP) / HP 25	[5]	2:1	60 Gb/s [12]	recent results, relaxed specifications
Decision Circuit	22 Gb/s (RUB + Siemems) / B6HF	[6]	CPM =180°		
Demultiplexer	32 Gb/s (RUB + Siemems) / B6HF	[6]	1:2	60 Gb/s [13]	
Static Frequency Divider	19 GHz (RUB + Siemens) / B6HF	[7]	2:1, 2 Outputs (0° and 90°)	37 GHz [14,15]	
Laser/Modulator Driver	14 Gb/s (RUB + Siemens) / B6HF	[8]	Swing ΔV_Q= 3.6 V (50 Ω)	23 Gb/s [16]	40 Gb/s [18] 50 Gb/s [19]
Preamplifier (Transimp. Type)	13 Gb/s (RUB + ANT + Siemens) / B6HF	[9]	Transimp. Z_{TI} = 615 Ω f_C = 10 GHz Noise $\bar{j}_N$ =10.5 pA/$\sqrt{Hz}$	20 Gb/s [17]	40 Gb/s [20]
Main Amplifier, AGC	13 Gb/s (RUB + ANT + Motorola) /Advanced MOSAIC V	[10]	S_{21} = 37 dB, f_C = 10 GHz Dynamic Range = 41 dB	——	
Main Amplifier, limiting	15 Gb/s (RUB + ANT + Siemens) / B6HF	[11]	S_{21} = 52 dB (max.) Dynamic Range = 46 dB	——	

Table 1 shows the state-of-the-art for the circuits in the signal path of an optical-fiber link, with respect to maximum speed. The ICs were designed in cooperation between the author's group at Ruhr-University Bochum (RUB) and several industrial partners (given in parentheses). The maximum data rates achieved with present Si-bipolar *production* technologies (mostly with the B6HF of Siemens,[21] cf. Sec. 3) are given in the second column and some further specifications in the third column. Note especially the high output voltage swing ΔV_Q of the modulator driver, the high transimpedance Z_{TI} and low equivalent input noise current density $\bar{j}_N$ of the preamplifier (averaged over the bandwidth), and the high gain S_{21} and high dynamic range of the (AGC and limiting) main amplifiers. The 3-dB cut-off frequency f_C of the linear amplifiers is about 10 GHz.

In the 4th and 5th columns, the maximum data rates of ultra-fast ICs are given, which have been developed in cooperation between RUB and Siemens and have been fabricated using a SiGe laboratory technology by Siemens[22] (cf. Sec. 3). These data rates are record values for Si-based technologies and some of them even for all kinds of semiconductor technologies. The ICs in the 4th column were derived from the circuits realized in production technologies and have, therefore, specifications which are similar to those given in the 3rd column (for details see Sec. 5). The ICs in the 5th column have recently been developed in order to complete the chipset for a 40 Gb/s experimental TDM link[1,2] (cf. the German R&D program Photonik II). However, to achieve 40 Gb/s with these most speed-critical circuits, the other specifications must be relaxed, as discussed in Secs. 6 and 7.

It should be noted that all the results given in Table 1 were measured on *mounted* chips (rather than on wafer only). These chip modules can, therefore, easily be used in practical links.[b] Moreover, it is worthwhile to note, that for the development of all the circuits the same design principles were applied, which already have been described in detail in Ref. 23. This is an important precondition for a fair comparison of the different circuits, required for the conclusions given below.

Some of the ICs in Table 1 and their measuring results are shortly presented in Secs. 5 and 7. Further details about these circuits and more information on the other ones can be taken from the references in Table 1 (given in brackets).

From the results of Table 1 we learn that there is enough speed margin to fabricate *all* ICs (even the speed-critical ones) in the signal path of 10 Gb/s systems in available Si-bipolar production technologies, as long as the circuits are carefully designed. However, the situation looks quite different for 20 Gb/s systems. Here, advanced laboratory technologies are required to just achieve this data rate with the speed-critical circuits (especially amplifiers and modulator driver), as long as the other specifications are not substantially relaxed. The impact of this experience on the potential application of Si-based ICs in 40 Gb/s systems will be discussed in Secs. 6 and 7.

---

[b]Successful application in 10, 20, and 40 Gb/s links has already been demonstrated, e.g., in European and German R & D programs (RACE and Photonik II,[1,2] respectively).

3. Si-Based Bipolar Technologies for High-Speed ICs

Today, the fastest bipolar technologies are based on self-aligned double-polysilicon processes. The main advantage of double-polysilicon compared to single-polysilicon technologies is the lower base resistance. A typical example of an advanced production technology is the B6HF of Siemens.[21] It is characterized by an f_T of about 27 GHz, low base resistance, 0.8 μm lithography (resulting in an effective emitter stripe width of $b_E = 0.4$ μm), three metallization layers, and a rather conservative isolation technique. Parameters for a typical transistor with an emitter length of 10 μm are given in Table 2. Most of the circuits presented in the second column of Table 1 are realized in this production technology.

Table 2. Transistor data of the laboratory SiGe-bipolar technology applied and their comparison with those of an advanced production technology (CBEB configuration with 10 μm effective emitter length). The junction capacitances are zero-bias values and j_{CK} is given for $V_{CE} = 1$ V.

Technol. (Siemens)	b_E [μm]	f_T [GHz]	τ_F [ps]	r_B [Ω]	C_{EB} [fF]	C_{CB} [fF]	C_{CS} [fF]	j_{CK} [mA/μm^2]
Prod. Techn. (B6HF)	0.4	27	4.5	51	37	18	51	≈ 0.75
Labor. Techn. (SiGe)	0.3	72	1.7	38	28	19	20	≈ 2

For the high-speed electronics in 20 or even 40 Gb/s systems, production technologies are no longer sufficient. Instead, advanced laboratory technologies are required, at least for the speed-critical circuits. Several promising approaches for the worldwide fastest Si-based bipolar IC technologies have been reported: Pure silicon technology with very steep and shallow emitter and base profiles,[24] SiGe HBT technology ("true" HBT) with its high speed potential,[25] and SiGe drift-transistor technology.[22,26,27] (For a rough comparison of the different SiGe technologies see, e.g., Ref. 28.)

For the choice of the best technology, the designer should not consider the transistors' cut-off frequencies f_T and f_{max} only, but also the specific basic transistor parameters (e.g., r_B and C_{CB} related to the emitter length), the current carrying capability of the transistors (high admissible collector current density j_{CK}), and the number of available metallization layers.[23,28] Usually, a high j_{CK} should be aimed at with respect to a good high-speed driving capability of a transistor stage (often dominated by $j_{CK} A_E / C_{CB}$, A_E = emitter area). However, the corresponding technological measures reduce the breakdown voltage, which, therefore, must be carefully considered in circuit design.[23]

For the circuits in the last two columns of Table 1, a SiGe laboratory technology of Siemens was used.[22] It is a self-aligned double-polysilicon technology with 0.6 μm lithography (resulting in an effective emitter stripe width of $b_E = 0.3$ μm) and three metallization layers. A gradient of the Ge concentration in the epitaxial base causes an accelerating drift field, which reduces the transit time τ_f and thus increases f_T ($= 72$ GHz) and $f_{max}(= 74$ GHz). The speed potential is further increased by the high admissible collector current density j_{CK} (≈ 2 mA/μm^2).[c] The parameters for a transistor with a 10 μm long emitter are given in Table 2 and compared there with the data of a B6HF transistor, which has the same emitter length but a lower maximum collector current (3 mA instead of 6 mA, mainly due to the different values for j_{CK}).

4. High-Speed Circuit Principles

In this section, some elementary circuits frequently used in ICs for optical-fiber applications are shortly recapitulated, in order to simplify the description of the circuits in Secs. 5 and 7.

4.1. *Digital circuits*

Figure 2 shows the circuit diagram of a 2:1 MUX, as an example for a simple but frequently used digital circuit. It consists mainly of the MUX core and two data input buffers.[d]

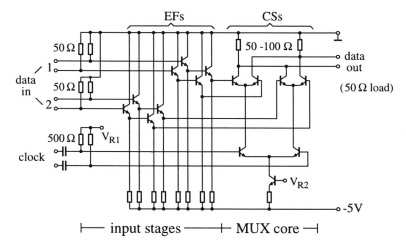

Fig. 2. Circuit diagram of a high-speed 2:1 MUX. The bias voltages V_{R1}, V_{R2} are generated on the chip. The output voltage swing can be varied via an external potentiometer. The on-chip output resistances are optional 50, 75, and 100 Ω.

From this diagram most of the circuit principles recommended for high-speed ICs can be taken, which shall be summarized briefly:

- E^2CL *instead of ECL*. This means that emitter followers (EF) are used at the data inputs and limiting differential stages, called current switches (CS), at the out-

---

[c]This value is not limited by reliability problems — in contrast to III-V HBT technologies, where for this reason the maximum collector current density is usually chosen lower than 1 mA/μm^2.
[d]Sometimes, also a buffer for the *clock* inputs is used to improve matching of the driving clock lines. For this, we often apply the circuit diagram proposed in Ref. 6. But also in this case the external blocking capacitors at the clock input are mostly required.

put. Among other advantages, this circuit concept shows substantially improved behavior in a transmission line environment compared to ECL (e.g., better line matching and steeper pulse edges).

- *Series gating.* Besides high-speed logic combination (e.g., of data and clock signals) this concept enables differential operation.
- *Differential operation.* This principle results in many important advantages, e.g., summarized in Ref. 23. One of them is a low voltage swing.
- *Low voltage swing.* This measure increases speed (especially of the CS) and reduces power consumption. Optimum values for *internal differential swings* of the data pulses are usually about $2 \times 200 \text{ mV}_{p\text{-}p} = 400 \text{ mV}_{p\text{-}p}$.
- *Cascaded emitter followers.* At a high operating speed, the decoupling capability (impedance transformation) of EFs is rather limited due to the reduced effective current gain of the transistors. Thus, two or even three cascaded EF pairs are often required.[e] Moreover, by this measure, V_{CE} of the CS transistors and thus the maximum collector current density j_{CK} are increased.[23] As a consequence, the size of the CS transistors can be reduced, further reducing C_{CB} (together with C_{EB} and C_{CS}) and thus usually increasing operating speed.
- *On-chip matching* of external transmission lines, especially at the data inputs. Compared to off-chip matching, the return loss is substantially improved and the potential instability of EF inputs can be reliably suppressed.[23] Usually, line matching at the IC inputs is easier compared to the outputs, due to the smaller transistor size and thus lower input capacitance. Nevertheless, on-chip output resistors are recommended to further reduce double-reflections and thus time jitter.[29] However, they are often chosen higher than the characteristic line impedance, in order to limit the required current through the output CS (at the given voltage swing).

4.2. Broadband amplifiers

Amplifiers in broadband communications must be operated down to very low frequencies (e.g., down to about 30 kHz in optical-fiber links). Thus, in monolithic integrated amplifiers d.c. coupling is necessary. Moreover, it is highly recommended to apply the principle of strong impedance mismatching between succeeding stages. As a typical example, a chain of alternating transadmittance stage (TAS), transimpedance stage (TIS), and emitter followers (EFs) is shown in Fig. 3.[f] Due to this mismatching, the transfer functions of the different stages (e.g., transadmittance and transimpedance) remain approximately constant up to comparatively high frequencies, resulting in a high cut-off frequency. This is because of the substantially reduced influence of the strongly frequency-dependent input and output

---

[e]However, it should be mentioned that the use of cascaded EFs requires careful circuit design.[23]
[f]Sometimes, the cut-off frequency of the gain magnitude is increased by using an EF in the feedback path of the TIS. This measure has not only advantages but also disadvantages. For example, the group delay can be deteriorated, even resulting in a reduction of the maximum usable operating speed.

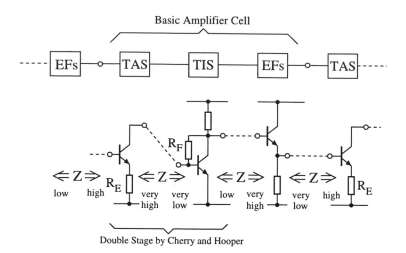

Fig. 3. Principle of strong mismatching in broadband amplifiers (Z is the input/output impedance of a single transistor stage).

impedances of the different stages. As another advantage of mismatching, all nodes in the circuit show a very low impedance, which reduces the influence of parasitic capacitances and thus further increases the bandwidth. At high frequencies, where mismatching is degraded, conjugate complex input and output impedances are observed at the interfaces between the different stages. This fact can be used to further increase the bandwidth; however, very careful circuit design is mandatory.[30]

In Fig. 3, the original double-stage proposed by Cherry and Hooper,[31] consisting of a TAS and a TIS, is extended by (one to three) EFs between the output of the TIS and the input of the TAS. These EFs are required for improving the insufficient high-frequency mismatching at this interface, for level shifting, and for gain peaking near the upper frequency limit.

In the receivers of optical-fiber links (without an optical amplifier in front of the photodiode) a low-noise preamplifier behind the photodiode is required, succeeded by a high-gain main amplifier. Both circuits must be designed for a wide dynamic range. They are often realized on separate chips connected by 50-Ω transmission lines.[32] In the preamplifier, the first stage in the chain of mismatched stages is generally a TIS in order to guarantee broadband mismatching between the driving (high-impedance) photodiode and the amplifier input. This stage is followed by EFs and an output buffer (TAS) with external 50-Ω loading.[9,30] As in the case of digital circuits, using EFs at the output is not recommended.

The main amplifier must have a high gain at 50-Ω input impedance. Therefore, often on-chip matching and a single EF pair are used at the input, followed by several basic amplifier cells (cf. Fig. 3) and a TAS output buffer.[10,11] For this amplifier, which must have a constant output amplitude (i.e., independent of the input signal amplitude), two solutions are possible: automatic-gain-control (AGC) or limiting amplifier. If any is possible, the differential operation should be preferred.

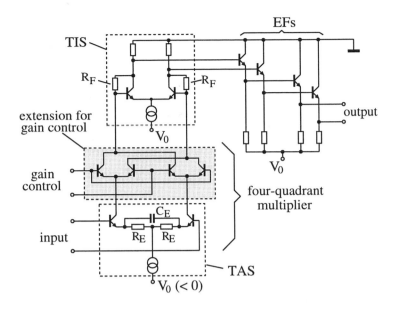

Fig. 4. Differential broadband amplifier cell. The shaded part is required for AGC amplifiers only.

A differential version of the basic amplifier cell in Fig. 3 is shown in Fig. 4. Without the shaded circuit block it has a constant gain at small input amplitudes, but it operates also properly in the limiting range and is, therefore, well-suited for limiting amplifiers, too (but often with $R_E = 0$).[11] For AGC amplifiers, the gain of the amplifier cell must be variable. To our experience, the best results for such amplifiers with respect to flat gain response and high cut-off frequency *over a wide dynamic range* are obtained if the TAS is extended to a four-quadrant multiplier by the shaded circuit block in Fig. 4.[g] In this example, the gain is controlled by a d.c. voltage at the upper input.

For compensating the gain drop at high frequencies, on-chip peaking capacitors C_E are often shunted to the emitter feedback resistors R_E of the TAS, as shown in Fig. 4. If realized as pn-junction capacitances, C_E can be varied by adjusting its bias voltage via an external potentiometer. In this way, the frequency response of the amplifier's transfer function can be optimized even after chip mounting.[h]

Finally, it should be noted that a high cut-off frequency of the gain magnitude does not guarantee a proper high-speed operation of the amplifier.[30] In publications, a proof of the additionally required high cut-off frequency of the group delay is often missing. Reliable operation in digital transmission systems can be best demonstrated by eye diagrams, which are, therefore, preferred in this paper.

---

[g]This especially holds if such an amplifier cell is combined with another kind of gain-controllable cell, but of similar circuit configuration.[10]
[h]For other peaking measures see, e.g., Ref. 30.

4.3. *Utilization of bond inductances for improving circuit performance*

An obvious disadvantage of present Si technologies, compared to compound semi-conductor technologies, is that cost-effective on-chip inductors with sufficiently high Q factor are usually not yet available at very high frequencies. Thus we have to think about improvement of circuit performance by applying bond wires as inductors. Some examples which clearly demonstrate the necessity of this measure in certain applications shall be shortly summarized (for details see Ref. 23).

Bond inductors are favorably used at the input of low-noise transimpedance preamplifiers, mainly to increase the signal-to-noise ratio, but also to improve frequency response of the amplifier's transfer function.[9,30] Moreover, they are an efficient measure to substantially reduce rise and fall times of the output pulse edges of driver circuits, thus increasing the data rate.[8] Also transmission-line matching can further be improved by use of a bond inductor.[23] Good matching up to high frequencies, especially at the data inputs of an IC, is a must in high-speed links, since otherwise double-reflections can drastically increase time jitter.[29] Simulations on the base of a laboratory SiGe-bipolar technology showed that there is a good chance of achieving a sufficiently low input return loss even up to at least 30 GHz by applying a matching network at the (50-Ω) IC input. It consists of the bond inductance and a small additional capacitance at the end of the driving transmission line.[23]

5. Examples for Practical Circuits with Typical Specifications

The circuits discussed in this section have already been listed in columns 2 to 4 of Table 1, together with some specifications and references. For a fair comparison, we distinguish again between circuits fabricated in production technologies and more recent circuits fabricated in the SiGe laboratory technology discussed in Sec. 3. Most of the ICs in production technologies have been designed for long-haul 10 Gb/s TDM systems and, thus, their specifications are "typical" for this application. Apart from operating speed, the SiGe ICs in the 4th column have similar specifications, because they were derived from the ICs fabricated in production technologies. This is in contrast to the speed-critical ICs for a 40 Gb/s system, listed in the 5th column of Table 1 and discussed later on in Sec. 7.

5.1. *Circuits in production technologies*

Here only a few examples are selected which stand for the speed-critical circuits in 10 Gb/s systems.

5.1.1. *Laser/modulator driver*[8]

Figure 5 shows the (simplified) circuit diagram of a laser/modulator driver developed for the transmitter of a 10 Gb/s optical-fiber link. It was fabricated with the Siemens production technology B6HF (cf. Sec. 3). Its single-ended output is able

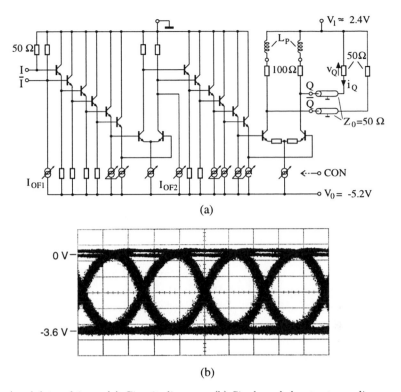

(a)

(b)

Fig. 5. Laser/modulator driver. (a) Circuit diagram. (b) Single-ended output eye diagram (v_Q) at 14 Gb/s (29 ps/div.).

to drive both a laser diode directly or an external modulator. While in the first case the output current swing ΔI_Q has to be varied within a wide range (here: 15 mA$_{p\text{-}p}$ $\leq \Delta I_Q \leq$ 60 mA$_{p\text{-}p}$), a comparatively high but constant output voltage swing ΔV_Q (here: maximum 3.6 V$_{p\text{-}p}$) is required in the second case. The circuit consists of two current switches, which (for the reasons mentioned in Sec. 4.1) are each driven by three emitter-follower pairs. For reducing double-reflections, again on-chip input and output resistors of 50 and 100 Ω, respectively, are used.

The basic problems of such a driver circuit and their solutions have already been discussed in Ref. 8 and shall, therefore, only briefly be listed here:

- *Asymmetrical pulse shape due to the single-ended output.* As a consequence, the eye diagrams can be substantially degraded. However, by use of two adjustable offset voltages (generated by the currents I_{OF1} and I_{OF2} in Fig. 5(a)), a quite symmetrical pulse shape can be obtained.
- *Large output time constant due to the high voltage swing* (approximately: $\tau_Q \propto \Delta V_Q$). Here, the only practicable way to reliably achieve the required quality of the output eye diagrams at 10 Gb/s and above was to steepen the pulse edges by use of bond inductors L_P ($\approx$ 2 nH) in series to the on-chip output resistors, as mentioned in Sec. 4.3.

- *Wide output current range for direct laser modulation.* For this, in addition to the current source of the output stage, the currents through several other stages have to be properly adjusted via a single (common) control input CON, in order to get clear eye diagrams.
- *Exceeding the breakdown voltage BV_{CEO} (≈ 3.7 V) of the output transistors.* This is a consequence of the contradicting demands on high output voltage swing and high collector current density in the on-state ($\approx j_{CK}$, cf. Sec. 3). The resulting maximum V_{CE} of 5.5 V can, however, be tolerated in the present case, due to the low output impedance of the driving source (emitter followers) and, moreover, due to balasting resistors connected in series to each emitter finger of the multi-emitter output transistors (which are also useful to avoid thermal problems).

The measured (single-ended) output eye diagram of the driver circuit at its maximum data rate of 14 Gb/s is shown in Fig. 5(b). In this case, the current switched by the output transistors is as high as 108 mA, resulting in an output voltage swing of $\Delta V_Q = 3.6$ $V_{p\text{-}p}$ (across an external load of $R_L = 50$ Ω) and a total on-chip power consumption of 2.5 W. For these measurements, as well as for those presented in the next section, a pseudo-random bit-pattern generator (PRBG) was applied (cf. also Sec. 5.2.1).

5.1.2. *High-gain main amplifiers*[10,11]

Based on the amplifier cell discussed in Sec. 4 (Fig. 4), a main amplifier with automatic gain control (AGC) was developed for a 10 Gb/s optical-fiber link and fabricated in an advanced version of Motorola's production technology MOSAIC V. The IC consumes 850 mW at a single supply voltage of -6.5 V. The block diagram of the d.c. coupled and fully differential circuit is given in Fig. 6(a). (For detailed circuit diagrams of most of these circuit blocks see Ref. 10.) The high-frequency part consists of a high-ohmic input buffer IB, two gain-controllable amplifier cells A1 (cf. Fig. 4) and A2 (modified compared to A1), a constant-gain cell A3, and two separated 50-Ω output buffers OB for driving the succeeding decision circuit and the clock extraction circuit separately.

The driving transmission line is well matched by a 50-Ω input termination circuit IT which is combined with an offset-control network OC. For gain control, the output signal voltage of A3 is detected by a peak-detector D and compared with the nominal voltage (REF). The voltage difference V_{GC} is amplified by a high-gain d.c. amplifier GC which generates the two different gain-control voltages for A1 and A2.

Aspects for the design of this amplifier have already been discussed in Ref. 10. Besides the high data rate (≥ 10 Gb/s) and the high gain ($S_{21} = 37$ dB), also the demands on high input dynamic range (41 dB) and sufficiently low equivalent noise voltage density at the input (2.5 nV/$\sqrt{\text{Hz}}$) had to be met. Figure 6(b) shows the measured frequency response of the gain magnitude ($|S_{21}|$) with the low-frequency gain as a parameter. For this measurement, the automatic gain control had to be

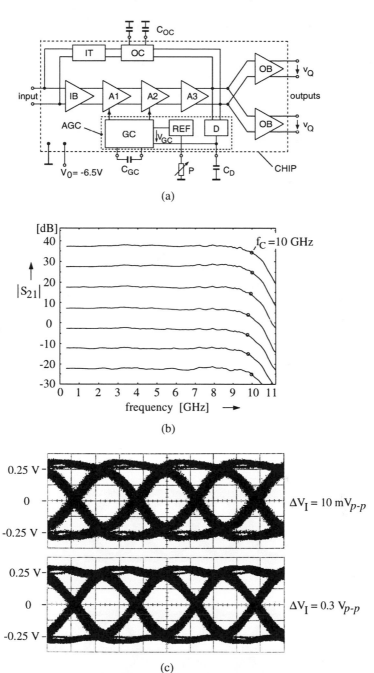

(a)

(b)

(c)

Fig. 6. Broadband AGC amplifier with two outputs. (a) Block diagram. (b) Gain vs. frequency characteristics for a wide range of gain. (c) Output eye diagrams (v_Q) at 13 Gb/s for two quite different input voltage swings: $\Delta V_I = 10$ mV$_{p\text{-}p}$ and 300 mV$_{p\text{-}p}$. Time scale: 30 ps/div. The (differential) output voltage swing in both cases is $\Delta V_Q = 500$ mV$_{p\text{-}p}$.

switched off and, instead, the voltage V_{GC} (and thus the gain) was adjusted via an external voltage source. Despite the extremely wide gain range (60 dB!), the characteristics remain flat and the 3-dB cut-off frequency f_C is nearly constant ($\approx$ 10 GHz). Moreover, clear output eye diagrams were measured even above 10 Gb/s at a constant output voltage swing of 500 mV$_{p\text{-}p}$ within the total input dynamic range. Figure 6(c) shows the (differential) output eye diagrams at the maximum data rate of 13 Gb/s for two quite different input voltage swings ΔV_I. Note, that there is nearly no difference between both diagrams.

The design of an AGC amplifier for such a high gain and wide dynamic range, operating at the speed limit of a given technology, is a very demanding task. As a consequence, if linear operation is not required, limiting amplifiers should be preferred for the following reasons: Easier design, higher operating speed, lower power consumption, lower supply voltage, and fewer external components. In addition, its input sensitivity is slightly higher due to the lower equivalent input noise. These advantages were demonstrated by the development of a differential limiting amplifier which was fabricated in the B6HF technology.[11] The block diagram is similar to that in Fig. 6a but without needing the automatic-gain-control part (AGC). Again, the amplifier cell in Fig. 4 is used, without the shaded block for gain variation.

Figure 7 shows the measured (differential) output eye diagrams at the maximum data rate of 15 Gb/s and with the same output voltage swing ($\Delta V_Q = 0.8$ V) for two quite different input voltage swings ΔV_I. Again, both diagrams nearly agree. The power consumption is 720 mW at a single supply voltage of -5 V. The input dynamic range is 46 dB and the small-signal gain $S_{21} = 52$ dB. For further specifications see Ref. 11.

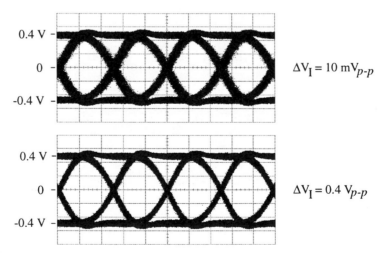

Fig. 7. Output eye diagrams of a limiting main amplifier at 15 Gb/s for two quite different input voltage swings: $\Delta V_I = 10$ mV$_{p\text{-}p}$ and 400 mV$_{p\text{-}p}$. Time scale: 25 ps/div. The (differential) output voltage swing in both cases is $\Delta V_Q = 800$ mV$_{p\text{-}p}$.

5.2. *Circuits in SiGe laboratory technology*

In this section, examples from the 4th column of Table 1 are presented: 60 Gb/s
MUX and DEMUX as well as the speed-critical circuits, 20 Gb/s preamplifier and
23 Gb/s modulator driver. As mentioned before, all these circuits were derived
from ICs which have already been fabricated in production technologies. Therefore, not only the specifications but also the circuit diagrams and the basic layout
configurations are very similar to these former ICs. Of course, the circuit components, especially the transistor geometries, were carefully adjusted to the new
technology and even more care was taken to optimize on-chip wiring in order to
avoid instabilities and ringing.[23]

 All these circuits have been successfully used in an experimental 20 Gb/s optical-
fiber link.[1] The experimental results presented are record values for silicon and some
of them even for any semiconductor IC technology.

5.2.1. *Mounting and measurement techniques*[12,23]

Besides the preamplifier, for which a ceramic substrate was used, all the ICs discussed in this section were mounted in a simple and inexpensive way, which shall be
presented now. It proved to be suited up to a data rate of 60 Gb/s. Figure 8 shows
schematically the chip and the surrounding part of the measurement socket, which
consists of a 254 μm thick teflon substrate (PTFE) with low permittivity ($\varepsilon_r = 2.2$)
soldered on a brass block.[23] The chip is glued into a recess in this socket and then
conventionally ultrasonic bonded. By this measure, the chip surface is at about the
same level as the microstrip line, so that the bond-wire lengths and thus the bond
inductances are reduced (about 0.3 nH for a single wire). Due to the differential
operation, the effective (odd-mode) bond inductance is further reduced by keeping
the bond wires in parallel and close together.

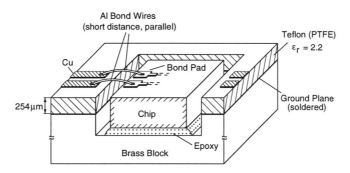

Fig. 8. Scheme of the simple mounting technique used for most of the high-speed ICs. The bond
wires are shown for a differential signal line.

 A photograph of the complete measurement socket for the 60 Gb/s MUX is
shown in Fig. 9.[12] At the data inputs (30 Gb/s) and the clock input (30 GHz)
K-connectors are used. At the output, the interface between coaxial cable and

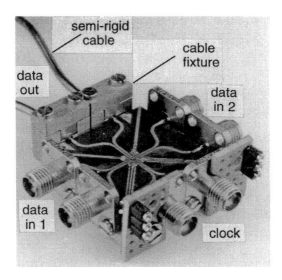

Fig. 9. Measurement socket of the 60 Gb/s MUX. Size of substrate is about 3 cm × 3 cm.

microstrip line is even more critical due to the extremely high data rate of 60 Gb/s. The required low reflection coefficient demands low discontinuities in the electro-magnetic field at this interface. For this, a semi-rigid cable, whose diameter fits the substrate thickness, was directly connected to the measurement socket by means of a simple fixture (see Fig. 9): The semi-rigid cable is clamped between an upper and lower brass block, with a small overlap of the outer conductor in order to guarantee a well defined ground contact between cable and microstrip ground plane. The centre conductor of the cable is soldered to the microstrip line for reliability reasons. Similarly good results were obtained with K-connectors if the assembly is slightly modified compared to the instructions of the manufacturer.

Due to the high operating speed of the developed ICs, commercial pseudo-random bit-pattern generators (PRBG) are not available. For example, the 60 Gb/s MUX must be driven by two pulse sequences of 30 Gb/s data rate and the DEMUX by a single 60 Gb/s sequence. Therefore, we had to build a measurement setup by ourselves using our own circuits.

The block diagram of the measurement setup for the MUX is shown in Fig. 10. The two PRBGs, each with two independent outputs, generate four uncorrelated pseudo-random bit sequences (PRBS) at 7.5 Gb/s with a word length of $2^{15} - 1$ bits,[33] which are multiplexed in a first MUX stage. Then, the data stream is split up by 6-dB power splitters and delayed in one path by $T_0 = m \times bitwidth$. The time delay T_0 is introduced in order to get an only weakly correlated second bit sequence in a simple way. Several values of m between 5 and 100 are proven to be without any significant influence on the output eye diagram of the MUX under test. It is, therefore, expected, that the two test sequences well approximate the results obtained by uncorrelated PRBSs.

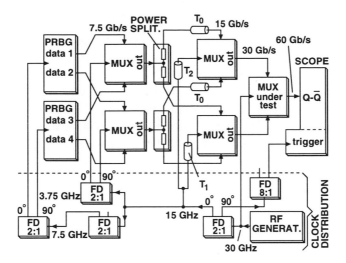

Fig. 10. Block diagram of the measurement setup for the MUX under test, shown for an output data rate of 60 Gb/s. FD: frequency divider, PRBG: pseudo-random bit-pattern generator.

The PRBGs and the two MUXs in the first stage are each clocked by a separate static frequency divider (FD) with two outputs, shifted in phase against each other by 90°. As a consequence, the transients of the two data input signals of the succeeding MUXs in the first and second stage are shifted by half a bitwidth, which is a precondition for optimal multiplexing. In the first MUX stage, the clock phases are adjusted to be ±45° out of the centre of the MUX's clock-phase-margin (CPM) range.[34,i] This measure, which simplifies the measurement setup, is possible because at this data rate the two MUXs still have nearly the ideal CPM of 180°. Similar clock timing is applied to the MUXs in the second stage. Here, the (ideal) 90° difference in their clock phases is adjusted by the delay lines T_1 and T_2. Fine adjustment can be obtained by varying T_1 or T_2, in order to optimize the output eye diagram of the MUX under test.

To minimize the influence of trigger jitter (caused by the measurement setup) on the time jitter of the MUX under test, a stable RF generator frequency with low phase noise and a trigger signal with steep pulse edges are mandatory. If the line lengths in the data paths correspond to the line lengths in the clock distribution and if the delay lines T_1 and T_2 are replaced by an additional frequency divider, the data rate of this setup can easily be changed by varying the RF generator frequency and the delay time T_0 (variable delay line).

The measurement setup of the MUX in Fig. 10 is also used for the 60 Gb/s DEMUX. For this, the data input of the DEMUX is driven by the "MUX under test".

---

[i]Note that the phase difference of the two MUX outputs is fixed at 90°, given by the frequency divider.

5.2.2. *60 Gb/s MUX*[12]

The circuit diagram of the MUX has already been discussed in Sec. 4, Fig. 2. As recommended by us in former papers, a separate output buffer is not applied, in order to obtain maximum operating speed. The output stage was designed to switch 10 mA, resulting in a differential output voltage swing of 0.5 V_{p-p} at 50-Ω on-chip matching and 50 Ω external load. Design aspects are given in Ref. 35, and, more generalized, in Ref. 23. The power consumption is 300 mW at a single supply voltage of -5 V. For a chip photograph of the MUX see Ref. 35.

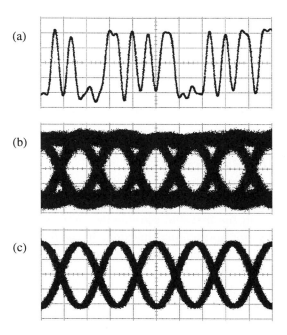

Fig. 11. Measurement results of the MUX at 60 Gb/s output data rate. (a) Section of output pulse sequence (50 ps/div., 125 mV/div.). (b) Output eye diagram (10 ps/div., 125 mV/div.). (c) Clock signal and its inversion (10 ps/div., 400 mV/div.).

Figure 11 shows the measurement results of the mounted MUX chip at an output data rate of 60 Gb/s (clock frequency 30 GHz): A section of the pulse sequence at the (differential) output is given in Fig. 11(a) and the output eye diagram in Fig. 11(b). The rise and fall times (20 to 80%) of the pulse edges are about 7 ps, including the degradation caused by the measurement setup and the sampling scope (HP 54124). Thus, the real pulse edges of the circuit are even steeper. The "eyes" in Fig. 11(b) are sufficiently open although the time jitter is considerably degraded by the measurement equipment, especially by the scope. This is demonstrated by the oscillogram in Fig. 11(c) showing the clock signal of the MUX and its inversion. Here, the scope is triggered in the same way as in the measurement of the output eye diagram of the MUX and, therefore, the time jitter in Fig. 11(c) stands for the trigger jitter. From this result we suppose that the inherent time jitter of the MUX is substantially smaller than that taken from Fig. 11(b).

To the best of the author's knowledge, the 60 Gb/s achieved is the highest data rate ever generated by a mounted IC in any semiconductor technology. It should be mentioned that even at this data rate the MUX can be used, to a certain extent, for the retiming of data streams, since its CPM is still $130°$.

5.2.3. *60 Gb/s DEMUX*[13]

DEMUXs are key components in communication systems and measurement equipments. In addition to the intrinsic demultiplexer function, the excellent retiming capability combined with a high input sensitivity predestinates a DEMUX as a decision circuit in the receiver of very-high-speed optical-fiber transmission systems. This is of special interest if the operating speed of a single master-slave D-type flipflop (MS-D-FF) is no longer sufficient for the intended application (cf. Sec. 6).

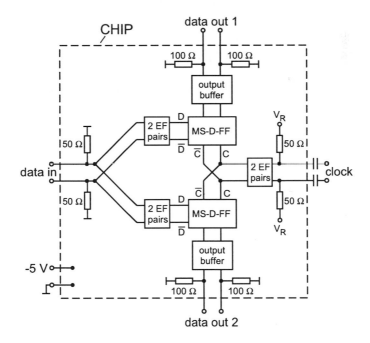

Fig. 12. Block diagram of a high-speed 1:2 DEMUX. V_R (< 0) is an on-chip generated bias voltage.

Figure 12 shows the block diagram of a 1:2 DEMUX which consists of several basic circuits:[6] The DEMUX core with two MS-D-FFs operating in parallel, the data and clock input stages, and two output buffers. These circuit blocks are, due to the modular concept applied, partly taken from other circuits and are, therefore, not all optimized with respect to the DEMUX performance.[6,36] This especially holds for the output buffers, which were designed for a high-speed decision circuit consisting of a *single* MS-D-FF and are, therefore, oversized for a DEMUX with its halved data rate at the outputs. Especially their power consumption is unnecessarily high (about 50% of the total DEMUX power consumption). Circuit diagrams of all the building blocks have already been presented and discussed in Refs. 6 and 36 and shall, therefore, not be recapitulated here. Even the chip layout is nearly unchanged compared to the former versions.

The DEMUX operates at a single supply voltage of -5 V and consumes 1.2 W in total (including the 0.6 W for both output buffers). On-chip output resistances of 50 Ω or 100 Ω can be used optionally. For 100 Ω, the differential output voltage swing at 50 Ω external load is 640 mV$_{p\text{-}p}$.

The chip was mounted as described in Sec. 5.2.1 and driven by the bit-pattern generator of Fig. 10 with the 60 Gb/s MUX as the last stage. As in the case of the MUX, the clock frequency is 30 GHz. An experimental result is shown in Fig. 13(a). The eye diagram of the driving bit pattern (top) is intentionally degraded in order to demonstrate the excellent retiming capability of the DEMUX even at 60 Gb/s. The eye diagram of the output signal (bottom), measured for an on-chip output resistance of 100 Ω, is well opened and shows only small time jitter. The CPM for this measurement is estimated to be about 180° at a swing of the *differential* sinusoidal clock signal of 2×0.75 V$_{p\text{-}p} = 1.5$ V$_{p\text{-}p}$.

The high operating speed achieved (60 Gb/s input, 30 Gb/s output) is a record for DEMUXs in any IC technology. By a further measurement it was shown that the maximum operating speed of the driving MUX and the driven DEMUX under test can be increased up to 70 Gb/s. For this, both chips were mounted close together on a common measurement socket with the MUX output directly bonded to the DEMUX input.

Due to the intended application of the IC as a DEMUX with additional decision function in the receiver of a 40 Gb/s system (cf. Sec. 6), the CPM versus differential input data swing was measured at this data rate. The results in Fig. 13b demonstrate the high sensitivity and excellent retiming capability of this circuit. It should be pointed out that, to the best of the author's knowledge, single MS-D-FFs with sufficiently high CPM at 40 Gb/s (i.e., at 40 GHz clock frequency) have not yet been reported. Therefore, if in the receiver of a 40 Gb/s TDM system a decision circuit without DEMUX function is desired, the DEMUX must be succeeded by the MUX described before. The function of the MUX, which also has a high retiming capability at this data rate, is here to multiplex the two 20 Gb/s output signals of the DEMUX to the original 40 Gb/s data stream.[37,38]

5.2.4. *20 Gb/s low-noise high-gain preamplifier*[17]

The simplified circuit diagram of the amplifier shown in Fig. 14(a) is similar to the 13 Gb/s and 10 Gb/s preamplifiers presented in Refs. 9 and 30, respectively. It consists of a transimpedance input stage, three emitter followers, and a trans-

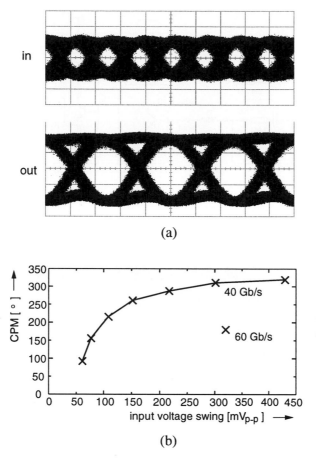

(a)

(b)

Fig. 13. Measurement results of the regenerating DEMUX. (a) Eye diagrams at 60 Gb/s input data rate (160 mV/div., 13 ps/div.), top: intentionally deteriorated input signal (60 Gb/s), bottom: regenerated output signal (30 Gb/s), CPM ≈ 180°. (b) Signal regeneration capability at 40 Gb/s input data rate: clock phase margin vs. input data swing. For comparison, the measurement result from (a) at 60 Gb/s is inserted.

admittance output stage which drives an external 50-Ω load (cf. Sec. 4.2). C_1 and C_2 are external decoupling capacitors and C_E is an on-chip peaking capacitor realized by an emitter and collector junction capacitance connected in parallel. The diodes in series to several collectors drop V_{CE} below the critical breakdown voltage. The power consumption of the IC is 145 mW at a single supply voltage of -6 V and the total transimpedance is 800 Ω (58 dBΩ) at 50 Ω external load.

The shunt feedback resistance R_F of the transimpedance stage was chosen as high as 900 Ω because at that time we aimed at increasing the amplifier's sensitivity as far as possible. A high value of R_F not only reduces the noise generated by this resistor but also increases the gain of the first stage and thus reduces the noise contribution of the succeeding stages related to the input. At high frequencies, the signal-to-noise ratio and the frequency response of the transfer function are further

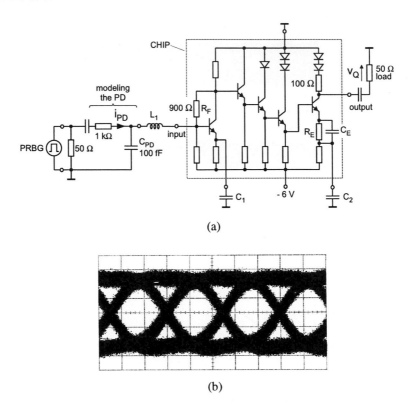

(a)

(b)

Fig. 14. 20 Gb/s transimpedance preamplifier. (a) Simplified circuit diagram and measurement conditions. (b) Single-ended output eye diagram (v_Q) at 20 Gb/s (20 ps/div.). The output voltage swing is 200 mV$_{p-p}$ at 0.25 mA$_{p-p}$ input current swing.

improved by optimization of the bond inductance between photodiode (PD) and input bond pad (here $L_1 \approx 1$ nH), as already mentioned in Sec. 4.3.[9,30] The equivalent input noise current density $\bar{j}_N$ was calculated by SPICE simulation to be about 12 pA/$\sqrt{\text{Hz}}$, averaged over the total bandwidth of the amplifier.[j] This noise is much lower than that reported for other high-gain preamplifiers (in any semiconductor technology) with comparable speed. To the best of the author's knowledge such a high R_F has never been used in high-speed transimpedance amplifiers, since it reduces the bandwidth by increasing the dominating time constant $R_F C_{CB}$ (C_{CB} is the collector-base capacitance of the first transistor).[30] In order to still achieve 20 Gb/s, a third emitter follower has been added compared to our former designs, which shifts the drop of the amplifier's gain to higher frequencies.

For measurement, the IC was mounted on a measurement socket with ceramic substrate and driven by a 20 Gb/s pseudo-random bit-pattern generator (PRBG). The length of the pulse sequence is $2^{15} - 1$ bits. For signal transmission, the ratio

[j]To our experience with other broadband amplifiers, the measured noise proved to be even slightly lower than the SPICE-simulated one (probably due to the lacking correlation between the noise sources in the SPICE transistor model).

of the output voltage v_Q of the amplifier and the *intrinsic* PD current i_{PD} is of main interest. Therefore, in this pure electrical measurement the PD is modeled by a simple equivalent circuit, realized on the substrate of the measurement socket, which is inserted between the PRBG and the amplifier chip (see Fig. 14(a)).[k]

Figure 14(b) shows the measured clear eye diagram of the output signal across 50 Ω external load. In this example, the output voltage swing is $\Delta V_Q = 200$ mV$_{p\text{-}p}$ and thus just below its maximum value (given by the signal amplitude limitation). This corresponds to the present application of the amplifier in a 20 Gb/s optical-fiber link with a comparatively high input current swing of 0.25 mA$_{p\text{-}p}$ (due to a preceeding optical amplifier),[1] but is in contrast to the initially intended application with very low input signal currents, for which this low-noise amplifier was designed.

5.2.5. *23 Gb/s modulator driver*[16,17]

The circuit diagram is very similar to that described in Sec. 5.1.1, Fig. 5, and also its power consumption is nearly the same (2.6 W). As the main differences, only two (instead of three) emitter-follower pairs are used at the input, and the currents through the emitter followers are no longer controllable (since the application is now restricted to modulator driving with constant voltage swing).

The circuit was mounted as described in Sec. 5.2.1 (but now using inexpensive SMA connectors) and driven by the PRBG mentioned in Sec. 5.2.4. Figure 15 shows very similar eye diagrams for the single-ended and differential output voltage at 23 Gb/s. The single-ended output swing across the external 50-Ω load is $\Delta V_Q = 3.5$ V$_{p\text{-}p}$ (differential 7 V$_{p\text{-}p}$),[1] and the corresponding current switched by the output stage is as high as 112 mA (at an on-chip output resistance of 84 Ω).[16]

6. Some System and Circuit Aspects of 40 Gb/s TDM Links

From the results in Sec. 5.2 and Table 1, 4th column, we may conclude that one of the worldwide best SiGe laboratory technologies is just good enough to achieve about 20 Gb/s with the most speed-critical circuits. Therefore, the question arises: "Is SiGe at all a potential candidate for the complete electronics in 40 Gb/s TDM systems?"[39]

Of course, we can expect that the IC technologies will be further improved. However, in the author's opinion the resulting improvement in circuit speed for this kind of broadband application will be by far below the desired factor of two. Therefore, additional measures are required to achieve our goal of using only Si-based ICs in 40 Gb/s systems.

---

[k]Recent measurements with real photodiodes, which show as clear 20 Gb/s eye diagrams as the pure electrical measurements, demonstrate that this simple PD modeling leads to reliable results (cf. also Ref. 30).

[1]These values are substantially higher than those reported in Ref. 17. This is because the initially occuring stability problems (ringing) at high output voltage swing, which proved to be mainly caused by ground-metallization parasitics, have now been removed by a redesign of the metallization masks, based on careful resimulations.

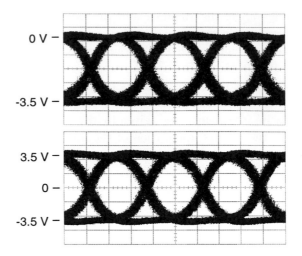

Fig. 15. Output eye diagrams of the modulator driver at 23 Gb/s (17 ps/div.). Top: single-ended signal, $\Delta V_Q = 3.5$ $V_{p\text{-}p}$. Bottom: differential signal, $\Delta V_Q = 7$ $V_{p\text{-}p}$.

Two of the most promising measures are:[39]

- Elimination of speed-critical circuits in the links as far as possible.
- Relaxed specifications for the remaining speed-critical circuits (in favor of speed).

These measures, which will be discussed now, are applied in a 40 Gb/s TDM system of a German R&D program, called Photonik II.[1,39] The corresponding optical-fiber link is shown in Fig. 16. There are several differences to the concept in Fig. 1, where all data rates and frequencies had to be increased by a factor of four if it should be applied to a 40 Gb/s system. As a main simplification compared to Fig. 1, the decision function is now performed by the first stage of the two-stage 1:4 DEMUX. As already demonstrated by Fig. 13(b), such a 1:2 DEMUX stands out for a high operating speed at excellent retiming capability, and, moreover, for a high input sensitivity. As a consequence, speed-critical single MS-D-FFs (clocked at 40 GHz) for both 40 Gb/s decision and 40 GHz frequency division are no longer needed. Moreover, clock extraction is now reduced to 20 GHz, substantially relaxing the demands on this circuitry. An adequate proposal has already been experimentally verified[40] and will shortly be recapitulated at the end of this section.

Due to the use of a gain-controlled optical amplifier (OA) in the receiver, the specifications of the electronic amplifier can be substantially relaxed:[m] Gain and dynamic range can be reduced considerably and the input sensitivity (mainly determined by the equivalent input noise current density) is no longer a severe restricting condition. As a further consequence, the total amplifier can be realized on a single chip, thus reducing costs and avoiding critical interfaces. However, despite

---

[m]For example, compared to the specifications of a 10 Gb/s system without an OA in the receiver (cf. Table 1, 3rd column).

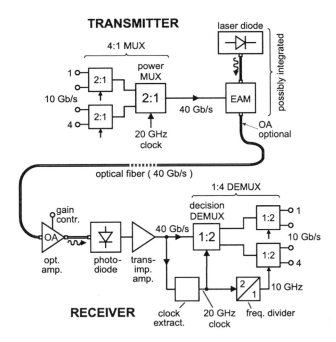

Fig. 16. Scheme of a 40 Gb/s optical-fiber TDM link. The loop for the automatic gain control of the optical amplifier (OA) is not shown.

of the relaxed specifications, the design of the 40 Gb/s transimpedance amplifier still remains a demanding task.

On the transmitting side, as another speed-critical circuit the 40 Gb/s modulator driver was eliminated in the scheme of Fig. 16. Instead, the driver function is now performed by a power version of the last 2:1 MUX ("power MUX"), which stands out for steeper pulse edges compared to conventional driver concepts. As a further advantage, nearly jitter free output signals are generated due to the inherent retiming capability of a MUX.[34,n]

Improvements of the opto-electronic components are another important precondition to relax the specifications of the speed-critical circuits. Today, photodiodes with a total capacitance as low as 50 fF and an intrinsic cut-off frequency of about 50 GHz are available. A severe problem is the high voltage swing required for driving the modulator, which limits the speed of the driver circuit.[8] Electroabsorption modulators (EAM) for 2 V swing and below are under development. If a symmetrical EAM configuration is used, as in the present case,[1] the output swing of the driver circuit can be halved (e.g., 1 V_{p-p} at each of the complementary outputs, resulting in a *differential* swing of 2 V_{p-p}). But even then it is not easy to achieve 40 Gb/s with standard circuit concepts. This is why the driver function is performed here by a power MUX.

---

[n]Therefore, a retiming MS-D-FF, sometimes used in front of the driver IC (but today not available with sufficient CPM at 40 Gb/s), is not required.

As already mentioned, the TDM system in Fig. 16 requires a clock frequency which is only half the bit frequency, i.e., a 20 GHz clock is extracted from the transmitted 40 Gb/s data stream. (This is in contrast to the system in Fig. 1 where clock and bit frequency have the same value.) Therefore, it is no problem to realize the IC components of a clock extraction circuit for a 40 Gb/s system in advanced Si bipolar technologies.

As an example, Fig. 17 shows the block diagram of a clock extraction unit which has already been successfully verified.[40] Here, an anticorrelation method is applied to generate a d.c. voltage, depending on the phase difference between clock and data signal, which drives a 20 GHz voltage-controlled oscillator (VCO). For this, the 40 Gb/s data stream drives, in addition to the DEMUX, a MS-D-FF, which is also clocked with 20 GHz but delayed by half the input bit width T_B (i.e., by 12.5 ps). Thus, if the clock frequency has the right value, the MS-D-FF is clocked at the data pulse edges. Its output signal is compared with the signals at both DEMUX outputs by use of two EXOR gates. The difference between their output signals, averaged by a loop filter, is the desired phase-detector voltage for driving the VCO. In this way, a phase-locked loop (PLL) is formed.

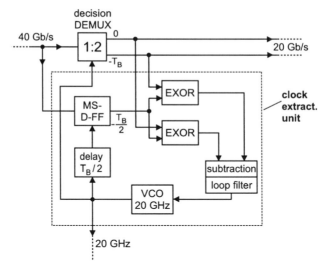

Fig. 17. Extraction of a 20 GHz clock signal from a 40 Gb/s data stream. In addition to the 40 Gb/s signal, the two 20 Gb/s output signals of the DEMUX are required as input data for the clock extraction unit (connections not drawn in Fig. 16). Compared to the upper output of the DEMUX, its lower output and the MS-D-FF output are delayed by T_B and $T_B/2$, respectively ($T_B = 25$ ps).

Presently, this clock extraction unit is realized as a multi-chip version, using ICs which have been fabricated in an advanced Si-bipolar laboratory technology.[36] This technology is slower than the SiGe technology applied for the circuits described in Secs. 5 and 7. However, the only circuits which have to process the maximum data rate of 40 Gb/s are the DEMUX (consisting of two MS-D-FFs, as shown in Fig. 12) and a single MS-D-FF. Since both circuits are clocked at 20 GHz only, they can easily be realized (even in a Si-bipolar technology) with an excellent input sensitivity and retiming capability, as demonstrated by the DEMUX results in Sec. 5.2.3 (cf. Fig. 13(b)).

In conclusion, only two speed-critical circuits remain in the 40 Gb/s system of Fig. 16: The transimpedance amplifier in the receiver and the power MUX (modulator driver) in the transmitter. Both circuits will be presented in the next section.

7. Design and Experimental Results of the Speed-Critical Circuits in a 40 Gb/s Link

The speed-critical circuits for the 40 Gb/s system were carefully designed, using the circuit concepts shortly presented in Secs. 4 and 5 and the design considerations discussed in Ref. 23. For the simulations, improved transistor and substrate (coupling) modeling[23,41] was applied, and the on-chip wiring parasitics as well as the mounting parasitics were considered very carefully. Moreover, accurate models for the photodiode (see Fig. 18) and the EAM,[o] respectively, were used. The circuits will now briefly be described, using simplified block and circuit diagrams.

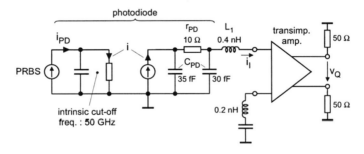

Fig. 18. Equivalent circuit of the photodiode used for the design of the transimpedance amplifier. Note that the transfer function to be optimized is v_Q/i_{PD}, where i_{PD} is proportional to the light intensity.

7.1. *Transimpedance Amplifier* [20]

The amplifier was designed for the data of the photodiode (PD) given in Fig. 18. The (differential) output voltage swing $\Delta V_Q = 0.6$ V$_{p-p}$ (at 50 Ω external load) is high enough to drive the succeeding circuits. It has to be constant for the (single-ended) input current swing ΔI_I generated by the PD, ranging from 0.3 to 0.6 mA$_{p-p}$. Thus the maximum large-signal transimpedance in the limiting mode is $Z_{TI} = \Delta V_Q/\Delta I_I = 2$ kΩ (while the small-signal transimpedance proved to be 10 kΩ). The equivalent input noise current density must be below 25 pA/$\sqrt{\text{Hz}}$ (simulated

---

[o]The EAM model used and its parameters are similar to those given in Ref. 1. However, in this reference there is a printing error in the equivalent circuit of Fig. 2: the n-contact of the EAM is not (as shown) grounded.

here: 21 pA/$\sqrt{\text{Hz}}$). The total power consumption is 800 mW at a single supply voltage of -6.5 V.

To reduce costs and avoid critical interfaces, the amplifier was not separated into a preamplifier and a main amplifier; instead, the total amplifier was realized on a single chip. Of course, due to the high gain, crosstalk on the chip must be simulated carefully during design, in order to guarantee stable operation and to avoid ringing.

Figure 19(a) shows the block diagram of the transimpedance amplifier with three basic amplifier cells discussed below. In contrast to the single-ended operating 20 Gb/s transimpedance preamplifier in Fig. 14(a), a fully differential configuration is chosen. Besides other advantages, this measure helps to reduce performance degradation caused by substrate coupling as well as by on-chip wiring and mounting parasitics.

Careful investigation of different amplifier concepts showed us that, at the high gain required, 40 Gb/s can hardly be reliably achieved with completely linear operation. Therefore, only the first amplifier cell operates in the linear and the others in the limiting mode. As another advantage of limiting operation, the constant output voltage swing is obtained without needing automatic gain control. An important design aspect for the division into a linear and a limiting part is as follows. The group delay of the first amplifier cell should deviate only weakly from the constant (low-frequency) value up to high frequencies. This is because this deviation would result in time jitter which, usually, cannot be substantially improved by the succeeding cells. However, a drop in the gain of the first cell with increasing frequency can be compensated by amplitude limitation in the succeeding cells. Therefore, the first cell is allowed to have a comparatively low cut-off frequency for the magnitude of the gain (here about 24 GHz), but not for the phase.

The signal input of the first amplifier cell is d.c. coupled to the PD, resulting in an offset current which depends on the signal amplitude of the input current i_I. Therefore, automatic offset control, using a d.c. amplifier and a low-pass (LP) filter, is required. It feeds an equivalent d.c. current I_{OS} into the other (complementary) input terminal of the first stage, which is decoupled by an off-chip capacitor. While the bond inductance to this capacitor has to be as low as possible, the input bond inductance L_1 is here chosen mainly with respect to optimum frequency response (cf. also Fig. 14). There is also the possibility to adjust an additional contribution to the input offset current, which is independent of the signal amplitude (not shown in Fig. 19(a)). It can be used to optimize the switching threshold, thus minimizing the bit error rate of the receiver (especially if optical amplifiers are applied).

The simplified circuit diagram of the linear and limiting amplifier cells in Fig. 19(a) is shown in Fig. 19(b). It is very similar to the amplifier cell discussed in Sec. 4, Fig. 4, but without the shaded block for gain control. In addition, three (instead of two) EF pairs are required, and the TAS is extended to a cascode configuration by use of a grounded-base stage (GBS) at its output, mainly to mitigate potential transistor breakdown problems. The bias voltage of the latter stage, V_B, is generated on chip. A resistor in series to V_B is highly recommended to avoid potential ringing, caused by on-chip wiring parasitics.[18]

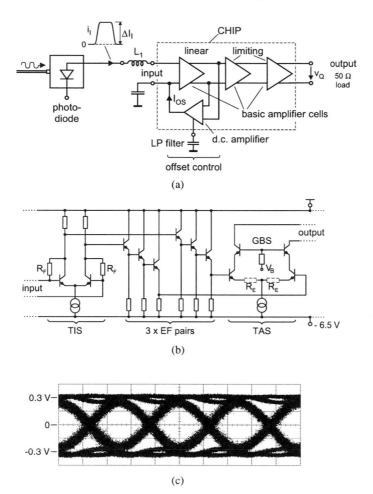

Fig. 19. Limiting transimpedance amplifier for a 40 Gb/s link. (a) Simplified block diagram. (b) Basic amplifier cell. (c) Output eye diagram (v_Q) at 40 Gb/s (10 ps/div.), $\Delta V_Q = 0.6$ V$_{p\text{-}p}$. For clearness, the diodes for V_{CE} reduction (cf. Fig. 14(a)) are not shown in (b).

A discussion of the slight differences in circuit diagram (e.g., with or without R_E) and dimensioning of the three amplifier cells is beyond the scope of this paper. At the output of the third cell (output stage), which must drive a (differential) 50-Ω transmission line, on-chip output resistors (60 Ω) are used to reduce double reflections. This means that the corresponding output TAS has to switch a current of 12 mA at 40 Gb/s.

The amplifier chip (size: 0.9 mm × 1.4 mm) was mounted and measured in the same way as discussed for the preamplifier in Sec. 5.2.4. For the pure electrical measurement, the simplified PD model of Fig. 14(a) was used again, but with the total PD capacitance now reduced to 65 fF. The validity of this simplification (compared to the accurate PD model in Fig. 18 used for circuit design) was confirmed by network simulations.

Figure 19(c) shows the measured clear output eye diagram of the differential output voltage v_Q at 40 Gb/s, for the lower limit of the input current swing, $\Delta I_I = 0.3$ mA$_{p\text{-}p}$. Even better signal quality (at the same output voltage swing of $\Delta V_Q = 0.6$ V$_{p\text{-}p}$) was observed at the upper specified limit of input current swing, $\Delta I_I = 0.6$ mA$_{p\text{-}p}$.

7.2. *Power MUX as a modulator driver*[19]

As discussed in Sec. 6, a symmetrical EAM configuration is used in the present 40 Gb/s system. Thus, the driving circuit has to generate a differential voltage swing of about $\Delta V_Q = 2$ V$_{p\text{-}p}$. The mainly capacitive load, represented by the EAM, may result in problems caused by double reflections, if driver and EAM are connected via a transmission line. Therefore, we preferred — in contrast to the usual practice — to bond the driver outputs directly to the EAM chip. High-speed performance is further improved and mounting costs are reduced by implementing the load resistors R_Q as well as the EAM biasing network on the driver chip.[18] Note that the nominal value of R_Q is chosen as low as 25 Ω in order to reduce the output time constant, which suffers under the comparatively high capacitive loading of the output nodes (including the EAM capacitance of about 100 fF).

For the reasons mentioned in Sec. 6, a power MUX was preferred to a conventional driver concept. Its simplified block diagram is shown in Fig. 20(a). Details of the circuit and its output stage are given in Ref. 19 and 18, respectively, together with design aspects. Again, a completely differential operation is applied. Besides the much higher switched output current (up to 50 mA), the main differences compared to the 60 Gb/s MUX in Fig. 2 are as follows. At the (50-Ω) data inputs, more powerful buffer stages are used, which each consists of a current switch with two EF pairs at input and three at output. Using only two or three EF pairs as a buffer would result in large input transistors and thus in a high input capacitance (in parallel to the 50-Ω on-chip resistors), which would severely degrade transmission line matching. Moreover, comparatively large signal amplitudes would be required at the data inputs. At the output of the intrinsic power MUX a (differential) grounded-base stage (similar to Fig. 19(b)) is used, in order to mitigate the transistor breakdown problems and to increase the operating speed.[P] Finally, in series to the load resistors R_Q a network is realized on the chip, which biases the EAM and, in addition, partly compensates its low-pass characteristic by peaking.[18] The EAM bias voltage V$_{Bias}$ can be varied between 0 V and -2 V via an external potentiometer, at a nominal value of -1 V.

To achieve the desired specifications, the driver circuit must be tailored to the EAM under consideration, for which an accurate model was available. For this, the circuit was optimized with respect to the signal voltage across the inner quantum wells of the EAM (and thus to the modulated light intensity) rather than to the

---

[P]This is because the output capacitance of the intrinsic power MUX is rather high due to the tied collectors of two comparatively large transistors at each output node (cf. also Fig. 2).

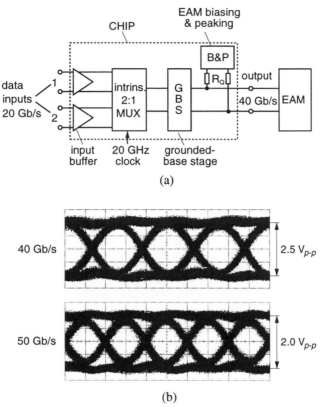

Fig. 20. Power MUX for driving the EAM in a 40 Gb/s link. (a) Simplified block diagram. (b) Output eye diagrams (v_Q) at 40 Gb/s (top) and 50 Gb/s (bottom). Time scale: 10 ps/div. The voltage scales are adjusted to the different voltage swings.

voltage at the EAM input terminal. To check whether the specifications of the fabricated driver IC are met, a restriction to pure electrical measurements is recommended, as already discussed in the case of the transimpedance amplifier. In principle, a simplified version of the equivalent circuit of the EAM could be realized on a ceramic substrate. However, measuring the signal voltage of interest would be a very complicated matter. Therefore, we tested another way to check the IC performance. For this, the driver outputs were a.c. coupled to the 50-Ω inputs of a 50-GHz sampling scope (HP 54124). In order to obtain the same voltage swing as in the case of the EAM load, the on-chip resistances R_Q must be increased from nominal 25 Ω to 50 Ω (yielding the same effective load of 25 Ω). This is achieved by ultrasonic cutting the interconnections of several shunted on-chip resistors. This measurement condition was simulated and the result was compared to both the simulation result with EAM load and the experimental result presented below. The good agreement between all three eye diagrams lets us expect that also in reality the measured eye diagrams well approximate those across the inner quantum wells of the EAM.

clock

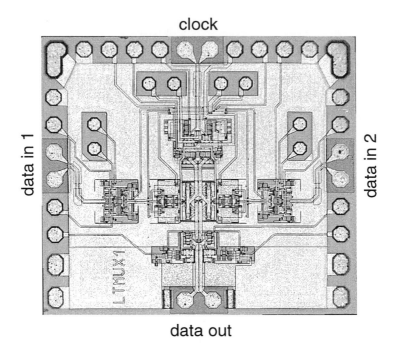

data in 1

data in 2

data out

Fig. 21. Chip photograph of the power MUX (1.0 mm × 1.1 mm).

The chip of the power MUX is shown in Fig. 21. It was mounted on the simple measurement socket with teflon substrate, described in Sec. 5.2.1., as well as on a special industrial measurement socket with ceramic substrate. The latter was designed with respect to a much lower thermal resistance, and was, therefore, preferred here. The power consumption was measured to be about 2 W.

The power MUX was driven by the same measurement setup as used for the 60 Gb/s MUX, shown in Fig. 10. The *differential* voltage swing at the data inputs was chosen as low as $2 \times 150 \text{ mV}_{p\text{-}p} = 300 \text{ mV}_{p\text{-}p}$ for all measurements, in order to demonstrate the comparatively high gain of the circuit. The clock input is driven by a *differential* sinusoidal signal with a voltage swing of $2 \times 0.5 \text{ V}_{p\text{-}p} = 1 \text{ V}_{p\text{-}p}$ at 40 Gb/s and $2 \times 0.75 \text{ V}_{p\text{-}p} = 1.5 \text{ V}_{p\text{-}p}$ at 50 Gb/s, respectively. Figure 20(b) (top) shows the measured eye diagram at the nominal data rate of 40 Gb/s with a differential output voltage swing of 2.5 $\text{V}_{p\text{-}p}$ (i.e., higher than the nominal value of 2 $\text{V}_{p\text{-}p}$). Well-opened "eyes" with small time jitter can be observed. The CPM measured at this data rate is 135°, confirming the retiming capability of the circuit. To demonstrate the high speed potential of the power MUX, the data rate is increased to 50 Gb/s. Even at this data rate, sufficiently clear eye diagrams are obtained at the nominal output swing of 2 $\text{V}_{p\text{-}p}$, as shown in Fig. 20(b) (bottom). Both eye diagrams in Fig. 20(b) are given for the nominal value of the EAM bias voltage $V_{Bias} = -1$ V, but no degradation of signal performance was observed within the adjustable bias range of 0 to -2 V.

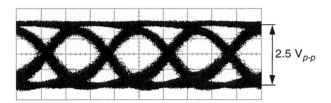

Fig. 22. Output eye diagram of a conventional EAM driver at 40 Gb/s (10 ps/div.) for comparison with the power MUX results in Fig. 20(b). For both ICs the same output stage is used.

In addition to the power MUX, a conventional EAM driver was designed and fabricated in the same technology. Its output stage, consisting of a grounded-base stage as well as an EAM biasing and peaking network, is the same as in the case of the power MUX (cf. Fig. 20(a)).[18] The measurement result in Fig. 22 still shows a clear eye diagram at 40 Gb/s and 2.5 $V_{p\text{-}p}$ (differential) output swing. However, as already expected from simulations, the measurements confirm that a data rate of 40 Gb/s is near to the speed limit of this circuit — in contrast to the power MUX (cf. Fig. 22 and Fig. 20(b), top). Therefore, the power MUX, which additionally stands out for its retiming capability, is confirmed to be the superior solution for this application.

8. Conclusions

It has been demonstrated that all ICs in 10 Gb/s long-haul optical-fiber TDM links can be fabricated in Si-bipolar production technologies, as long as the circuits are carefully designed. However, for the speed-critical ICs in 20 Gb/s systems present laboratory technologies are required, if the circuit specifications, apart from the data rate, must remain unchanged.

Recent measurement results even confirmed that all ICs in a 40 Gb/s TDM system (i.e., also the speed-critical ones) can be realized in advanced SiGe technologies. However, compared to ICs in 10 and 20 Gb/s systems, the other circuit specifications must be relaxed. This is possible by use of optical amplifiers and improved opto-electronic components as well as by system modifications, which further make possible elimination of some of the speed-critical circuits.

In all cases, even up to 60 Gb/s, good agreement between simulation and measurement results was observed. This, however, required the application of very advanced modeling, not only of the transistors but also of the mounting and on-chip metallization parasitics as well as of Si-substrate coupling.

It should be noted that all the experimental results presented are measured on *mounted* chips (using conventional wire bonding) rather than on-wafer only, and that most of the circuits have been applied in well operating experimental TDM links.

However, the results presented in this paper also show, that for the two speed-critical circuits in a TDM link, transimpedance amplifier and modulator driver,

40 Gb/s are near the speed limit of today's best Si-based laboratory technologies, even if advanced photodiodes and modulators are used. Therefore, for applications in future commercial systems, the SiGe bipolar technologies as well as the opto-electronic components should be further improved and, moreover, more accurate modeling tools must be commercially available. This is in order to pay for the potential spread in circuit performance, caused, e.g., by parameter spread in production and by design uncertainties.

The results and statements of this paper are focussed on *long-distance* transmission systems. However, they may also be helpful to judge the chance for an application of Si-based technologies in *short-distance* optical links with high data rates. The higher production volume for ICs in these applications might encourage the semiconductor companies to further push the development of high-speed Si-based bipolar technologies.

Acknowledgments

The work described in this paper ranged over a period of several years and a lot of experts were involved in design, fabrication, mounting, and measuring of the ICs. They are/were with the author's group at Ruhr-University Bochum and with his industrial partners, as Siemens AG, Munich, Bosch Telecom GmbH (formerly ANT Nachrichtentechnik), Backnang, and Hewlett Packard, Palo Alto. The author would like to acknowledge the excellent work of all these people and to thank them for the fruitful cooperation. Most of the ICs were fabricated by Siemens (cf. Table 1). Part of the work was financially supported by the European Union within the ESPRIT Project TIBIA and the RACE Project 1051, by the German Federal Ministry BMBF within the R&D program PHOTONIK II, and by our industrial partners mentioned above.

References

1. E. Gottwald, A. Felder, H.-M. Rein, M. Möller, M. Wurzer, B. Stegmüller, M. Plihal, J. Bauer, F. Auracher, W. Zirwas, U. Gaubatz, A. Ebberg, K. Drögemüller, C. Weiske, G. Fischer, P. Krummrich, W. Bogner, and U. Fischer, "Towards a 40 Gbit/s electrical time division multiplexed optical transmission system", in *Proc. ICCT '96*, Beijing, China, May 1996, pp. 60–63.
2. W. Bogner, E. Gottwald, A. Schöpflin, and C.-J. Weiske, "40 Gbit/s unrepeatered optical transmission over 148 km by electrical time division multiplexing and demultiplexing", *Electron. Lett.* **33** (1997) 2136–2137.
3. M. Nakamura, "Challenges in semiconductor technology for multi-megabit network services", in *Tech. Dig. ISSCC '98*, San Francisco, Feb. 1998, pp. 16–20.
4. J. Pietzsch, H. Burghardt, and P. Herger, "ATM switching network for Multigigabit/s throughput", in *Proc. ISS '97*, Toronto, Canada.
5. H.-M. Rein, J. Hauenschild, M. Möller, W. McFarland, D. Pettengill, and J. Doernberg, "30 Gbit/s Si multiplexer and demultiplexer IC's in silicon bipolar technologies", *Electron. Lett.* **28** (1992) 97–99.

6. J. Hauenschild, A. Felder, M. Kerber, H.-M. Rein, and L. Schmidt, "A 22 Gb/s decision circuit and a 32 Gb/s regenerating demultiplexer IC fabricated in silicon bipolar technology", in *Proc. IEEE BCTM '92*, Minneapolis, Oct. 1992, pp. 151–154.

7. A. Felder, J. Hauenschild, M. Kerber, and H.-M. Rein, "Static silicon frequency divider for low power consumption (4 mW, 10 GHz) and high speed (160 mW, 19 GHz)", in *Proc. IEEE BCTM '92*, Minneapolis, Oct. 1992, pp. 159–162.

8. H.-M. Rein, R. Schmid, P. Weger, T. Smith, T. Herzog, and R. Lachner, "A versatile Si-bipolar driver circuit with high output voltage swing for external and direct modulation of laser diodes in 10 Gb/s optical fiber links", *IEEE J. Solid-State Circuits* **29** (1994) 1014–1021.

9. M. Neuhäuser, H.-M. Rein, H. Wernz, and A. Felder, "A 13 Gbit/s Si bipolar preamplifier for optical front ends", *Electron. Lett.* **29** (1993) 492-493.

10. M. Möller, H.-M. Rein, and H. Wernz, "13 Gb/s Si-bipolar AGC amplifier with high gain and wide dynamic range for optical-fiber receivers", *IEEE J. Solid-State Circuits* **29** (1994) 815–822.

11. M. Möller, H.-M. Rein, and H. Wernz, "15 Gbit/s high-gain limiting amplifier fabricated in a Si-bipolar production technology", *Electron. Lett.* **30** (1994) 1519–1520.

12. M. Möller, H.-M. Rein, A. Felder, and T. F. Meister, "60 Gbit/s time-division multiplexer in SiGe-bipolar technology with special regard to mounting and measurement technique", *Electron. Lett.* **33** (1997) 679–680.

13. A. Felder, M. Möller, M. Wurzer, M. Rest, T. F. Meister, and H.-M. Rein, "60 Gbit/s regenerating demultiplexer in SiGe bipolar technology", *Electron. Lett.* **33** (1997) 1984–1985.

14. M. Wurzer, T. F. Meister, H. Schäfer, H. Knapp, J. Böck, R. Stengel, K. Aufinger, M. Franosch, M. Rest, M. Möller, H.-M. Rein, and A. Felder, "42 GHz static frequency divider in a Si/SiGe bipolar technology", in *Tech. Dig. ISSCC '97*, San Francisco, Feb. 1997, pp. 122–123.

15. M. Wurzer, private communications, Siemens AG, Munich, 1997.

16. R. Schmid, T. F. Meister, M. Rest, and H.-M. Rein, "23 Gbit/s SiGe modulator driver with 3.5 V single-ended output swing — Design aspects and measuring results", in *Proc. IEEE Workshop on MMIC Design, Packaging, and System Applications*, Freiburg (Germany), Oct. 1998, pp. 87–88.

17. R. Schmid, T. F. Meister, M. Neuhäuser, A. Felder, W. Bogner, M. Rest, J. Rupeter, and H.-M. Rein, "20 Gb/s transimpedance amplifier and modulator driver in SiGe-bipolar technology", *Electron. Lett.* **33** (1997) 1136–1137.

18. R. Schmid, T. F. Meister, M. Rest, and H.-M. Rein, "40 Gbit/s EAM driver IC in SiGe bipolar technology", *Electron. Lett.* **34** (1998) 1095–1097.

19. M. Möller, T. F. Meister, R. Schmid, J. Rupeter, M. Rest, A. Schöpflin, and H.-M. Rein, "SiGe retiming high-gain power MUX for directly driving an EAM up to 50 Gbit/s", *Electron. Lett.* **34** (1998) 1782–1784.

20. J. Müllrich, T. F. Meister, M. Rest, W. Bogner, A. Schöpflin, and H.-M. Rein, "40 Gbit/s transimpedance amplifier in SiGe bipolar technology for the receiver in optical-fibre TDM links", *Electron. Lett.* **34** (1998) 452–453.

21. H. Klose, R. Lachner, K. R. Schön, R. Mahnkopf, K. H. Malek, M. Kerber, H. Braun, A. v. Felde, J. Popp, O. Cohrs, E. Bertagnolli, and P. Sehrig, "B6HF: A 0.8 micron 25 GHz/25 ps bipolar technology for 'mobile radio' and 'ultra fast data link' IC-products", in *Proc. IEEE BCTM '93*, Minneapolis, Oct. 1993, pp. 125–127.

22. T. F. Meister, H. Schäfer, M. Franosch, W. Molzer, K. Aufinger, U. Scheeler, C. Walz, M. Stolz, S. Boguth, and I. Böck, "SiGe base bipolar technology with 74 GHz fmax and 11 ps gate delay", in *Tech. Dig. IEDM '95*, Washington, Dec. 1995, pp. 739–742.

23. H.-M. Rein and M. Möller, "Design considerations for very-high-speed Si-bipolar ICs operating up to 50 Gb/s", *IEEE J. Solid-State Circuits* **31** (1996) 1076–1090.
24. Y. Kyota, E. Ohue, T. Onai, K. Washio, M. Tanabe, and T. Inada, "Lamp-heated rapid vapor-phase doping technology for 100-GHz Si bipolar transistors", in *Proc. IEEE BCTM '96*, Minneapolis, Sept. 1996, pp. 173–176.
25. A. Schüppen, H. Dietrich, S. Gerlach, H. Höhnemann, J. Arndt, U. Seller, R. Götzfried, U. Erben, and H. Schumacher, "SiGe-technology and components for mobile communication systems", in *Proc. IEEE BCTM '96*, Minneapolis, Sept. 1996, pp. 130–133.
26. K. Washio, E. Ohue, K. Oda, M. Tanabe, H. Shimamoto, and T. Onai, "A selective-epitaxial SiGe HBT with SMI electrodes featuring 9.3-ps ECL-gate delay", in *Tech. Dig. IEDM '97*, San Francisco, Dec. 1997, pp. 785–789.
27. D. L. Harame, J. H. Comfort, J. D. Cressler, E. F. Crabbé, J. Y. C. Sun, B. S. Meyerson, and T. Tice, "Si/SiGe epitaxial-base transistors - Part I and II", *IEEE Trans. Electron Devices* **42** (1995) 455–482.
28. H.-M. Rein, "Very-high speed Si and SiGe bipolar ICs", in *Proc. ESSDERC '95*, The Hague, The Netherlands, Sept. 1997, pp. 45–56.
29. J. Hauenschild and H.-M. Rein, "Influence of transmission-line interconnections between Gbit/s IC's on time jitter and instabilities", *IEEE J. Solid-State Circuits* **25** (1990) 763–766.
30. M. Neuhäuser, H.-M. Rein, and H. Wernz, "Low-noise, high-gain Si-bipolar preamplifiers for 10 Gb/s optical-fiber links — design and realization", *IEEE J. Solid-State Circuits* **31** (1996) 24–29.
31. E. M. Cherry and D. E. Hooper, "The design of wide-band transistor feedback amplifier", *IEE Proc.* **110** (1963) 375–389.
32. M. Neuhäuser, M. Möller, H.-M. Rein, and H. Wernz, "Low-noise, high-transimpedance Si-bipolar AGC amplifier for 10 Gb/s optical-fiber links", *IEEE Photonics Technol. Lett.* **7** (1995) 549–551.
33. M. Bussmann, U. Langmann, B. Hillery, and W. W. Brown, "A 12.5 Gb/s Si-bipolar IC for PRBS generation and bit error detection up to 25 Gb/s", in *Tech. Dig. ISSCC '93*, San Francisco, Feb. 1993, pp. 152–153.
34. J. Hauenschild, H.-M. Rein, W. McFarland, and D. Pettengill, "Demonstration of the retiming capability of a silicon bipolar time-division multiplexer operating up to 24 Gbit/s", *Electron. Lett.* **27** (1991) 978–979.
35. M. Möller, H.-M. Rein, A. Felder, J. Popp, and J. Böck, "50 Gbit/s time-division multiplexer in Si-bipolar technology", *Electron. Lett.* **31** (1995) 1431–1433.
36. A. Felder, M. Möller, J. Popp, J. Böck, and H.-M. Rein, "46 Gb/s DEMUX, 50 Gb/s MUX, and 30 GHz static frequency divider in silicon bipolar technology", *IEEE J. Solid-State Circuits* **31** (1995) 481–486.
37. C. Clawin, U. Langmann, and B. Bosch, "Silicon bipolar decision circuit handling bit rates up to 5 Gbit/s", *J. Lightwave Technol.* **5** (1987) 537–541.
38. H.-M. Rein, J. Hauenschild, W. McFarland, and D. Pettengill, "23 Gbit/s Si bipolar decision circuit consisting of 24 Gbit/s MUX and DEMUX ICs", *Electron Lett.* **27** (1991) 974–975.
39. H.-M. Rein, E. Gottwald and T. F. Meister, "Si-bipolar — A potential candidate for high-speed electronics in 20 and 40 Gb/s TDM systems?", in *Proc. OSA Spring Topical Meetings '97: Ultrafast Electronics and Optoelectronics*, Incline Village, Nevada, March 1997, invited paper UTUA1-1, pp. 118–120, and in OSA Trends in Optics and Photonics Series **13**; Ultrafast Electronics and Optoelectronics, Optical Society of America, 1997, pp. 124–128.

40. W. Bogner, "40 Gb/s decision, demultiplexer and clock recovery circuit for ultra high speed optical transmission systems", Technical Report, Siemens AG, Munich, 1997, published in the internet: http://w2.siemens.de/oen/.
41. M. Pfost and H.-M. Rein, "Modeling and measurement of substrate coupling in Si-bipolar IC's up to 40 GHz", *IEEE J. Solid-State Circuits* **34** (1998) 582–591.

International Journal of High Speed Electronics and Systems, Vol. 9, No. 2 (1998) 385–397

LOW TRANSIMPEDANCE-FLUCTUATION DESIGN FOR 10-GHz Si-BIPOLAR PREAMPLIFIER IN 10-Gb/s OPTICAL TRANSMISSION SYSTEMS

TORU MASUDA, RYOJI TAKEYARI and KATSUYOSHI WASHIO

ULSI Research Department, Central Research Laboratory, Hitachi, Ltd.,
1-280 Higashi-koigakubo, Kokubunji, Tokyo 185-8601, Japan

KENICHI OHHATA

Musashino Office, Hitachi Device Engineering, Co. Ltd.,
1-280 Higashi-koigakubo, Kokubunji, Tokyo 185-8601, Japan

KAZUO IMAI

Device Development Center, Hitachi, Ltd.,
6-16-3 Shin-machi, Ome, Tokyo 198-8512, Japan

Design considerations concerning Si preamplifiers for 10-Gb/s optical transmission systems are discussed. A preamplifier with 300-Ω transimpedance was designed by focusing on attaining low transimpedance fluctuation in the frequency response despite bias variation caused by the photo current. To ensure good design accuracy, we optimized the current density of the transistor and the open-loop voltage gain using measured results to obtain the desired bandwidth. We also developed a low-loss pad structure that has U-grooves to improve the bandwidth. The U-grooves increase the pad parasitic resistance and reduce signal-loss from pad to Si substrate. A preamplifier IC fabricated using a Si bipolar technology with f_T of 35-GHz provided a bandwidth of 10.2 GHz, transimpedance fluctuation within 0.5 dB, an input dynamic range of up to 1.6 mA$_{p-p}$, and a low averaged input noise current density of 11.5 pA/$\sqrt{\text{Hz}}$.

1. Introduction

Optical-fiber transmission systems with a data rate of 10 Gb/s are now being used in practical applications. The key integrated circuits (ICs) in the transmitters and receivers of these systems were originally developed based on compound semiconductor technologies.[1,2] However, with the recent improvements in Si technology, most of these have been replaced with Si ICs which offer lower cost, easier fabrication, higher integration, and lower power consumption.[3,4] A preamplifier, which is a key IC in the optical receiver, requires high transimpedance and low noise characteristics with sufficient bandwidth in the frequency response, because it largely determines the maximum bit-rate and the S/N (signal-to-noise ratio) of the receiver. Moreover, low transimpedance fluctuation (the deviation of the gain within the frequency range from DC to 8 GHz) is needed to suppress jitter as measured by an eye diagram. These requirements must be satisfied even if the bias variation in the preamplifier is caused by photo current swing. To adopt a preamplifier IC for practical use, modeling of the mounting parasitic elements has to be included in the design as well as optimization of the transistor operating point. Pad patterns are used to connect ICs, and as the operating speed increases, the parasitic elements of the pads affect IC performance. To achieve sufficient bandwidth and transimpedance, the leading preamplifier so far reported for 10 Gb/s operation needs to have its transimpedance fluctuation adjusted by an external potentiometer.[3] Furthermore, in that report, there was no description of the variation in the frequency response relative to the magnitude of the photo current swing or of the behavior of the pad parasitic elements.

In this paper, we described a 300-Ω transimpedance preamplifier using 35-GHz Si bipolar technology. We focus especially on design considerations aimed at reducing the influence of the signal-pad parasitic elements and on the optimization of the operating point to suppress the variation in the frequency response caused by the photo current swing.

2. Circuit Configuration

Figure 1 shows the schematic of the preamplifier discussed in this paper. It consists of a transimpedance amplifier stage and an emitter follower as an output buffer. The transimpedance stage consists of an emitter common amplifier (T_1, R_L) and a shunt feedback loop through T_2, D_1, and R_F. The T_2 and an emitter follower (T_4) are biased by the current source circuit. Resistor R_O was introduced to set the output impedance to 50 Ω. The V_{R_L} bias circuit is designed to supply a stable bias voltage for R_L. Because the transimpedance fluctuation in the frequency response (ΔZ_T) needs to be suppressed to reduce jitter, V_{R_L} should be independent of the V_{CC} variation because the amplitude of the emitter common amplifier is sensitive to variation in the collector current of T_1 and this amplitude dominates the frequency response of the preamplifier. V_{R_L} is defined from the ground level according to circuit elements D_7-D_9, R_7-R_9, T_6, and T_7 and can be precisely adjusted by tuning resistors R_7-R_9. In the current source circuit, R_{10}-R_{11} and T_8-T_9 determine the base node voltage of T_3 and T_5. To prevent interference, the ground pattern of the current source circuit was isolated from other grounds. Capacitors C_1-C_4 were introduced to stabilize the constant voltages in the preamplifier IC. Diodes $(D_2$-$D_6)$ were set to reduce the collector-emitter voltage of T_2 and T_4 to below the breakdown voltage. We used a 0.15-pF capacitor (C_{PD}) as an equivalent model of a pin photodiode. The capacitance is the total parasitic capacitance when the preamplifier IC is mounted on a ceramic substrate.

With this preamplifier design, we mainly used a small signal frequency response to investigate the preamplifier performance, because a practical input signal will be very small and it is easy to introduce noise behavior. The transient response can

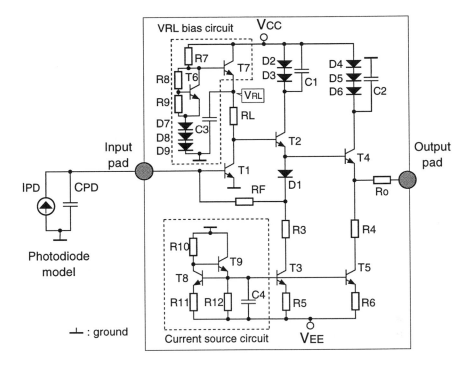

Fig. 1. Schematic of preamplifier.

be predicted from a small signal frequency response when the preamplifier is held in the linear operating range.

3. Improvement in Bandwidth

Wire bonding is a common and inexpensive mounting technique, but the large parasitic inductance degrades the IC performance and has to be considered in the circuit design to decrease the likelihood of high Q resonance. Furthermore, the wire length must be decreased to reduce the parasitic inductance by that which the IC performance connected with a substrate satisfy the specifications. However, the parasitic inductance cannot be reduced since the wire must have enough length to compensate for tolerance between the substrate and chip. Therefore, we used a flip chip bonding (FCB) technique to connect the preamplifier IC to a ceramic substrate. In this technique, the flipped IC and the substrate are connected using Au bonding bumps. It is suitable for mounting high-speed ICs over 10-Gb/s optical transmission systems, because the bonding has little effect on the IC performance. The parasitic inductance between the IC and the ceramic substrate can be reduced to as low as 0.15 nH by using FCB technology.[5] However, the pad-pattern area required for FCB is about four times as large as that for wire bonding. Thus we have to consider the influence of the pad metallization in the circuit design. The parasitic capacitance of the input pad in particular affects the preamplifier performance, because the time constant at the input node of the preamplifier largely determines the frequency response. For example, increasing the parasitic capacitance of a pad decreases the bandwidth and causes the frequency response to become uneven. When using compound semiconductor technology, the parasitic capacitance of the pad is easy to model because the substrate is an insulator and parasitic capacitance mainly occurs between metallization components. On the other hand, in Si technology, the parasitic capacitance of a pad is greater and it occurs between the pad metallization and the Si substrate which has a finite impedance. Thus, the characteristics of the Si substrate need to be considered in pad modeling. The value of the parasitic element also depends on the pad construction. Therefore, the influence of the pad has to be considered in high-speed Si IC design.

Figure 2 shows the influence of pad parasitic resistance and capacitance on the simulated bandwidth of a preamplifier. The input and output pad of the preamplifier were modeled as a capacitor ($Cpad$) and a resistor ($Rpad$) connected in series. The parasitic capacitance of the photodiode was represented by a 0.15-pF capacitor. The bandwidth decreased with an increase in $Cpad$. The dependence of the bandwidth on $Cpad$ was especially strong when $Rpad$ was below 300 Ω, and the bandwidth was also sensitive to $Rpad$. In this case, because the pad was taken as a peaking capacitor, the time-constant increased and the bandwidth decreased with an increase in $Cpad$. To achieve a bandwidth of more than 10 GHz, $Cpad$ must be less than 100 fF. However, decreasing the area of pad metallization to reduce $Cpad$ is difficult. On the other hand, when $Rpad$ is above 100 Ω, the bandwidth increased with $Rpad$ because the signal loss to the Si substrate via the pad parasitic element was reduced as $Rpad$ increases. Moreover, the $Cpad$ dependence and the $Rpad$ dependence of the bandwidth weakened as $Rpad$ increased. That is, if $Rpad$ is large enough, the bandwidth will increase and converge on a certain value,

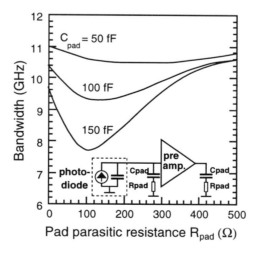

Fig. 2. Influence of pad parasitic resistance and capacitance.

regardless of the value of *Cpad*. When the cut-off frequency of the pad parasitic elements ($= 1/(2\pi \times Cpad \times Rpad)$) is close to 10 GHz, the bandwidth is minimized because the signal loss to the Si substrate increases. Accordingly, we should increase *Rpad* to more than 340 Ω to obtain a bandwidth of more than 10 GHz without the need to reduce *Cpad*. For example, when *Cpad* is 150 fF, increasing *Rpad* from 100 Ω to 340 Ω improves the bandwidth from 7.7 GHz to 10.0 GHz.

Figure 3(a) is a simple cross-section of a conventional pad metallization structure and an equivalent circuit extracted from measured scattering parameters. A Si-on-insulator (SOI) wafer was used for the fabrication. The diameter of the pad was 135 μm. The pad metallization was on a 3.1-μm-thick SiO_2 layer. The 1.5-μm-thick n^- silicon layer ($\rho = 10 \ \Omega$ cm) below the SiO_2 layer was biased to ground. *Cpad* was 150 fF and *Rpad* was 20 Ω. A bandwidth of 9.2 GHz was expected based on Fig. 2. To improve the bandwidth without reducing the area of pad metallization, meshed

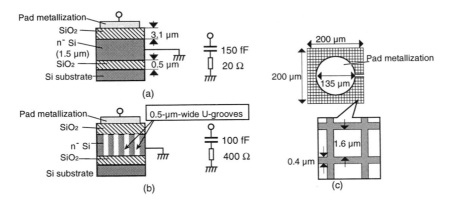

Fig. 3. Cross-section and equivalent circuit of pad structure. (a) conventional, (b) low loss structure with U-grooves, (c) top view of pad layout.

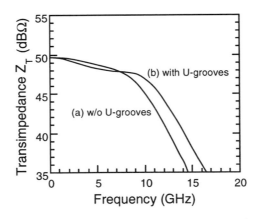

Fig. 4. Frequency response of transimpedance (a) without U-grooves ($Cpad = 150$ fF, $Rpad = 20 \ \Omega$) (b) with U-grooves ($Cpad = 100$ fF, $Rpad = 400 \ \Omega$).

U-grooves were made in a 200×200-μm area under the input and output pads and filled with oxide silicon (Figs. 3(b) and (c)). The 0.4-μm-wide grooves were spaced 1.6 μm apart. The insertion of silicon oxide in the n^- silicon layer increased the pad parasitic resistance in this low loss structure. We extracted the pad equivalent circuit, and found that $Rpad$ increased from 20 Ω to 400 Ω and $Cpad$ decreased from 150 fF to 100 fF when the U-grooves were used. Therefore, a bandwidth of 10.5 GHz can be expected. Figure 4 shows the calculated frequency response of the preamplifier. While the bandwidth was improved by using the low-loss structure, the transimpedance fluctuation (ΔZ_T) was degraded by -1.6 dB at 5 GHz. The methods used to improve ΔZ_T are discussed in the next section.

4. Optimization of the Frequency Response

The most important role of a preamplifier is the faithful regeneration of the input current waveform. Achieving a well-opened output eye diagram requires a flat transimpedance gain in the frequency response within the frequency range from DC to 8 GHz in 10-Gb/s systems. Moreover, the frequency response should be insensitive to variation in the magnitude of the photo current to avoid waveform distortion. To satisfy these requirements, the bias condition variation with the photo current must be considered in the design of the transimpedance gain stage. However, since the bandwidth was close to the cut-off frequency of transistor $f_T (= 35$ GHz) used in this preamplifier and peaking is caused by phase delay in the transconductance of transistors and parasitic elements of integrated circuit components, we could not predict the frequency response with sufficient accuracy with our circuit simulator. Therefore, we investigated the relation between the bias condition and frequency response by measuring TEG (test element group) circuits. These TEG circuits allow the voltage of the base node of T_7 (Fig. 1) to be varied by adjusting the external voltage supply, and the load resistance R_L to be changed.

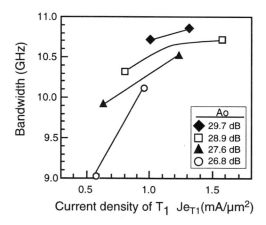

Fig. 5. Bandwidth versus current density of T_1.

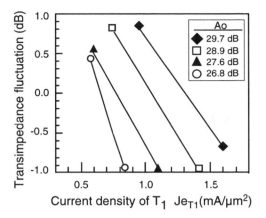

Fig. 6. Transimpedance fluctuations versus current density of T_1.

Figures 5 and 6 show the dependence of the bandwidth and the transimpedance fluctuation (ΔZ_T) on the collector current density of T_1 (Je_{T_1}). These characteristics were obtained by measuring the TEG circuits. The collector current of T_1 ($I_{C_{T_1}}$) was normalized by the current density and the emitter area was 4.2 μm^2. A maximum cut-off frequency of 35 GHz was achieved with a current density of 1.3 mA/μm^2. The bandwidth increased with Je_{T_1} when the open-loop voltage gain (A_O) was constant. Similarly, the bandwidth increased with A_O. ΔZ_T decreased with an increase in Je_{T_1} when A_O is constant, and increased with A_O. Here, ΔZ_T was defined as the deviation of the transimpedance from 50 MHz to 8 GHz. It is evident that careful optimization is needed to simultaneously obtain a bandwidth of over 10 GHz and a low ΔZ_T.

Next, design aspects for this optimization are briefly discussed for the preamplifier shown in Fig. 1. The transimpedance Z_T is approximately given by

$$Z_T \approx \frac{Z_{TO}}{1 + \dfrac{s}{p_1} + \dfrac{s^2}{p_1 \cdot p_2}} , \qquad (1)$$

where

$$p_1 \approx \frac{A_O}{R_F \cdot C_T} \qquad (2)$$

and

$$p_2 \approx \frac{1}{R_L \cdot C_{TC}} \qquad (3)$$

are the dominant pole p_1 and non-dominant pole p_2, Z_{TO} is the DC transimpedance gain, C_T is the total parasitic capacitance at the base node of T_1, and C_{TC} is the total parasitic capacitance at the collector node of T_1. The value of A_O is given by

$$A_O \approx gm_{T_1} \cdot R_L \approx \frac{I_{C_{T_1}}}{V_T} \cdot R_L \qquad (4)$$

so A_O is determined by the product of load resistor R_L and the transconductance of T_1 (gm_{T_1}) and gm_{T_1} is proportional to $I_{C_{T_1}}$. The feedback resistor R_F was designed to be 640 Ω to achieve a transimpedance gain of 300 Ω. The ratio of the poles $(= p_2/p_1)$ is a measure of the peaking level, which increased with a decrease in the ratio. These equations show the following:

(1) When A_O is fixed, the resistance of R_L decreases as Je_{T_1} increases, so p_2 and the ratio of the poles increase with Je_{T_1}. Therefore, the bandwidth increases and ΔZ_T decreases with an increase in Je_{T_1}.

(2) If Je_{T_1} is constant, p_1 increases with A_O, but p_2 simultaneously decreases as R_L increases, thus the ratio of the poles decreases. Consequently, the bandwidth and ΔZ_T increase with A_O.

These characteristics agree with the measured results shown in Figs. 5 and 6.

Next, the variation in the bias condition with the input photo current (I_{PD}) is considered because Je_{T_1} and A_O change with the swing of I_{PD}. However, to reduce output waveform distortion, same-frequency responses are required over the entire range of I_{PD} derived from system specifications. Therefore, we have to optimize the operating point of the transistor to reduce the dependence of the frequency response on the I_{PD} swing. A rise in I_{PD} will increase the voltage drop at R_F (which is approximately $I_{PD} \times R_F$) and the same voltage drop will occur at the collector node of T_1. Since V_{R_L} is set to a constant voltage, $I_{C_{T_1}}$ increases proportionally with the ratio $(I_{PD} \times R_F)/R_L$ and A_O increases in proportion to I_{PD}. Because the bandwidth increases with Je_{T_1} and A_O, the bandwidth is minimized when I_{PD} is 0. Therefore, the operating point of T_1 should be designed for the situation when I_{PD} is 0. Moreover, the collector current should be minimized within a range that allows both a bandwidth of more than 10 GHz and a low ΔZ_T, because, if Je_{T_1} becomes too large due to an increase in I_{PD}, Je_{T_1} would exceed the current density of 1.3 mA/μm^2 at which the maximum cut-off frequency can be achieved. An excessive current density would sharply decrease the operating speed of T_1 and the bandwidth. To avoid this, Je_{T_1} should be as small as possible.

Figure 7(a) shows the bandwidth dependence on A_O when Je_{T_1} is set to 0.8 mA/μm^2. When A_O is larger than 26.7 dB, a bandwidth of more than 10 GHz can be achieved. In the ΔZ_T dependence on A_O shown in Fig. 7(b), if A_O is within the range from 27.0 to 28.2 dB, ΔZ_T within 0.5 dB can be achieved when Je_{T_1} is set to 0.8 mA/μm^2. Therefore, to achieve low ΔZ_T, we designed A_O to be 27.6 dB. In this case, R_L was designed to be 370 Ω, and a bandwidth of 10.2 GHz and a low ΔZ_T of -0.1 dB can be predicted.

The photo-current dependence of the bandwidth and ΔZ_T are shown in Fig. 8. These characteristics were estimated on the basis of the operating point when I_{PD} is 0 $(Je_{T_1} = 0.8 \text{ mA}/\mu\text{m}^2, A_O = 27.6 \text{ dB})$. Since Je_{T_1} and A_O can be calculated when I_{PD} is 0.5 mA or 1.0 mA, the bandwidth and ΔZ_T can be predicted based on the measured results shown in Figs. 5 and 6. As a result, in a range of I_{PD} from 0 to 1.0 mA, a bandwidth of more than 10 GHz and a low ΔZ_T within -0.1 dB were estimated.

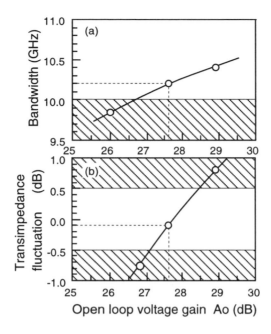

Fig. 7. Optimization between the open-loop voltage gain and (a) bandwidth, and (b) transimpedance fluctuation, when $Je_{T_1} = 0.8$ mA/μm^2.

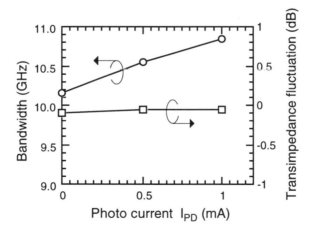

Fig. 8. Predicted photo-current dependence of bandwidth and transimpedance fluctuation.

5. Experimental Results for the Preamplifier

We fabricated a preamplifier IC using Si bipolar technology with a cut-off frequency of 35 GHz.[6] The transistor parameters are shown in Table 1, and a photomicrograph of the preamplifier IC is shown in Fig. 9. The round bumps for flip-chip bonding were laid in such a way that thermal resistance would be minimized, because when

Table 1. Transistor parameters.

Emitter area	0.3×2.8 μm
h_{FE}	450
R_B	230 Ω
R_E	10 Ω
R_C	37 Ω
C_E	8.7 Ω
C_C	6.4 Ω
C_S	8.0 Ω
f_T	35 GHz (@$V_{CE} = 2$ V)

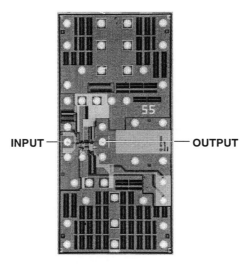

Fig. 9. Chip photomicrograph (1.85×3.90 mm).

using an SOI wafer, considerable thermal radiation through the bumps is required. The active area where the circuit elements were laid out was only 300×300 μm. The chip area was 1.85×3.90 mm. MOS-capacitors were arranged over the entire preamplifier IC to improve the supply voltage stability. A 30-nm-thick SiO_2 layer was used as an insulator. The capacitance between V_{CC} and the ground was 270 pF, and the capacitance between V_{EE} and the ground of the current source circuit was 360 pF. To reduce the parasitic resistance in series to the capacitor, comb-shaped plates were used.

The measured frequency responses of the transimpedance and the equivalent input noise current density are shown in Fig. 10. The transimpedance was calculated from scattering parameters measured by using RF wafer probes, taking into account the 0.15-pF parasitic capacitance of the photodiode. Transimpedance of 300 Ω and a -3-dB bandwidth of 10.2 GHz were obtained when I_{PD} is 0 mA. To measure the equivalent input noise current density, both a pin photodiode and a preamplifier

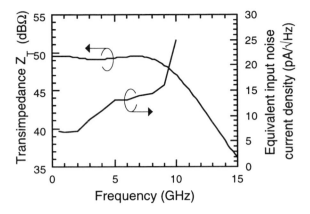

Fig. 10. Frequency response of transimpedance gain and equivalent input noise current density.

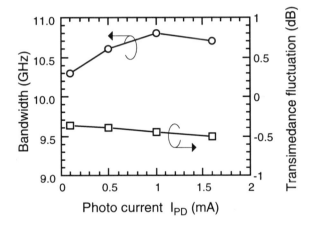

Fig. 11. Photo-current dependence of bandwidth and transimpedance fluctuation.

IC were mounted on a ceramic substrate. The averaged equivalent input noise current density up to 8 GHz was 11.5 pA/√Hz. Figure 11 shows the photo-current dependence of the bandwidth and the transimpedance fluctuation. In the range of I_{PD} from 0.1 to 1.6 mA, bandwidths of over 10 GHz are obtained and ΔZ_T is within a range from -0.5 to -0.4 dB. We attribute these results to the optimization of Je_{T_1} and A_O. ΔZ_T was also evaluated in the frequency range from 50 MHz to 8 GHz.

The estimated bandwidth dependence (Fig. 8) agreed well with the measured results, but ΔZ_T was slightly degraded compared to the estimated value. A well-opened output eye diagram was obtained when the preamplifier was driven by a pin photodiode (Fig. 12). In this case, the swing of the photo current was 0.85 mApp. The features of the preamplifier are summarized in Table 2. The power supply

Table 2. Features of preamplifier.

−3-dB Bandwidth	10.2 GHz
Transimpedance	300 Ω
Transimpedance fluctuation	−0.5 dB
Averaged equivalent input noise current density	11.5 pA/$\sqrt{\text{Hz}}$
Maximum input current swing	1.6 mA$_{p-p}$
Suupply voltage	+5.0 V/−5.2 V
Power consumption	258 mW

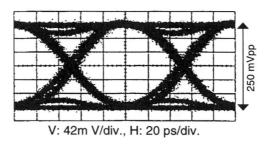

V: 42m V/div., H: 20 ps/div.

Fig. 12. Eye diagram at 10-Gb/s operation.

voltages V_{CC} and V_{EE} were 5.0 V and −5.2 V, respectively, and the power consumption was 258 mW.

6. Conclusion

We have developed a Si preamplifier for a 10-Gb/s optical receiver and have fabricated the preamplifier using Si bipolar technology. We optimized various operating parameters to achieve good gain flatness in the frequency response against bias variation caused by the photo current. To ensure good design accuracy, measured results from TEG circuits were used to simultaneously obtain sufficient bandwidth and low transimpedance fluctuation. Moreover, the low-loss pad structure made with U-grooves improved the bandwidth by increasing the pad parasitic resistance.

The fabricated preamplifier IC had a transimpedance gain of 300 Ω, a wide bandwidth of 10.2 GHz, a low transimpedance fluctuation of −0.5 dB with an input dynamic range of up to 1.6 mApp, and a low averaged equivalent input noise current density of 11.5 pA/$\sqrt{\text{Hz}}$.

Acknowledgments

We thank Dr. I. Imaizumi, Dr. K. Shimohigashi and Dr. K. Miyauchi for their encouragement and valuable suggestions. We also thank Mr. R. Kaneko for the measurements.

References

1. M. Miyashita *et al.*, "Ultra broadband GaAs MESFET preamplifier IC", *IEEE Trans. Microwave Theory and Techniques* **40**, 12, December 1992, pp. 2439–2443.
2. I. Amemiya *et al.*, "An AlGaAs/GaAs HBT preamp IC for high speed optical fiber communication systems", *1990 Autumn National Convention Record IEICE Japan,* October 1990, Paper B-775 (in Japanese).
3. M. Neuhauser, H.-M. Rein, and H. Wernz, "Low-noise, high-gain Si bipolar pre-amplifier for 10 Gb/s optical-fiber links–design and realization", *IEEE J. Solid-State Circuits* **31** (1996) 24–29.
4. H. Hamano *et al.*, "10 Gbit/s optical front end using Si-bipolar preamplifier IC, flipchip APD, and slant-end fibre", *Electron. Lett.*, 29th August 1991, 1602–1605.
5. T. Harada *et al.*, "Si bipolar multiplexer, demultiplexer, and prescaler ICs for 10 Gb/s SONET systems", *ISSCC 1993 Digest of Technical Papers*, Feb. 1993, pp. 154–155.
6. T. Hashimoto *et al.*, "Advanced process technology for a 40-GHz f_T self-aligned bipolar LSI", *1994 Bipolar/BiCMOS & Technology Meeting*, pp. 76–79.

International Journal of High Speed Electronics and Systems, Vol. 9, No. 2 (1998) 399–435
© World Scientific Publishing Company

20-40-Gbit/s-CLASS GaAs MESFET DIGITAL ICs FOR FUTURE OPTICAL FIBER COMMUNICATIONS SYSTEMS

TAIICHI OTSUJI, KOICHI MURATA, KOICHI NARAHARA,
KIMIKAZU SANO and EIICHI SANO
*NTT Optical Network Systems Laboratories, 1-1 Hikarino-oka,
Yokosuka, Kanagawa, 239-0847, Japan*

KIMIYOSHI YAMASAKI
*NTT System Electronics Laboratories, 3-1 Morinosato Wakamiya,
Atsugi, Kanagawa, 243-01, Japan*

This paper describes recent advances in high-speed digital IC design technologies based on GaAs MESFETs for future high-speed optical communications systems. We devised new types of a data selector and flip-flops, which are key elements in performing high-speed digital functions (signal multiplexing, decision, demultiplexing, and frequency conversion) in front-end transmitter/receiver systems. Incorporating these circuit design technologies with state-of-the-art 0.12-μm gate-length GaAs MESFET process, we developed a DC-to-44-Gbit/s 2:1 data multiplexer IC, a DC-to 22-Gbit/s static decision IC, and a 20-to-40-Gbit/s dynamic decision IC. The fabricated ICs demonstrated record speed performances for GaAs MESFETs. Although further operating speed margin is still required, the GaAs MESFET is a potential candidate for 20- to 40-Gbit/s class applications.

1. Introduction

There is an urgent need to expand transmission capacity because of the movement toward broadband integrated services digital network (B-ISDN) and multimedia services. 10-Gbit/s systems are commercially available. With the emergence of WDM (Wavelength-Division Multiplexing) and OTDM (Optical Time-Division Multiplexing) technologies, transmission throughput is exceeding terabits per second at an experimental level.[1-3] From the view point of practicality, there are many issues to be solved in photonic devices and optical fiber-based components. Advances in electronic integrated circuits (ICs) are the most promising way to expand the transmission throughput because of their maturity and cost effectiveness. The next generation of systems is targeting a single channel bit rate of 20 to 40-Gbit/s.

Many types of high-speed transistors have been developed using Si or compound semiconductors. Historically, Si bipolar transistors have been continuing the longest effort to improve the speed performance. However, compound semiconductors can potentially offer much higher speeds of operation than can Si bipolars. Hetero-structure semiconductor devices are the most attractive for breaking through the speed limit. Currently InP-based high-electron-mobility transistors (HEMTs) or hetero-bipolar transistors (HBTs) achieve a current-gain-cutoff frequency (fT) and/or maximum oscillation frequency (fmax) of beyond 200 GHz,[4–6] and a 40-Gbit/s class optical communication IC chip set using InP-based HEMTs has been experimentally demonstrated.[7–9] In terms of process ease and maturity, together with high-speed performance, GaAs metal-semiconductor field-effect transistors (MESFETs) are considered to be one of the best suited for 20- to 40-Gbit/s applications.

By shortening the transistor feature size (the gate length) and suppressing the short-channel effects, GaAs MESFETs have been improving their speed. 0.2-μm class gate-length MESFETs have provided commercial products of the 10-Gbit/s class logic IC families.[10] Recently, 0.15-μm class MESFETs having an fT and fmax approaching 100 GHz were successfully developed.[11] A record fT of 160 GHz has also been demonstrated with a 0.06-μm FET.[12]

The speed of GaAs MESFETs, in fact, is not sufficient to provide 20-Gbit/s static delay-flip-flop (D-FF) and 40-Gbit/s dynamic D-FF operation with enough speed margin. This is because even if we assume state-of-the-art process technology (low-energy ion implantation and i-line photolithography), further shrinkage of the FET's gate length to below 0.1 μm causes considerable short channel effect. This cannot improve the large-signal operation speed although an fT and fmax at a typical bias point must surely increase. This means the digital IC speed can no longer improve by further shrinking the device feature size. Therefore, circuital advances to provide faster operation are essential to develop 20- to 40-Gbit/s class GaAs MESFET ICs in the real world.

As data rates increase beyond 10-Gbit/s, the wavelength of the signal approaches the physical size of the IC chips, which gives rise to substantial difficulties in designing broadband ICs: (1) numerous parasitics; (2) cavity resonance; and (3) bandwidth limitation due to lumped-circuit treatment. These are the major reasons that transistor performance no longer directly reflects circuit and module performance.[9]

In terms of analog IC design, such as baseband amplifiers and mixers, the synthesis of distributed and lumped-circuit design techniques is an effective approach to overcoming the performance limit. Distributed amplifiers or traveling-wave amplifiers are commonly used in millimeter wave applications. To obtain a broadband operation from near 0 Hz to beyond 40 GHz, a frequency-dependent bias-termination network has been introduced to compensate for the loss at low frequencies due to the drain conductance of FETs.[13] Up to now, many types of over-40-Gbit/s-class analog ICs, including baseband amplifiers, frequency multipliers, and signal distributors, have been developed using InP-based 0.1-μm class HEMTs and 0.2-to-0.15-μm-class GaAs MESFETs.[14–18]

The digital ICs, however, stay at a relatively lower stage. Introduction of distributed design into digital ICs (especially for sequential circuits such as flip-flops) is really difficult because the propagation delay time along with critical signal path limits the circuit speed performance. Consequently, making full use of analog circuit design in digital ICs is the key to obtaining faster speeds.

This paper describes recent advances in 20- to 40-Gbit/s class high-speed digital IC design technologies based on GaAs MESFETs for future high-speed optical communications systems. Section 2 gives an overview of the fundamentals of optical communications ICs. Section 3 briefly describes state-of-the-art GaAs MESFET device technology. In Sec. 4, we discuss the relation between digital IC speed and device figure of merit. In Sec. 5, advanced high-speed digital IC design technology is described in detail. In Sec. 6, we actually fabricate several ICs and discuss their performances. In Sec. 7, future trends and subjects are discussed.

2. Fundamentals of Optical Communications ICs

Figure 1 depicts the basic transmitter and receiver configurations for lightwave communications systems. The functionality required for transmitter and receiver electronic ICs is basically the same (multiplexing, amplifying, retiming, regenerating and demultiplexing) for all TDM and WDM transmission systems with the small exception of extra blocks (e.g., exclusive-OR or wideband mixer) for coherent systems. The transmitter block consists of a laser diode (LD), an optical modulator (MOD), a modulator driver (DRV), and time-division multiplexers (MUXs). The receiver block consists of a photodiode (PD), a preamplifier (Pre), a baseband amplifier (Base), a decision circuit (DEC), time-division demultiplexers (DEMUXs), frequency dividers (DIVs), and a clock recovery circuit that includes a differentiator (DIF), a rectifier (REC), a resonator (RES), and a limiting amplifier (Limit). The transmitter and receiver ICs, except for the clock recovery block, require broadband operation from near DC to the maximum bit rate with good eye openings.

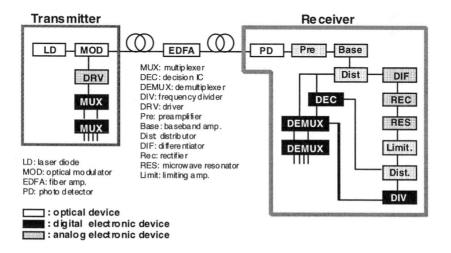

Fig. 1. Basic transmitter and receiver configuration for lightwave communications systems.

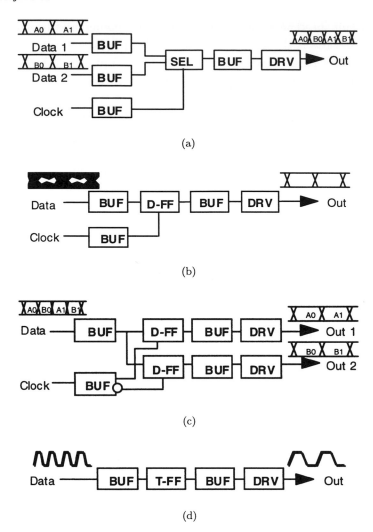

Fig. 2. Simplified circuit block diagrams of (a) MUX, (b) DEC, (c) DEMUX, and (d) DIV.

In terms of digital ICs, MUXs, DEC, DEMUXs, and DIVs are the key components. Typical and simplified circuit block diagrams are shown in Fig. 2. We see that each block has a common configuration where data and/or clock buffer(s) are/is followed by a core functional block whose outputs are followed by (an) output buffer(s). The whole core functional block can be configured with one or a combination of the following circuitry: a data selector; a delay flip-flop (D-FF); and a toggle-FF (T-FF). From the view point of the circuit's configuration, T-FFs are regarded as a part of D-FFs in which the output data is invertedly fed back to the data input. Furthermore, the D-FF is configured with a static circuit or with a dynamic circuit that can generally operate faster than a static one, in which however, lower bit-rate operation is sacrificed. The input and output buffers are so-called simple amplifier circuits and can be designed with a bandwidth wide enough to cover the core-circuit operation speed.

In general, the core functional blocks limit the operation speed because of their circuit complexity. Therefore, the main objective for the development of high-speed digital ICs is to improve the speed performance of data selectors and static/dynamic FFs. As is described later, wideband buffer circuit design also becomes important as the core digital circuits improve their speed by means of circuital advances.

For the purpose of high-speed operation, digital ICs are configured on the basis of a source-coupled FET Logic (SCFL) circuitry,[19] which is a sort of differential configuration compatible to the emitter-coupled logic (ECL) in bipolar transistors. The circuit diagram of a simplified SCFL D-type latch circuit is shown in Fig. 3. A so-called series-gated logic configuration (the first level for data and the second level for clock) is used to perform a combinational logic function. For sequential logic function in a D-FF and a T-FF, the feedback-loop is configured along the data-signal path. Based on these basic circuit configurations, advanced circuit design, including various extensions and/or transformations, will be discussed later to improve the speed performance.

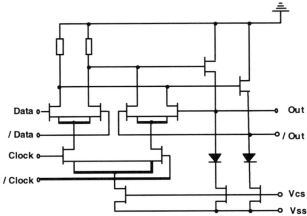

Fig. 3. Circuit diagram of an SCFL D-type latch.

3. Short-Gate-Length GaAs MESFET Process Technology

For higher operation speed for digital ICs, FETs demand higher current-gain cut-off frequency (fT) and larger transconductance (gm). The FET gate length has been shortened to 0.15 μm and shorter with suppressing the short-channel effects as much as possible. A cross-sectional view of the 0.1-μm-class gate-length GaAs MESFET we have developed is shown in Fig. 4. The two-step buried p-layer (BP and BP2) lightly-doped drain (LDD) structure is effective in suppressing the short channel effects.[20] The short gate-length (Lg) was obtained by using i-line photolithography and shrinking the resist size by O_2-RIE (reactive ion etching) and ECR (electron

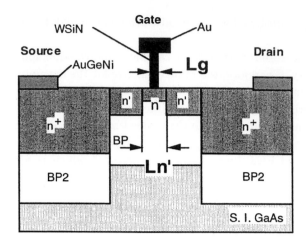

Fig. 4. Cross-sectional view of a BP-LDD GaAs MESFET.

cyclotron resonance) etching. The standard deviation of the gate-length is about 0.01 μm, which is small enough to give uniformity of the threshold voltage (Vth) for ICs. The T-shaped Au/WSiN gate structure is optimized to minimize the gate parasitic capacitance without increasing gate resistance very much. Offsets (Ln'-Lg), which separate the gate electrode from the n' layers, also help reduce the gate parasitic capacitance. These two techniques reduce the gate capacitance to 3/4 of that of conventional 0.2-μm gate-length MESFETs. The Vth, transconductance, fT, and fmax are 55 mV, 530 mS/mm, 100 GHz, and 101 GHz, respectively. Typical FET parameters for a 0.15-μm gate-length MESFET are summarized in Table 1.

Table 1. Transistor parameters of BP-LDD 0.15-μm gate-length GaAs MESFET.

Threshold voltage	Vth	55 mV
Transconductance	gm^*	530 mS/mm
Drain conductance	gd^*	36.4 mS/mm
Gate resistance	rg	12.85 Ω^{**}
Source resistance	rs	10.44 Ω^{**}
Drain resistance	rd	10.44 Ω^{**}
Gate-Source capacitance	Cgs^*	38.6 fF**
Gate-Drain capacitance	Cgd^*	10.7 fF**
Drain-Source capacitance	Cds^*	15.2 fF**
Current-gain cutoff freq.	fT^*	100 GHz
Maximum oscillation freq.	$fmax^*$	101 GHz

*The values are at Vds=1.5 V and Vgs=0.55 V.
**The values are for a gate width of 50 μm.

A record fT value of 163 GHz has been achieved for 0.06-μm gate-length MESFETs.[12] However, the short channel effects become so severe that the circuit speed performance is considerably degraded. One of the most typical aspects is the dull slope of fT (or transconductance) dependence on Vgs. Although the threshold voltage decreases in the negative bias region as the gate length is shortened, the Vgs point that gives the maximum fT (or the maximum transconductance gm$_{max}$) stays at a high level ($\sim$ 0.5 V). This, in turn, results in a dull DC transfer curve for large-signal operation of logic circuits if the circuit is designed to make full use of the device speed performance.

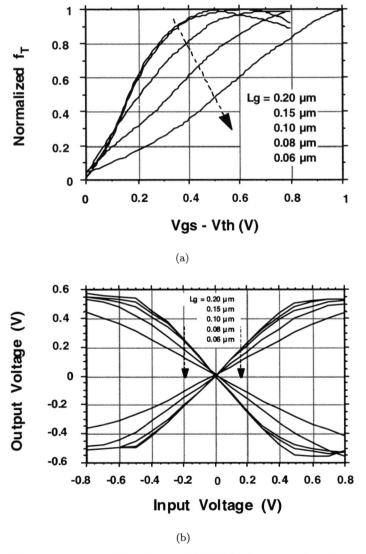

(a)

(b)

Fig. 5. Gate-bias dependencies of (a) fT and (b) DC-transfer curves for different gate-length MESFETs.

Typical fT versus Vgs is plotted for several FETs with different gate lengths in Fig. 5(a). Simulated DC transfer curves for SCFL buffer gates using those FETs are also plotted in Fig. 5(b) under the condition that the product of the gm_{max} and the load resistance R_L is constant. A good fT slope and sharp DC transfer are obtained for the devices having a gate length of no less than 0.10 μm. For these devices, the short channel effects are well suppressed. For devices with gate lengths shorter than 0.10 μm, the performances are drastically degraded by shortening the gate length. The degradations in fT slope and DC transfer correspond well to each other. From the above, a gate length of 0.10 μm is the critical condition under which the short channel effects can be well suppressed and the digital circuits can make full use of the device speed performance. Taking the process margin and repeatability into account, we decided to utilize 0.12- to 0.15-μm gate-length MESFETs for high-speed IC fabrications.

4. Relation between IC Speed Performance and FET Parameters

4.1. Analytical expression for FF toggle frequency

Figure 6 shows the trend in D-FF speed versus transistor fT for FETs. Previously reported IC speeds stay 1/5 to 1/3 of fT.[21] Static D-FFs (Master-slave D-FFs) stay around 1/5, and dynamic D-FFs (clocked-inverter D-FFs) around 1/3 of fT. The data selectors are also about 1/3 of fT. This indicates that a 40-Gbit/s static

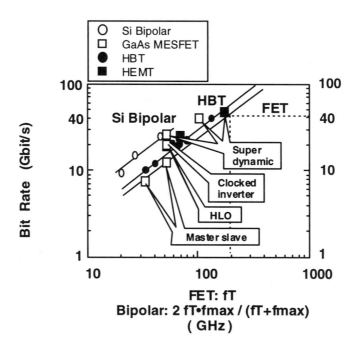

Fig. 6. Trend in D-FF speed versus device figure of merits.

D-FF operation needs 200-GHz FETs. It is impossible to develop such an ultra-high-speed GaAs MESFET. We must, therefore, clarify the device design criteria to maximize the IC speed performance as well as devise a new circuit design to improve operation speed. In that sense, a precise study of device figure of merits from the view point of digital IC operation speed is very important to relaxing the device requirements.

An analytical expression of the FF toggle frequency has been derived on the basis of the small-signal transfer function of an SCFL inverter gate.[22] IC speed can be roughly estimated with it but the expression does not give good agreement with either circuit simulation results using HSPICE, or with measured results. Why? This is because of the exclusion of large-signal behavior in the expression. In fact, the transconductance and gate-source capacitance have significant bias-dependence, and in Ref. 22, each parameter is approximated to a constant value (mean value for large-signal operation). It is difficult to analytically express all of the circuit behavior. As a result, we tried to derive an accurate semi-empirical formula for the FF speed versus transistor parameters on the basis of a mixture of sensitivity analysis and the analytical expression.[23]

The proposed formula for an intrinsic toggle frequency $f_{MS\text{-}TFF}$ (excluding the interconnection parasitics) is expressed as

$$
f_{MS\text{-}TFF}
$$

$$
= \frac{gm}{\left\{ \alpha \cdot Cgd \left(1 + \frac{X}{1+X\frac{gd}{gm}}\right)(1 + \eta \cdot rg) + \beta \cdot Cgs(1 + \eta \cdot rg) + \gamma \cdot Cds \right\} \left(1 + \left(\frac{1}{1+\lambda\frac{gd}{gm}}\right)\right)},
$$

$$(1)$$

where $f_{MS\text{-}TFF}$ is the intrinsic toggle frequency excluding interconnection parasitics, gm, gd, rg, Cgd, Cgs, and Cds refer to the transconductance, drain conductance, gate resistance, gate-drain capacitance, gate-source capacitance, and drain-source capacitance, respectively, α, β, γ, and λ are the fitting parameters, X indicates the product of the load resistance and the gm (corresponds to the logic swing). All the factors of the time constants which appear in the analytical expression in Ref. 22 are included, but the weighting of each time constant that contributes to the circuit speed is optimized so as to satisfy the sensitivity analysis.[23]

The actual maximum toggle frequency ($F_{TFF\,max}$) can be expressed by the sum of the inverse of the intrinsic toggle frequency $f_{MS\text{-}TFF}$ and extra-delay due to interconnection parasitics:

$$
F_{TFF\,max} = \left[f_{MS\text{-}TFF}^{-1} + \sum_{i=1}^{2} (\tau_{CLi} + \tau_{PDi}) \right]^{-1},
$$

$$(2)$$

where τ_{CLi} and τ_{PDi} are the CR delay and the propagation delay due to the ith interconnection line. Hence, we assumed:

(i) the interconnections from the master to the slave FF ($i = 1$) and from the slave to the master FF ($i = 2$) are dominant and others are negligible;

(ii) the ith interconnection-line capacitance C_{Li} is driven by the source follower circuit; and

(iii) the ith interconnection line has a length of L and a propagation velocity of V_P.

Thus, τ_{CLi} and τ_{PDi} can be approximated as follows:

$$\tau_{CLi} = \frac{C_{Li}}{gm\left(1 + 2\dfrac{gd}{gm}\right)} ; \tag{3}$$

$$\tau_{PDi} = L \cdot V_{Pi} . \tag{4}$$

First we measured the DC and AC characteristics of fabricated MESFETs having a gate length of 0.15 μm. Then FET model parameters for HSPICE were extracted for each device by using the optimizer function in HSPICE. A set of parameters for a nominal performance of FET (summarized in Table 1) was regarded as the reference. In order to prepare for substantial variations in the parameter set (within + or −40% of a nominal value for each parameter), many extra sets of parameters were prepared, in each of which the value of a specific parameter was artificially changed from the reference set.

τ_{CLi} and τ_{PDi} are extracted from an actually designed layout pattern on the basis of an electromagnetic simulator and a microwave circuit simulator. First, the interconnection lines are physically modeled onto the EM simulator to calculate their S parameters. Then, they are electrically modeled onto the circuit simulator and their electrical parameters are extracted so as to fit the circuit responses to the simulated S parameters.

By using each set of FET parameters, HSPICE simulation was performed to simulate $F_{TFF\,max}$. Then, the fitting parameters in Eq. (1) were optimized to minimize the sum of the errors between the calculated and simulated maximum toggle frequencies. The fitting results are shown in Fig. 7 and resultant values of the weighting coefficients are summarized in Table 2. A single set of the coefficients gives a very good coincidence over wide variations of the toggle frequencies. In order to investigate how widely and well the formula with the single set of extracted coefficients can estimate the T-FF speed of actually fabricated ICs, measured maximum toggle frequencies versus calculated ones were plotted for FETs having a gate length from 0.12 to 0.24 μm. The result is shown in Fig. 8. For wide variations of gate length, the formula gives good agreement with the measured results within a 12% error. This is very important to applying this formula to device optimization in practice.

For FETs with gate lengths shorter than 0.12 μm, the extracted coefficients in the formula would greatly overestimate the circuit speed. This means that the actual circuit performance would be more degraded due to the increase in short channel effects (see Fig. 5) than is extrapolated from the 0.24-to-0.12-μm device characteristics. For those shorter-gale-length FETs, the weighting coefficients in (1) should be re-fitted.

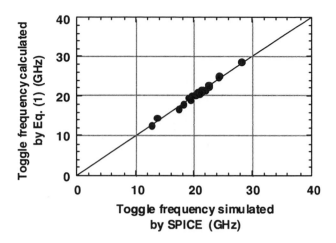

Fig. 7. Simulated maximum toggle frequency versus the calculated one using the proposed formula.

Table 2. Extracted values of weighting coefficients in Eq. (1).

α	7.35
β	1.40
γ	7.00
η	0.01
λ	−2.00

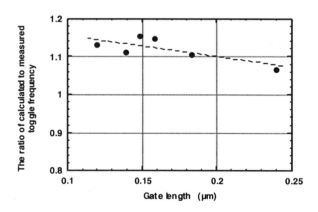

Fig. 8. Relation between the ratios of the calculated to the measured maximum toggle frequencies and the gate lengths of fabricated MESFETs.

4.2. *Device figure of merit for high-speed digital IC*

Equation (1) reveals that the Cgd is the most dominant parameter that limits IC speed. gm and gd/gm are also important parameters. The equation takes the similar

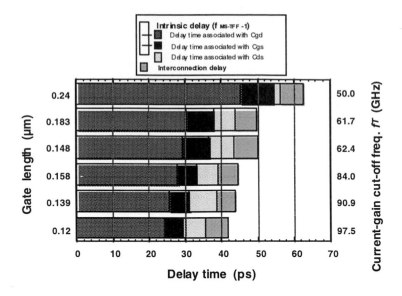

Fig. 9. The weight of switching-delay components of T-FFs. The intrinsic delay is classified in terms of the FET capacitances.

formations to the definition of fT; gm/(sum of capacitances). The IC speed is related to the fT, but not directly. In other words, FETs with lower fT may bring faster IC operation than those with higher fT if the Cgd is lower. Figure 9 expresses the weight of switching-delay components of the T-FF in terms of FET capacitances (Cgs, Cgd, and Cds) and the interconnection delay. It is clearly seen that the Cgd is the dominant factor that occupies about 50% of the total delay time for a 0.12-μm gate-length FET. High gm with small Cgd and Gds is the best measure for designing high-speed SCFL ICs.

Figure 9 also indicates that the interconnection parasitics become dominant as the transistor speed increases. This is because the minimum interconnection rules (line width and pitch) remain constant.

The circuit speed analysis using Eqs. (1)–(4) reveals that it is very difficult for conventional static (dynamic) FFs to achieve 20-Gbit/s (40-Gbit/s) operation even if the FET gate length approaches 0.1 μm. Consequently, in order to develop 20- to 40-Gbit/s class digital ICs, novel circuital advances are essential.

5. High-Speed Circuit Design Technology

As mentioned before, the fundamental elements (core elements) of the digital ICs for optical fiber communications are: (1) a data selector, (2) a D-FF, and (3) a T-FF. From the viewpoint of circuit configuration, T-FFs are regarded as a part of D-FFs in which the output data is invertedly fed back to the data input. Furthermore, the D-FF is configured with a static circuit or with a dynamic circuit. A dynamic D-FF can generally operate faster than a static D-FF but lower bit-rate operation is sacrificed. In terms of the circuit's configuration, we discuss the circuit design technologies for (1) data selectors, (2) static FFs, and (3) dynamic FFs. In the final part of this section, we discuss (4) wideband buffer circuit design which becomes critical as the core FF or selector gets faster. The FET parameters we assumed are those for the 0.15-μm gate-length GaAs MESFETs summarized in Table 1.

5.1. *Data selector*

Figure 10 shows the circuit diagram of a typical SCFL-type 2:1 data selector. The first-level differential pair, D1 and D2, is assigned to the input data reading, and the second-level is assigned to alternatively selecting the input data. Its operation principle is very simple and is as follows. When the input clock frequency (Hz) corresponds to the input data rate (bit/s), the input data is alternatively selected once at every clock cycle resulting in time-division multiplexing.

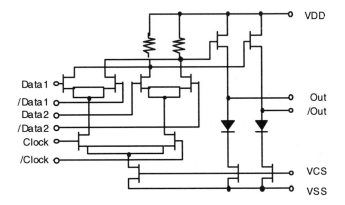

Fig. 10. Circuit diagram of a conventional SCFL 2:1 data selector.

It is seen from the above mentioned operation principle that, in the selector core, the highest frequency component at the output node is equal not to the input data frequency, but to the input clock frequency. Therefore, the AC characteristics along with the signal path from the clock input to the output is the most critical for the circuit speed. Taking this fact into account, a new high-speed data selector has recently been proposed.

Figure 11 shows the circuit diagram of the new data selector circuit.[24] A parallel feedback circuit[25] is added to the clock input differential gate, and inductors are connected to the load resistors in series. Figure 12 shows the calculated gain-bandwidth characteristics of the clock-to-data path for several types of selector circuit. The gain in the low frequency region is boosted by adding the parallel feedback circuit to the clock input gates. The feedback resistance RF is designed so as to maximize the DC gain of the circuit. Also, inductor peaking compensates for the losses of the gain in the high frequency region. Both the gain and bandwidth

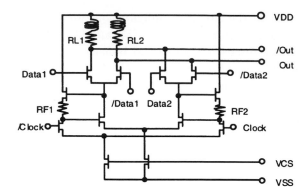

Fig. 11. Circuit diagram of the new 2:1 data selector.

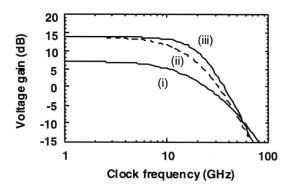

Fig. 12. Calculated gain-bandwidth characteristics of the clock-to-data path for several types of data selector. (i) conventional circuit, (ii) circuit including parallel-feedback only, and (iii) new circuit including both parallel-feedback and inductor peaking.

are improved simultaneously by applying two analog wideband design techniques to the selector circuit. The SPICE simulation shows that the new circuit improves its maximum operating speed by more than 10% over that of the conventional one.

5.2. *Static FF*

The most familiar conventional circuit is the master-slave D-FF (MS D-FF). Figure 13 shows a diagram of the circuit. In Fig. 13, MR and ML mean the reading and the latching source-coupled pair (SCP) circuit in the master FF, and SR and SL mean the reading and the latching circuit in the slave FF. The conventional circuit consists of a cascade of a simple series-gated master FF and slave FF, in each of which the reading and latching circuit are symmetrically configured.

When the clock is high, MR and SL are active, so the input data can be read and the output of the master FF is changed according to the input data. This operation cycle is regarded as the reading period. When the clock is turned off (low), ML

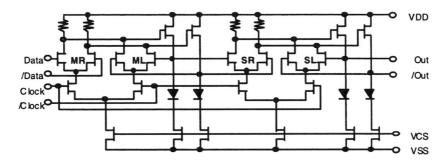

Fig. 13. Circuit diagram of a conventional SCFL MS D-FF.

and SR become active, so the master-FF output is latched and transferred to the slave-FF. At this moment, the output of the slave-FF, thus the output of the MS D-FF, is changed according to the latched data at the master-FF. According to this operation manner, the input data is retimed to the clock transition timing. Since the input data is stored by the latching circuit, the data never disappears even if the clock is stopped to maintain a low level. In this sense, the circuit can perform a static operation.

In order to improve the circuit speed, reducing the voltage swing without degrading the slew rate is effective, because it shortens the transition time. However, extra care must be taken to maintain the static operation. Unity gain along the critical path is the condition that guarantees static operation. A part-time shrinkage of the logic swing is effective to doing so. One example is the HLO (High-speed Latching Operation)-type D-FF, where the logic swing during the latching operation is somewhat reduced than that during the reading operation. Several types of such D-FFs have been proposed, including the HLO D-FF.[22,26-28] Furthermore, reducing the number of circuit elements is also effective to improving the speed because the interconnection parasitics can be reduced.

Figure 14 shows a new type of high-speed static D-FF we have recently developed.[29] It is hereafter called the High-speed Latching operation with Merged cell (HLM) D-FF. The HLM D-FF has two new features. One is the reduction of the number of circuit elements. The MR and the SL are coupled to create a common source, and the ML and the SR are also coupled. Furthermore, the second-level differential pairs in Fig. 13 are merged into a single pair (consisting of FET A and B). This configuration guarantees circuit operation equivalent to the conventional one, and is a kind of circuital degeneration which helps reduce the interconnection parasitics, resulting in faster operation.

The second feature of the HLM D-FF is the voltage bias levels fed to each first-level SCP. The bias level of the MR is set higher than that of other SCPs by providing the gate input of the MR from the anode of the upper diode of the source follower circuit. The gate inputs of the ML, SR, and SL, on the other hand, are

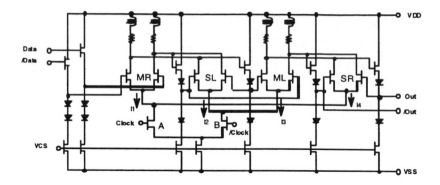

Fig. 14. Circuit diagram of the new static D-FF.

provided from that of the lower diode. As a result,

$$I1 > I2$$
$$I3 = I4 ,$$
(5)

where I1 (I2) is the drain current of FET A through the MR (SL) when FET A (B) is turned on (off), and I3 (I4) is the drain current of FET B through the ML (SR) when FET B (A) is turned on (off). Since the second-level differential pair is driven by a single current source, the following relation also holds;

$$I1 + I2 = I3 + I4 .$$
(6)

As a consequence,

$$I1 > I3 = I4 > I2 .$$
(7)

Since the logic swing of each SCP is given by the products of the load resistance and the currents, the output voltage of the SL is smaller than that of the MR. As is seen in Fig. 15, the transition time between the logical high and low levels is reduced in proportion to the reduction of the effective logic swing because the signal slew rate is not degraded in the circuit. The combination of the two features in the HLM D-FF reduces the voltage amplitude of data latching and results in higher speed operation than in a conventional circuit. Although the HLM D-FF has the drawback that the output amplitude shrinks somewhat intermittently, a static operation can be guaranteed if the output buffer/driver have enough DC gain to recover the voltage swing to the normal level.

Assuming a 0.15-μm-class gate-length GaAs MESFETs, HSPICE simulation indicates that the HLM D-FF can operate over 10% faster than the conventional one. As the transistor speed increases, the speed improvement is expected to be greater because the interconnection parasitics become more evident in limiting the circuit speed. If we assume a 200-GHz-class fT device with the present interconnection design rules, the HLM D-FF will be able to operate about 20% faster than a conventional one.

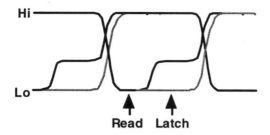

Fig. 15. A typical output waveform of the new D-FF.

5.3. *Dynamic FF*

The dynamic FF is a modified version of the static FF, in which the circuit can operate faster than the static one but the static operation is sacrificed. This means the dynamic FF requires a repetitive clock input with a specific clock rate to perform a normal operation.

The most popular dynamic D-FF is the clocked-inverter D-FF[22] as shown in Fig. 16. The circuit completely eliminates the latching circuit (ML and SL in Fig. 13) resulting in a cascade of two clocked inverters consisting of MR and SR each. This helps reduce the parasitic capacitance associated with the drain output nodes. Furthermore, when the clock is turned off, there is no driving current, so the logic low level instantaneously starts to move to the logic high level. If the clock is turned on before the logic low level exceeds the threshold bias level, the logic data can be restored to the normal logic level. Then, the circuit operates normally. Such low-level transition in the +direction of logic high-level shrinks the effective logic swing. These two factors: (1) reduction of the parasitic capacitance; and (2) shrinkage of the effective logic swing make the circuit operate faster than the conventional MS D-FF. The SPICE simulation shows that the clocked inverter D-FF can operate about 70% faster than the MS D-FF. The clocked inverter D-FF had been the fastest circuit until the super-dynamic D-FF was devised.

The new type of dynamic D-FF we have recently developed is the super-dynamic D-FF.[30] A diagram of the circuit is shown in Fig. 17. It is based on the HLO D-FF[22]

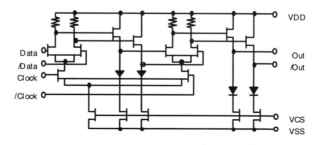

Fig. 16. Circuit diagram of a conventional SCFL clocked-inverter D-FF.

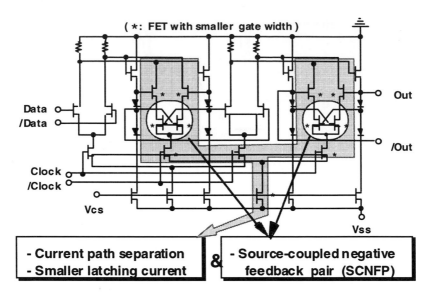

Fig. 17. Circuit diagram of the super-dynamic D-FF.

for GaAs LSCFL (low power source-coupled FET logic). Its circuit features are: (1) a series-gate connection to separate the current path of the reading and latching circuits; (2) a smaller latching current (I_{latch}) than the reading current (I_{read}); and (3) a source-coupled negative feedback pair (SCNFP) inserted in the first-level latching differential pair in a cascade manner. The first two features coincide with the HLO D-FF, but the last feature makes the circuit operation completely different from the HLO D-FF.

The operating waveforms at a very low speed are schematically shown in Fig. 18. Due to the negative feedback of the SCNFP, the complementary output levels are instantaneously forced to equalize at the middle of the logic swing at every transition between the latching and reading operations. Remember that the logic swing is smaller during the latching period than during the reading period. As a result, the equilibrium state is also different. This preserves the input data information for just a short time at every transition between reading and latching. This makes the FF operate dynamically. The SCNFPs drastically reduce the effective logic swing from $I_{read}R_L$ to $(I_{read} - I_{latch})R_L/2$ without any degradation of the signal transition slew rate, where R_L is the load resistance. In this figure, the original logic levels are shown by the dashed lines (designated as "Hi" and "Lo"). This is the key to increasing the FF speed performance. This FF operates normally at higher bit rates where the equilibrium state vanishes. The first trial sample of the super-dynamic decision IC was fabricated using production-level 0.2-μm gate-length GaAs MESFETs and demonstrated an excellent speed performance of 24-Gbit/s operation that corresponded to 45% of transistors' fT.[30]

Figure 19 shows simulated D-FF speed versus gate width ratio Wg_{latch}/Wg_{read} (corresponding to I_{latch}/I_{read}). For the simulation, the normal operation was judged

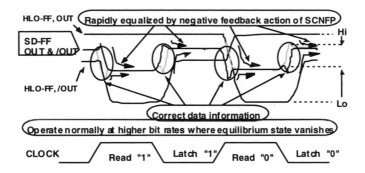

Fig. 18. Output waveform of the super-dynamic D-FF. To show the operation principle, clock rate is set low enough (mis-operating region).

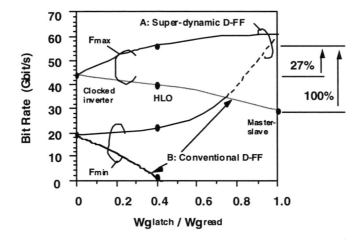

Fig. 19. Simulated super-dynamic D-FF speed versus gate-width ratio Wg_{latch}/Wg_{read}. Fmax: maximum operating bit rate; Fmin: minimum operating bit rate.

under the following conditions: (1) the output voltage swing is larger than 750 mV; and (2) the phase margin is larger than 150 degrees. The interconnection parasitics were ignored in the simulation. Wg_{read} was fixed at 40 μm, and R_L was optimized for each FF. Curve B shows the route of conventional FFs from the completely static (master slave) FF to the dynamic (clocked inverter) FF. The super dynamic FF (curve A) can increase its operating speed range by about 30% over that of the clocked inverter FF, and by 100% over that of the master-slave FF when the gate-width ratio is set at 0.4 (dotted on curve A). HSPICE simulation suggests that the maximum operating bit rate of the super-dynamic D-FF exceeds one half of the FETs' f_T.

Figure 20 shows the simulated FF speed versus the load resistance R_L with the parameter of Wg_{latch} when the Wg_{read} is set at 40 μm. The circuit has a wide tolerance for R_L. When the Wg_{latch} is set at 20 μm, a wide operation range from

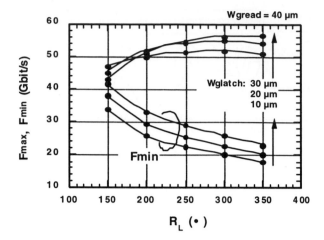

Fig. 20. Simulated super-dynamic D-FF speed versus load resistance R_L. W_{gread} is set at 40 μm.

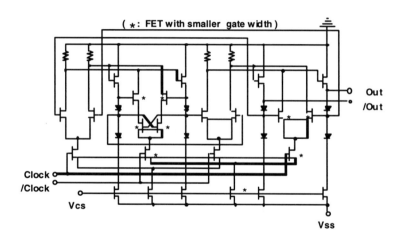

Fig. 21. Circuit diagram of the super-dynamic T-FF.

24 ± 1-Gbit/s to 50.5 ± 0.5-Gbit/s can be obtained for a large variation in R_L from 250 to 350 Ω.

Figure 21 shows a schematic of a super-dynamic T-FF. The SCNFP is introduced only in the master latching circuit so as to obtain sufficient voltage swing at the feedback data input nodes.

5.4. *Wideband data buffers*

As described in Sec. 2 with Fig. 2, digital ICs for optical fiber communications have a common configuration where data and/or clock buffer(s) are/is followed by a core functional block (such as an FF or a selector) whose outputs are followed by (an) output buffer(s). The input and output data buffers require a broadband operation with a high and flat gain from DC to the maximum operation frequency. These factors affect the input data sensitivity and output waveform quality. In particular, the input data buffer is the most critical because its data input is single-ended and a source follower level shifter must be incorporated prior to the buffer while the output buffer is pair-driven. When we assume a standard non-return-to-zero (NRZ) data format, these buffers are usually required to cover the bandwidth over approximately 70% of the input data bit rate. The conventional buffer circuit (based on a differential amplifier) cannot cover the operating range of the new data selector and the super-dynamic D-FF.

We, therefore, employed two types of wideband data buffers with them. One was a parallel feedback amplifier,[25] shown in Fig. 22(a) and the other was an inductor-peaking amplifier, shown in Fig. 22(b). For both circuits, capacitance peaking (Cpf) was also incorporated in the source followers (SFs), which compensates for the loss of the SFs at high frequencies. The parallel feedback amplifier can optimize its gain-bandwidth characteristics by means of the value of the load resistance RL and the feedback resistance RF. Meanwhile the inductor-peaking amplifier can do the same by means of the values of the load resistance RL and the inductor L.

The typical simulated gain-bandwidth characteristics of these buffers are shown in Fig. 23. In this simulation, the circuit configuration of a data input buffer

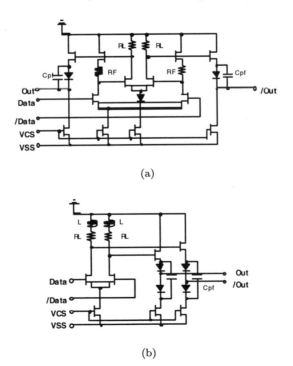

(a)

(b)

Fig. 22. Wideband data buffers: (a) the parallel-feedback amplifier; (b) the inductor-peaking amplifier.

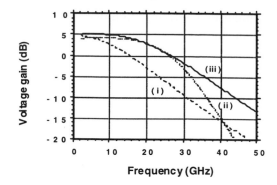

Fig. 23. Simulated gain-bandwidth characteristics. (i) conventional buffer, (ii) parallel-feedback buffer and (iii) inductor-peaking buffer.

(consisting of a series of a pair of input-stage source followers and two-stage amplifiers with a pair of output-stage source followers) was assumed. The circuit parameters used are as follows: $Wg = 40$ μm, $RL = 200$ Ω, $RF = 100$ Ω, $L = 0.5$ nH, and $Cpf = 2.0$ pF. Simulation results indicate that the -3-dB bandwidths for both buffers were improved by over 80% over those for a conventional buffer. The parallel feedbac amplifier achieves slightly wider bandwidth with flatter gain than those of the inductor peaking amplifier, although the roll-off is steeper.

For applications to 20 to 40-Gb/s class ICs, an inductor value of the order of subnano-henries is needed. It can be monolithically integrated using the metal interconnection layers, but requires a fairly large area of about 100–200 μm square for each. Thus, designers should select the circuit configuration taking both the required performance and the allowable layout area into account.

5.5. Wideband clock buffers

Clock buffers for D-FF and T-FF ICs require the highest bandwidth of all the digital ICs. As the core FF speed increases, the bandwidth of a conventional SCFL clock buffer can no longer cover the FF operating range. This causes not only a limit on circuit speed performance, but also serious degradation in the retiming capability of D-FFs.[30]

In the case of a static circuit that is subject to broadband operation from DC, the clock buffer as well as the data buffer must guarantee the DC operation. For this purpose, the wideband design techniques (including feedback and/or peaking) used in the data buffers are similarly utilized. However, the clock buffer needs a much wider bandwidth than that for the data buffer. For example, a 20-Gbit/s static decision IC or a 40-Gbit/s 2:1 data selector IC needs an input data buffer having a bandwidth from DC to at least 14 GHz and a clock input buffer having a bandwidth from DC to at least 20 GHz. The requirement for the clock buffer bandwidth is, therefore, the most critical issue. Assuming a 0.15-μm-class gate-length GaAs MESFET, HSPICE simulation indicates that DC-to-over-20-GHz bandwidth can be achieved by incorporating the parallel feedback and inductor peaking circuitry into a standard differential buffer gate. Consequently, a 20-Gbit/s static decision IC and a 40-Gbit/s 2:1 data selector IC can take this circuit configuration for the clock input buffer.

For use in 40-Gbit/s D-FF and T-FF, the clock buffer should cover a frequency range beyond 40 GHz. To further enhance the clock buffer bandwidth, a parallel-feedback differential buffer was first considered. The circuit diagram is shown in Fig. 24. A capacitively coupled resistive divider was newly introduced as a low-loss, passive RF level shifter instead of the source-follower.[30] This is because source-followers are a major cause of loss, which is due to relatively large drain conductance (gds). DC analysis gives the transfer loss as approximately gds/(gm + gds), where gm is the transconductance. This is very different from the emitter-follower in bipolar transistors.

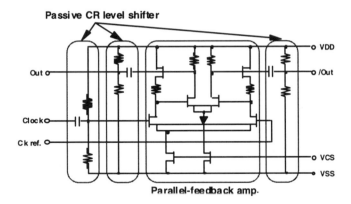

Fig. 24. A wideband clock buffer consisting of a cascade of an input-stage passive-level shifter, a parallel-feedback buffer and a pair of output-stage passive-level shifters.

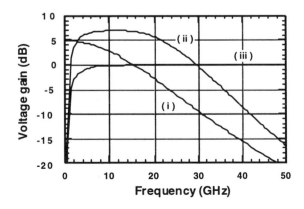

Fig. 25. Simulated gain-bandwidth characteristics for data input buffers. (i) conventional buffer, (ii) parallel-feedback buffer with passive level shifters, (iii) passive level shifter only.

Figure 25 shows simulated gain-bandwidth characteristics for several clock buffers. The conventional buffer (i) is the same as that in Fig. 23 (i). The parallel-feedback buffer basically has the same configuration as that for the data buffer shown in Fig. 23 (ii) except that the source followers are replaced by the passive level shifters. Compared to the parallel-feedback data buffer shown in Fig. 23, the gain drop with frequency is improved by replacing the source-follower with the passive level shifter resulting in a small-signal -3 dB-gain bandwidth of 2 to 34 GHz. However, it is insufficient to cover the 40-Gbit/s-class FF operation. One solution to cope with this is eliminating the differential amplifiers while keeping the passive level shifter. As shown in Fig. 25, a wide -3-dB bandwidth from 3 to beyond 50 GHz can be obtained, although the voltage gain is sacrificed. We adopted this simple level shifter as the clock input buffer in the 40-Gbit/s dynamic decision IC.

6. Experiments

On the basis of the circuit design technology described above, the following three ICs were actually designed, fabricated, and tested.

(i) a 2:1 data multiplexer (MUX) IC;
(ii) a static decision IC; and
(iii) a dynamic decision IC.

These ICs were fabricated using 0.12-μm gate-length GaAs MESFETs with standard double Au interconnection layers with a 2.5-μm thick Polyimide insulator in between. The device parameters extracted from the fabricated FETs are summarized in Table 3.

Table 3. Transistor parameters of BP-LDD 0.12-μm gate-length GaAs MESFET.

Threshold voltage	Vth	-45 mV
Transconductance	gm^*	481 mS/mm
Drain conductance	gd^*	31.4 mS/mm
Gate resistance	rg	15.7 Ω**
Source resistance	rs	9.55 Ω**
Drain resistance	rd	9.49 Ω**
Gate-Source capacitance	Cgs^*	47.3 fF**
Gate-Drain capacitance	Cgd^*	8.8 fF**
Drain-Source capacitance	Cds^*	10.7 fF**
Current-gain cut off freq.	$f_T{}^*$	98 GHz
Maximum oscillation freq.	$fmax^*$	97 GHz

*The values are at Vds = 1.5 V and Vgs = 0.55 V.
**The values are for a gate width of 50 μm.

6.1. *2:1 MUX IC*

Figure 26 shows the block diagram of the fabricated 2:1 MUX IC.[24] The target maximum speed was 44-Gbit/s. The IC has single data/clock inputs and differential outputs, which are respectively connected to impedance-matched 50-Ω and 100-Ω termination resistors to obtain clear eye patterns. The parallel-feedback-type new selector was adopted for the core multiplexer block. The data input buffer consists of a normal differential buffer and source followers with capacitive peaking. The clock buffer and the data output buffer consist of inductor-peaking amplifiers and source followers with capacitive peaking (see Fig. 22(b)). The supply voltage is −4.5 V and dissipation power is 0.95 W. A microphotograph of the chip is shown in Fig. 27. The chip is 2×2.5 mm^2. For comparison, we also fabricated a conventional MUX IC that consists of a conventional selector core and the same input/output buffers as used for the new MUX IC.

The fabricated ICs were tested on a wafer. The measurement setup is shown in Fig. 28. It consisted of clock distribution, data generation, signal probing, and

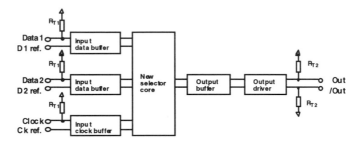

Fig. 26. Circuit block diagram of the 2:1 MUX IC.

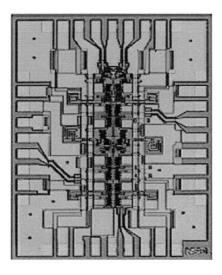

Fig. 27. Microphoto of the 2:1 MUX IC chip. 2 mm × 2.5 mm.

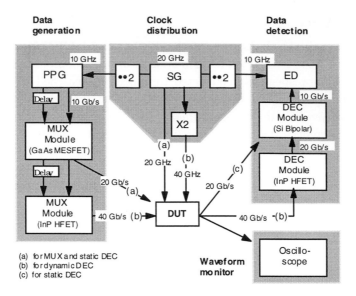

Fig. 28. Measurement setup.

monitoring blocks. (The data detection block in Fig. 28 was used in a subsequent measurement for the decision ICs.) In the clock distribution block, the quality of the clock synchronization becomes more important as the data rate increases. We used a single 20-GHz synthesized signal source (SG) as a master clock source, and 10-GHz clocks to the pulse-pattern generator (PPG) and the error detector (ED) were generated from the master clock by using a frequency divider.

In the data generation block, complementary pairs of a fundamental pseudo-random bit stream (PRBS) of PN $2^{23} - 1$ up to 12-Gbit/s were generated from the PPG. These were duplexed with appropriate delay against each other by a GaAs MESFET MUX unit.[31]

To probe the high-speed IC on a wafer, we utilized dedicated multiple contact probe heads having two DC pads and two 40-GHz-bandwidth RF pads. For monitoring the waveforms, a 50-GHz bandwidth digitizing oscilloscope HP 54123-T was used. 40-GHz bandwidth flexible coaxial cables with lengths as short as was practical were used for RF signal connection between the measurement instruments and the IC under test.

Figure 29 shows the measured operating waveforms of the new and conventional MUX ICs at 40 and 44-Gbit/s. Good eye openings with an output voltage of 850 mVp-p were obtained at 40-Gbit/s for the new MUX IC, whereas the conventional IC barely opened its output eye. At 44-Gbit/s, the new MUX still maintains a good eye opening, whereas the conventional one can no longer operate normally. As is expected, the proposed new data selector circuit enhanced the circuit speed by about 10% over the conventional one.

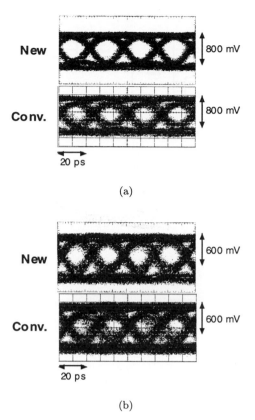

(a)

(b)

Fig. 29. Measured eye diagrams of the new and conventional MUX ICs (a) at 40-Gbit/s and (b) at 44-Gbit/s. Upper: new; lower: conventional.

6.2. *Static decision IC*

Figure 30 shows a block diagram of the fabricated static decision IC.[29] The target maximum speed was 20-Gbit/s with sufficient input sensitivity and clock phase margin. The input/output interfaces are the same as those of the MUX IC. The HLM D-FF was adopted for the core D-FF block. The data/clock buffers were designed with similar configurations to the MUX IC to cover the frequency range needed for 20-Gbit/s operation. The supply voltage is -4.5 V and dissipation power is 1.08 W. A microphotograph of the chip is shown in Fig. 31. The chip is 2×2.5 mm^2. For comparison, we also fabricated a conventional static decision IC that consists of a conventional MS D-FF as the core and the same input/output buffers as used for the new static decision IC.

The fabricated ICs were tested on a wafer. The measurement setup was basically the same as for the 2:1 MUX IC, which is shown in Fig. 28, except for inclusion of the data detection block where the output of the static decision IC with a duplexed data rate up to 22-Gbit/s was demultiplexed to a fundamental data rate up to 11-Gbit/s by a Si bipolar decision IC and then fed to the ED.

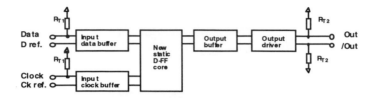

Fig. 30. Circuit block diagram of the static decision IC.

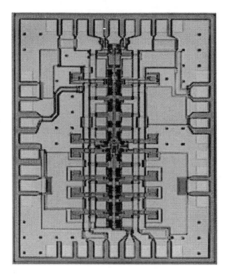

Fig. 31. Microphoto of the static decision IC. 2 mm × 2.5 mm.

Figure 32 shows the measured input and output eye diagrams of the static decision IC at 22-Gbit/s. To evaluate the retiming performance of the IC, in this figure, the data input timing was swept by 20 ps (corresponding to a phase of ~170 deg.) against the clock input timing. The output exhibited good retiming performance at this rate. The output voltage was 750 mV$_{p-p}$ and slightly lower than the designed value (900 mV$_{p-p}$), which is due to the slightly lower sheet resistance of the fabricated resistors.

Error-free operation of the new decision IC was confirmed from under 1 to 22-Gbit/s. Figure 33 shows the bit-rate dependence of the input sensitivity and the phase margin of the new and conventional decision ICs. The input sensitivity and phase margin of the new (conventional) decision IC at 20-Gbit/s operation were 180 mV (310 mV) and 211 deg. (72 deg.), respectively. The implementation of the new static D-FF greatly improves the input sensitivity (71% higher) and phase margin (194% wider) at 20-Gbit/s, and increased the maximum operation speed by 10% more than that of a conventional one, which is acceptable for real applications to 20-Gbit/s systems.

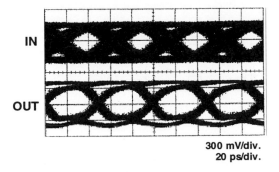

300 mV/div.
20 ps/div.

Fig. 32. Measured eye diagrams of the new static decision IC at 22-Gbit/s showing retiming performance. Upper: input; lower: output. The input data timing was swept by 20 ps against the clock input timing.

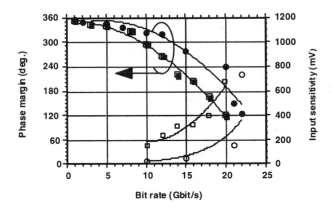

Fig. 33. Measured input data sensitivity and clock phase margin of the new static decision IC (circles) in comparison with a conventional one (squares).

6.3. *Dynamic decision IC*

Figure 34 shows a block diagram of the fabricated dynamic decision IC.[33] The objective operating range is from 20 to 40-Gbit/s with sufficient input sensitivity and clock phase margin. The IC has single data/clock inputs and differential outputs. The super-dynamic D-FF was adopted for the core D-FF block. The data input buffer consists of a two-stage parallel feedback amplifier with capacitive-peaking source followers (see Fig. 22(a)). The clock buffer, which requires 20-to-40-GHz bandwidth, was designed as a simple capacitively-coupled passive level shifter (see Fig. 24). The data output buffer consists of inductor-peaking amplifiers and source followers with capacitive peaking (see Fig. 22(b)). The supply voltage is −4.5 V and dissipation power is 0.98 W. A microphotograph of the chip is shown in Fig. 35. The chip is 2×2.5 mm^2.

The fabricated ICs were tested on a wafer. The measurement setup is basically the same as for the 2:1 MUX IC except for the inclusion of the data detection block,

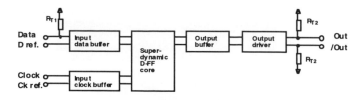

Fig. 34. Circuit block diagram of the super-dynamic decision IC.

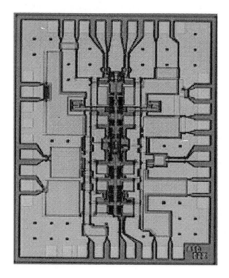

Fig. 35. Microphoto of the super-dynamic decision IC. 2 mm × 2.5 mm.

which is shown in Fig. 28. The 40-GHz clock to the decision IC was generated from the master clock source of 20 GHz using a frequency doubler. For the data generation block, complementary pairs of a PRBS of PN $2^{23} - 1$ up to 12-Gbit/s from the PPG were quadruplexed by a GaAs MESFET MUX[31] unit and an InP HEMT MUX module.[32]

In the data detection block, the output of the MUX IC with a quadruplexed data rate up to 44-Gbit/s was first demultiplexed to a half rate by an InP HEMT DEMUX module,[32] and then to the fundamental data rate by a Si bipolar decision IC and then fed to the ED. The output of the dynamic decision IC with a duplexed data rate up to 22-Gbit/s was directly demultiplexed to the fundamental data rate up to 11-Gbit/s by a Si bipolar decision IC and then fed to the ED.

The fabricated super-dynamic decision IC operated from under 20-Gbit/s to 40-Gbit/s. Figure 36 shows the input and output eye diagrams at 40-Gbit/s. Good eye opening with a 1.1 V_{p-p} swing was obtained. Figure 36 also shows 20-Gbit/s and 10-Gbit/s demultiplexed signals in 40-Gbit/s operation. We confirmed stable error-free operation at this bit rate. The input sensitivity and phase margin at

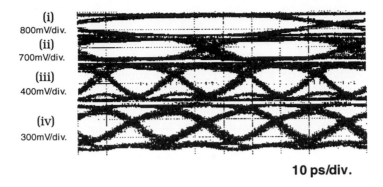

Fig. 36. Measured eye diagrams of the super-dynamic decision IC at 40-Gbit/s error-free operation. (i) 10-Gbit/s demultiplexed signal, (ii) 20-Gbit/s demultiplexed signal, (iii) 40-Gbit/s input signal, and (iv) 40-Gbit/s output signal.

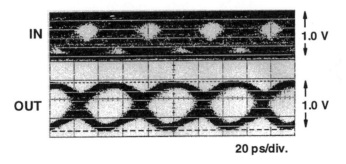

Fig. 37. Measured eye diagrams of the super-dynamic decision IC at a lower-limit bit rate of 20-Gbit/s. Upper: input; lower: output. To show the retiming performance, the data input timing is swept by 220 deg.

40-Gbit/s (PN $2^{23} - 1$) were 400 mV and 27 deg., respectively. When the fundamental 10-Gbit/s PRBS is shortened to PN $2^{15} - 1$, the phase margin was increased to 57 deg. The maximum bit rate of 40-Gbit/s corresponds to approximately 40% of the transistor fT, which is slightly lower than is expected from the super-dynamic D-FF core circuit operation. It is inferred that the newly designed data/clock buffers in the decision IC barely satisfied the required bandwidth for 40-Gbit/s operation, but still limit the circuit speed.

Figure 37 shows the input and output eye diagrams at 20-Gbit/s, which is close to the lower speed limit. To show the retiming performance, the input data timing was swept by 220 deg. Because of the dynamic circuit operation, low and high levels cannot be kept constant and are periodically forced to the center level at every transition timing. The measured phase margin at this bit rate was 254 deg., which is wide enough. Except for that dynamic operation, the output exhibits a clear eye-opening. Therefore, this circuit works well as a retimer circuit even at this low bit rate.

7. Future Trends and Subjects

Table 4 summarizes the obtained performances of the newly-designed digital ICs. The fabricated ICs achieved 22-Gbit/s operation for the static decision IC and 40-Gbit/s operation for the 2:1 MUX IC and the super-dynamic decision IC, all of which are record performances for GaAs MESFET ICs. For 40-Gbit/s operation in particular, however, we must enhance the circuit speed margin so that the process variations and harsh operating environments in the real world can be tolerated.

Table 4. Obtained performances for fabricated ICs.

	2:1 MUX IC	Static Decision IC	Super-dynamic Decision IC
Chip size	2 mm × 2.5 mm	2 mm × 2.5 mm	2 mm × 2.5 mm
Power supply	−4.5 V	−4.5 V	−4.5 V
Power dissipation	0.95 W	1.08 W	0.98 W
Interface	SCFL	SCFL	SCFL
Operating Speed	DC to 44-Gbit/s	DC to 22-Gbit/s	20 to 40-Gbit/s
Input data sensitivity	————	180 mV @ 20-Gbit/s	400 mV @ 40-Gbit/s
Phase Margin	————	211 deg. @ 20-Gbit/s	27 deg. @ 40-Gbit/s

From the viewpoint of process technology, there still may be some chance to improve FET performance. As mentioned in Sec. 3, a gate length of 0.10 μm is the critical condition that can well suppress the short channel effects, and under which digital ICs can make full use of device speed performance if we assume state-of-the-art MESFET process technology. If the gate length can be shortened from 0.12 μm, the present case, to 0.10 μm with good repeatability, an fT of 120 GHz is expected, which is almost 20% higher than the present case and thus the speed margin is gained.

One serious problem is the degradation of IC speed by interconnection parasitics. As Y. Umeda suggested in Ref. 34, the following two critically limit the circuit speed:

(i) the interconnection propagation delay; and
(ii) waveform distortion due to multiple reflection caused by the mismatch between the logic-gate-output impedance and the characteristic impedance of the interconnection line.

In the case of the 40-Gbit/s super-dynamic decision IC, the interconnection delay occupies about 25% of the total switching time that determines the maximum speed. Therefore, advanced interconnection technology with high propagation velocity and impedance controlability is a key to improving the IC speed. A multiple-layered interconnection structure with low-permittivity insulation layers is a possible solution.

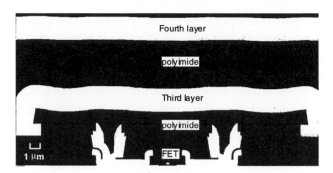

Fig. 38. Cross-sectional view of multilayer interconnection structure calculated by process simulator.

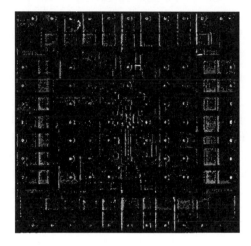

Fig. 39. Microphoto of the super-dynamic decision IC using multilayer interconnection technology. Most FETs and signal lines are concealed by the top-layer ground plane. The chip size is 2 mm × 2.5 mm.

One example we tried is the introduction of an impedance-controlled multilayer interconnection technology with relatively thick (10 μm) polyimide insulators.[12] Figure 38 shows a cross-sectional view. The top metal layer acts as a ground plane. Ground impedance is kept very low by using many connecting bumps on the plane. The impedance-controlled inverted-microstrip line (IMSL) is constructed with a third-layer strip and a top-layer ground plane. We designed another super-dynamic decision IC using 0.13-μm GaAs MESFETs incorporating this interconnection process. A microphoto of the chip is shown in Fig. 39. Most FETs and signal lines are concealed by covering the top-layer ground plane. Unfortunately, the data/clock buffers were designed with conventional circuitry, so the maximum speed was limited to 32-Gbit/s.[12,35]

The IMSL has relatively lower characteristic impedance that matches the source-follower-output impedance well and higher propagation velocity than those for standard lines without the top-layer ground. However, the IMSL includes larger coupling capacitance than normal lines without the top-layer ground, which significantly limits the circuit speed when the line length is far shorter than the signal wavelength. Thus, there is a trade off between the distributed line parameters and the lumped parameters. Designers should carefully design the interconnections taking these facts into account. Of course, many other types of smart interconnection structures should be considered.

Packaging is another key technology for real applications of high-speed ICs. We have recently developed an ultrabroadband package called "chip-size cavity package" that minimizes the inner cavity so as to shift undesirable cavity resonances out of the transmission band.[36,37] The digital-type package can accommodate up to 6 RF ports.[37] We have confirmed stable operation of this package up to 52-Gbit/s by using an InP-based HEMT MUX IC.[38] This is also thought to be best suited to the GaAs MESFET ICs. Now we are trying to mount the fabricated chips on to the packages.

8. Conclusions

We discussed recent advances in very-high-speed digital IC design technologies based on GaAs MESFETs for future high-speed optical communications systems. We devised new types of a data selector and static/dynamic flip-flops, which are key elements to performing high-speed digital functions (signal multiplexing, decision, demultiplexing, and frequency conversion) of front-end transmitter/receiver systems. Incorporating this circuit design technology together with state-of-the-art 0.12-μm gate-length GaAs MESFET process, we successfully developed a DC-to-44-Gbit/s 2:1 data multiplexer IC, a DC-to 22-Gbit/s static decision IC, and a 20-to-40-Gbit/s dynamic decision IC. The fabricated ICs demonstrated record speed performances (44-Gbit/s data multiplex and 40-Gbit/s error-free decision) for GaAs MESFETs. Although the present status is in the early development stage and the operating speed margin is not sufficient, further device/circuit optimization together with such new technologies as the high-speed interconnection and broadband packaging we've proposed will lead 40-Gbit/s GaAs MESFET ICs to the practical use.

Acknowledgements

The authors would like to acknowledge Shoji Horiguchi, Hiroshi Yoshimura, Kiyoshi Nosu, Katsushi Iwashita, and Yukio Akazawa for their direction and encouragement. They also thank Mikio Yoneyama, Yuhki Imai, Satoshi Yamaguchi, Shunji Kimura, Yasuro Yamane, Makoto Hirano, Masami Tokumitsu, Kou Inoue, Takashi Sugitani, Kazumi Nishimura, and Kiyomitsu Onodera for their extensive contributions throughout this work.

References

1. H. Onaka, H. Miyata, G. Ishikawa, K. Otsuka, H. Ooi, Y. Kai, S. Kinoshita, M. Seino, H. Nishimoto, and T. Chikama, "1.1 Tb/s WDM transmission over a 150 km, 1.3 μm zero-dispersion single-mode fiber", in *OFC '96 Tech. Dig.*, Feb. 1996, PD19-1.
2. Y. Yano, T. Ono, K. Fukuchi, T. Ito, H. Yamazaki, M. Yamaguchi, and K. Emura, "2.6 Terabit/s WDM transmission experiment using optical duobinary coding", in *ECOC '96 Tech. Dig.*, Feb. 1996, pp. 5.3–5.6.
3. T. Morioka, H. Takara, S. Kawanishi, O. Kamatani, K. Takiguchi, K. Uchiyama, M. Saruwatari, H. Takahashi, M. Yamada, T. Kanamori, and H. Ono, "100-Gbit/s × 10 channel OTDM/WDM transmission using a single supercontinuum WDM source", in *OFC '96 Tech. Dig.*, Feb. 1996, PD21-1.
4. L. D. Nguyen, A. S. Brown, M. A. Thompson, and L. M. Jelloian, "50-nm self-aligned-gate pseudomorphic AlInAs/GaInAs high electron mobility transistors", *IEEE Trans. Electron Devices* **39** (1992) 2007–2014.
5. T. Enoki, Y. Umeda, K. Osafune, H. Ito, and Y. Ishii, "Ultra-high-speed InAlAs/InGaAs HEMT ICs using pn-level-shift diodes", in *IEDM Tech. Dig.*, Dec. 1995, pp. 193–196.
6. S. Yamahata, K. Kurishima, H. Ito, and Y. Matsuoka, "Over-220-GHz-fT-and-fmax InP/InGaAs double-heterojunction bipolar transistors with a new hexagonal-shaped emitter", in *GaAs IC Symp. Tech. Dig.*, Oct. 1995, pp. 163–166.
7. T. Otsuji, Y. Imai, E. Sano, S. Kimura, S. Yamaguchi, M. Yoneyama, T. Enoki, and Y. Ishii, "40-Gb/s ICs for future lightwave communications systems", *IEEE J. Solid-State Circuits* **32** (1997) 1363–1370.
8. K. Hagimoto, M. Yoneyama, A. Sano, A. Hirano, T. Kataoka, T. Otsuji, K. Sato, and K. Noguchi, "Limitations and challenges of single carrier full 40-Gbit/s repeater system based on optical equalization and new circuit design", in *OFC '97 Tech. Dig.*, Feb. 1997, pp. 242–243.
9. M. Yoneyama, A. Sano, H. Hagimoto, T. Otsuji, K. Murata, Y. Imai, S. Yamaguchi, T. Enoki, and E. Sano, "Optical repeater circuits using 40-Gbit/s InAlAs/InGaAs HEMTs Digital IC chipset", in *IEEE MTT-S Int. Microwave Symp. Tech. Dig.*, May 1997, pp. 461–464.
10. M. Togashi, M. Ohhata, K. Murata, H. Kindo, M. Ino, M. Suzuki, and Y. Yamane, "10-Gb/s GaAs MESFET IC's", in *IEEE GaAs IC Symp. Tech. Dig.*, Oct. 1990, pp. 49–52.
11. K. Nishimura, K. Onodera, S. Aoyama, M. Tokumitsu, and K. Yamasaki, "High performance 0.1 μm self-aligned-gate GaAs MESFET technology", in *Proc. European Solid-State Device Res. Conf. (ESSDERC) '96*, 1996, pp. 865–868.
12. M. Tokumitsu, M. Hirano, T. Otsuji, S. Yamaguchi, and K. Yamasaki, "A 0.1-μm self-aligned-gate GaAs MESFET with multilayer interconnection structure for ultra-high-speed ICs", in *IEDM Tech. Dig.*, Dec. 1996, pp. 211–214.
13. S. Kimura and Y. Imai, "DC-to-40-GHz GaAs MESFET distributed baseband amplifier IC", in *Asia-Pacific Microwave Conf. Tech. Dig.*, 1994, pp. 249–252.
14. S. Kimura, Y. Imai, Y. Umeda, and T. Enoki, "A DC-to-50-GHz InAlAs/InGaAs HEMT distributed baseband amplifier using a new loss compensation technique", in *GaAs IC Symp. Tech. Dig.*, Oct. 1994, pp. 96–99.
15. S. Kimura, Y. Imai, and Y. Miyamoto, "Novel distributed baseband amplifying techniques for 40-Gbit/s optical communication", in *GaAs IC Symp. Tech. Dig.*, Oct. 1995, pp. 193–196.
16. S. Kimura, Y. Imai, Y. Umeda, and T. Enoki, "0-90 GHz InAlAs/InGaAs/InP HEMT distributed baseband amplifier IC", *Electron. Lett.* **31** (1995) 1430–1431.

17. Y. Imai, S. Kimura, Y. Umeda, and T. Enoki, "A DC to 38-GHz distributed analog multiplier using InP HEMT's", *IEEE Microwave Guided-Wave Lett.* **4** (1994) 399–402.

18. Y. Imai, S. Kimura, T. Enoki, and Y. Umeda, "A DC-to-100-GHz InP HEMT 1:2 distributor IC using distributed amplification", *IEEE Microwave Guided Wave Lett.* **6** (1996) 256–258.

19. T. Takada and M. Ohhata, "A new interfaceing method 'SCFL-interfacing' for ultra high-speed logic IC's", in *GaAs IC Symp. Tech. Dig.*, Oct. 1990, pp. 211–214.

20. M. Tokumitsu, M. Hirano, K. Murata, Y. Imai, and K. Yamasaki, "A 0.1 μm GaAs MESFET technology for ultra-high-speed digital and analog ICs", in *IEEE MTT-S Int. Microwave Symp. Tech. Dig.*, May 1994, pp. 1629–1632.

21. E. Sano and K. Murata, "An analytical delay expression for source-coupled FET logic (SCFL) inverters", *IEEE Trans. Electron Devices* **42** (1995) 785–786.

22. K. Murata, T. Otsuji, M. Ohhata, M. Togashi, E. Sano, and M. Suzuki, "A novel high-speed latching operation flip-flop (HLO-FF) circuit and its application to a 19-Gb/s decision circuit using 0.2 μm GaAs MESFET", *IEEE J. Solid-State Circuits* **30** (1994) 1101–1108.

23. K. Murata and T. Otsuji, "An analytical toggle frequency expression for source-coupled FET logic (SCFL) frequency dividers", *IEICE Trans. Electron* **E81-C** (1998) 1106–1111.

24. K. Sano and K. Murata, "A 44-Gb/s delector IC fabricated with 0.12-μm GaAs MESFETs", *Electron. Lett.* **33** (1997) 1375–1376.

25. N. Ishihara, O. Nakajima, H. Ichino, and Y. Yamauchi, "9 GHz bandwidth, 8-20 dB controllable-gain monolithic amplifier using AlGaAs/GaAs HBT technology", *IEE Electron. Lett.* **25** (1989) 1317–1318.

26. K. Murata, M. Ohhata, M. Togashi, and M. Suzuki, "20-Gb/s GaAs MESFET multiplexer IC using a novel T-type flip-flop circuit", *IEE Electron. Lett.* **28** (1992) 2090–2091.

27. T. Seshita, Y. Ikeda, H. Wakimoto, K. Ishida, T. Terada, T. Matsunaga, T. Suzuki, Y. Kitaura, and N. Uchitomi, "A 20 GHz 8 bit multiplexer IC implemented with 0.5 μm WNx/W-gate GaAs MESFET's", *IEEE J. Solid-State Circuits* **29** (1994) 1583–1588.

28. K. Ishii, H. Ichino, M. Togashi, Y. Kobayashi, and C. Yamaguchi, "Very-high-speed Si bipolar static frequency dividers with new T-type flip-flops", *IEEE J. Solid-State Circuits* **30** (1995) 19–24.

29. K. Narahara, T. Otsuji, and M. Tokumitsu, "A 20-Gbit/s static decision IC made with a novel D-type flip-flop", submitted to *IEICE Trans. Electron.*

30. T. Otsuji, M. Yoneyama, K. Murata, and E. Sano, "A super-dynamic flip-flop circuit for broadband applications up to 24-Gb/s utilizing production-level GaAs MESFETs", *IEEE J. Solid-State Circuits* **32** (1997) 1357–1362.

31. M. Ohhata, M. Togashi, K. Murata and S. Yamaguchi, "25-Gbit/s selector module using 0.2 μm GaAs MESFET technology", *IEE Electron. Lett.* **29** (1993) 950–951.

32. T. Otsuji, M. Yoneyama, Y. Imai, S. Yamaguchi, T. Enoki, Y. Umeda, and E. Sano, "46-Gbit/s multiplexer and 40-Gbit/s demultiplexer IC modules using InAlAs/InGaAs/InP HEMTs", *IEE Electron. Lett.* **32** (1996) 685–686.

33. K. Murata, T. Otsuji, M. Yoneyama, and M. Tokumitsu, "A 40-Gbit/s super-dynamic decision IC using 0.15-μm GaAs MESFETs", in *IEEE MTT-S Int. Microwave Symp. Tech. Dig.*, May 1997, pp. 465–468.

34. Y. Umeda, K. Osafune, T. Enoki, H. Ito, and Y. Ishii, "SCFL static frequency divider using InAlAs/InGaAs/InP HEMTs", in *Proc. 1995 European Microwave Conf.*, 1995, pp. 222–228.

35. T. Otsuji, K. Murata, M. Tokumitsu, and S. Sugitani, "32-Gbit/s super-dynamic decision IC using 0.13 μm GaAs MESFETs with mutilayer-interconnection structure", *IEE Electron. Lett.* **33** (1997) 480–482.

36. T. Shibata, S. Kimura, H. Kimura, Y. Imai, Y. Umeda, and Y. Akazawa, "A design technique for a 60 GHz bandwidth distributed baseband amplifier IC module", *IEEE J. Solid-State Circuits* **29** (1994) 1537–1543.

37. S. Yamaguchi, Y. Imai, S. Kimura, and H. Tsunetsugu, "New module structure using flip-chip technology for high-speed optical communication ICs", in *IEEE MTT-S Int. Microwave Symp. Tech. Dig.*, May 1996, pp. 243–246.

38. T. Otsuji, M. Yoneyama, Y. Imai, T. Enoki, and Y. Umeda, "A 64-Gbit/s 2:1 multiplexer IC using InAlAs/InGaAs/InP HEMTs", *IEE Electron. Lett.* **33** (1997) 1488–1489.

International Journal of High Speed Electronics and Systems, Vol. 9, No. 2 (1998) 437–472

20–40 Gbit/s GaAs-HEMT CHIP SET FOR OPTICAL DATA RECEIVER

Z. LAO,* M. LANG, V. HURM, Z. WANG, A. THIEDE,
M. SCHLECHTWEG, W. BRONNER, G. KAUFEL, K. KÖHLER,
A. HÜLSMANN, B. RAYNOR and T. JAKOBUS
Fraunhofer-Institute of Applied Solid-State Physics
D-79108 Freiburg, Germany

Using our 0.2 and 0.3 μm AlGaAs/GaAs/AlGaAs quantum well HEMT technology, we have developed a chip set for 20–40 Gbit/s fiber-optical digital transmission systems. In this paper we describe nine analog and digital receiver ICs: a 22 GHz high-gain transimpedance amplifier, a 20 Gbit/s OEIC front-end optical receiver, a 25 Gbit/s automatic-gain-control amplifier, a limiting amplifier with a differential gain of 26 dB and a bandwidth of 27.7 GHz, a 20–40 Gbit/s clock recovery, a 20 Gbit/s low-power Master-Slave-D-Flipflop with 24 mW power dissipation, a parallel data decision and a 1:4 demultiplexer, both for bit rates of 40 Gbit/s, and a 30 GHz static frequency divider, respectively. All chips were characterized on wafers with 50 Ω coplanar test probes.

1. Introduction

The progress in fiber optical digital communications has had a revolutionary impact on telecommunication and data transmission in the last two decades. The introduction of SONET (Synchronous Optical Network) and SDH (Synchronous Digital Hierarchy) has been promoted by the advances in ultra-high speed fiber-optical digital communications. The data rate of the highest level defined so far (SONET OC-192 or SDH STM-64) is about 10 Gbit/s. This successful transmission in such time division multiplexing (TDM) optical fiber communication systems is based on high-speed circuit technology. Systems operating at 10 Gbit/s are now already under development using different technologies to fabricate the high-speed ICs. Next generation systems using SONET/SDH are expected to operate at data rates of 20 Gbit/s or 40 Gbit/s[1] since the capacity of optical fibers is in the range of Tbit/s. Based on these 20 Gbit/s or 40 Gbit/s TDM systems, the transmission capacity can be further increased by wavelength division multiplexing (WDM). Although a few 20 to 40 Gbit/s ICs, which are mostly the digital circuits, have been developed by using different technologies,[2–4] the high-frequency, high-speed performance and the OEIC capability[5–7] make our AlGaAs/GaAs/AlGaAs HEMT (high electron

---

*Now with Rockwell Semiconductor Systems, Newbury Park, CA.

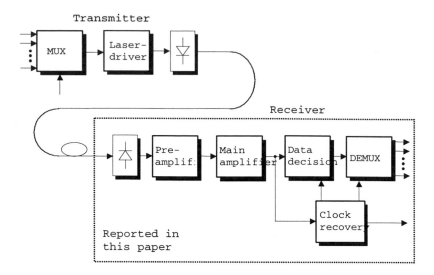

Fig. 1. Basic transmitter and receiver configuration for lightwave communication systems.

mobility transistor) technology especially attractive for large-scale or even mono-lithic integration of 20–40 Gbit/s optical transmitter and receiver modules.

A possible system configuration is shown in the block diagram of Fig. 1. At the transmitter, a multiplexer (MUX) combines parallel low-speed input signals into a single high-speed data stream. A laser driver or a modulator driver circuit amplifies the high-speed electrical signal to drive a laser diode or an optical modu-lator allowing the information to be transmitted through the fiber. At the receiver, a preamplifier circuit transforms the photocurrent detected by a photodiode into voltage signals. Following to the preamplifier IC a main-amplifier (AGC amplifier or a limiting amplifier) keeps the output voltage constant, i.e., independent of the in-put signal level. A data decision circuit makes a threshold decision and regenerates the data signal in time and amplitude. A clock recovery circuit extracts the clock from the data to operate the data decision and demultiplexer (DEMUX) circuits. A DEMUX splits the high-speed data stream into parallel low-speed signals.

On the transmitter side, a 40 Gbit/s AlGaAs/GaAs HBT 4:1 multiplexer[9] and a 20 Gbit/s SiGe HBT modulator driver[10] have been recently reported. We have designed a 45 Gbit/s 2:1 multiplexer and a 30 Gbit/s modulator driver using AlGaAs/GaAs/AlGaAs HEMTs in Refs. 11 and 12, respectively. In this paper we report on a complete chip set of 20–40 Gbit/s analog and digital ICs for the optical data receiver: a 22 GHz high-gain transimpedance amplifier, a 20 Gbit/s OEIC of frond-end optical receiver, a 25 Gbit/s automatic-gain-control amplifier, a limiting amplifier with a differential gain of 26 dB and a bandwidth of 27.7 GHz, a 20–40 Gbit/s clock recovery, a 20 Gbit/s low-power Master-Slave-D-Flipflop with 24 mW power dissipation, a parallel data decision (1:2 demultiplexer), a 1:4 de-multiplexer both for bit rates of 40 Gbit/s, and a 30 GHz static frequency divider, respectively. All are manufactured using our 0.2 μm and 0.3 μm gate-length en-hancement and depletion AlGaAs/GaAs/AlGaAs HEMT technology.

2. Technology

The fabrication of integrated circuits is based on a state-of-the-art III/V-semi-conductor process technology. For the active devices we have used three key technologies.[13] Firstly, molecular beam epitaxy (MBE) is employed for growing the epitaxial layers of our AlGaAs/GaAs/AlGaAs quantum well HEMT structure on 3″ GaAs wafers. Secondly, we define the gates of the transistors using electron beam lithography which enables gate lengths down to 0.1 μm. Because e-beam lithogra-phy is time consuming, we have developed a mix and match technology combining electron beam lithography with optical lithography using the e-beam lithography for writing the gates only. Thirdly, we perform a dry etch process (RIE) which stops precisely on an epitaxially grown $Al_{0.3}Ga_{0.7}As$ etch stop layer. Figure 2 shows the HEMT layer structure. NiCr thin film resistors are used with a sheet resistance of 50 Ω per square. Capacitors are realized as MIM structures using a dielectric layer with an average value for the capacitance per area of 225 pF/mm^2. The first interconnect layer consists of evaporated gold with a thickness of 310 nm. The critical dimension for this layer is 2 μm. The second interconnect layer is realized by electroplated gold with a thickness of 3 μm and a critical dimension of 3 μm. This layer is used to fabricate airbridges with a height of 1.4 μm and a thickness of 2.8 μm. Furthermore, inductors can be fabricated using airbridge technology.

The following mean values for the 0.2 μm gate length enhancement (E-mode) and depletion (D-mode) HEMTs, as shown in Table 1, were obtained: threshold voltages V_T = 0.05 V and -0.7 V, maximum transconductances g_m = 600 and 500 mS/mm, transit frequencies f_T = 60 and 50 GHz, respectively.

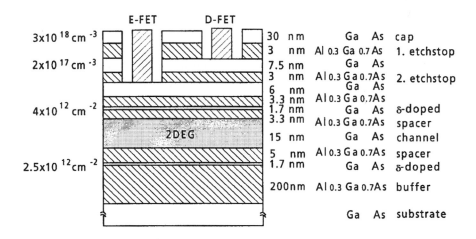

Fig. 2. HEMT layer structure for E- and D-FETs.

Table 1. Main parameters of the enhancement and depletion HEMTs.

Transistor	E-HEMT	D-HEMT
Gate length (μm)	0.2	0.2
I_{dsmax} (mA/mm)	200	180
	($V_{gs} = 0.6$)	($V_{gs} = 0$)
V_T (V)	0.05	-0.7
R_s (Ωmm)	0.6	0.6
g_m (mS/mm)	600	500
f_T (GHz)	60	50

3. General Circuit Techniques

High-speed ICs in broadband communication systems target not only operation at high-speed but also down to very low frequencies. The required low frequency limit of the circuits is determined by the section length in the bit stream. The transfer section overhead (frames) comes out every 125 μs (SDH standard), which corresponds to a frequency of 8 kHz. For monolithic circuits the necessary DC blocking must be provided off-wafer or by on-wafer DC-nulling techniques.

At low data bit rates some ICs were designed using direct coupled FET logic (DCFL) to keep power consumption low. To achieve maximum available operating speed, the fastest circuit principles must be used, combined with a very careful circuit design. The circuits presented in this paper are exclusively designed in source coupled FET logic (SCFL). This is important for ultra high speed circuit design to lower voltage swing and reduce common mode distortions. Low voltage swing, in turn, means increased speed and reduced power consumption. The values for internal differential swings are about 1 $V_{\mathrm{p-p}}$–1.5 $V_{\mathrm{p-p}}$. Figure 3 depicts a typical SCFL D-latch which is frequently used in high-speed digital circuits. The vertical series gating (for instance, track pair, latch pair and the clock pair in Fig. 3) is another advantage of using SCFL to build complex logic functions. Multiple source follower pairs are used for level shifting, impedance transformation and increasing the drain-gate voltage of the following current-switch transistors to lower their drain-gate capacitances. On-chip 50 Ω resistors were used at inputs to match the transmission lines and suppress the potential instability of source-follower inputs. 100 Ω–200 Ω resistors are suitable for direct ac coupling of the clock signal. Open drain buffers were chosen at the low-speed output, for instance for frequency dividers, to keep power dissipation low and to preserve high-speed performance. At high-speed outputs 100 Ω resistors were used for better matching over a wide frequency range.[14] The size of transistors was chosen based on driving and loading requirements between different stages. All transistors operate around maximum g_m for a given current.

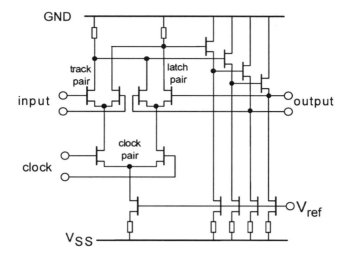

Fig. 3. A basic cell of D-latch in SCFL.

The layout strategy is to reduce the length of interconnections to a minimum, in particular the feedback lines in the static dividers. As far as is practical, interconnections carrying high frequency signals are kept short and formed using airbridges. These have very small parasitic capacitances and minimal ohmic resistances. All differential signal lines were arranged symmetrically to avoid differences in transit-times. Large ground contacts and separate ground and supply contacts were applied for analog and digital signal circuits. On-chip decoupling capacitances are placed between the voltage sources and ground as bypass capacitors for stabilizing the power supply. The design of the circuit has been optimized using the simulator HSPICE. Our transistor models are implemented in the HSPICE simulator and are commercially available.

4. Front-Ends of Optical Receivers

In optical-fibre links, low-noise, high gain (> 60 dBΩ [35]), high sensitivity and wide dynamic range preamplifiers are used as front-ends of receivers to convert photodiode current to voltage. Three kinds of amplifiers, i.e., high input impedance (HZ), transimpedance (TZ) and traveling wave (TW) or distributed amplifiers, can be used as preamplifiers. The HZ amplifier has the lowest cutoff frequency owing to the high input impedance combined with the photodiode capacitance, while the TW amplifier has the highest cutoff frequency even near $2/3$ f_T of utilized transistors within the circuits. However, for SDH applications the low frequency limit of the TW amplifiers must be extended down below 8 kHz by frequency compensation, and optical fiber amplifiers should be used due to low impedance gain of the circuits. Moreover, the interface to standard SCFL or ECL (emitter coupled logic) inputs and outputs can be difficult. The TZ amplifier in contrast to the TW amplifier has higher impedance gain, higher linearity and higher dynamic range, both due to the feedback configuration. For a typical TZ amplifier the input referred current-noise spectral-density is given by:

$$S_{nF}(f) = \frac{4kT}{R_F} + \left[4kT\Gamma g_m + \frac{4kT}{R_C}\right]\left(\frac{2\pi f C_{TF}}{g_m}\right)^2,$$

(1)

where R_F = feedback resistor, R_C = drain resistor in the first stage, C_{TF} = the sum of MSM photodiode capacitance and the HEMT input capacitance including the Miller capacitance, and Γ is the HEMT excess noise factor.[15] This equation shows that a smaller Γ and C_{TF} with larger g_m and R_F lead to higher sensitivity. Here Γ and g_m depend on the performance of the HEMTs, but the C_{TF} and R_F can be optimized by circuit design. The gain and the cutoff frequency of a transimpedance amplifier is represented by

$$Z_T = \frac{V}{I} = \frac{R_0 - AR_F}{1 + A + j2\pi C_{TF}(R_F + R_0)}$$

and

$$f_{-3\ dB} = \frac{1 + A}{2\pi C_{TF}(R_F + R_0)},$$

(2)

where A = the open-loop gain of the circuit, R_0 = is the output impedance of the source follower in the feedback loop. These equations show that the transimpedance gain is equal to the feedback resistor R_F when the open-loop gain A is very high, and that the circuit with a smaller input capacitance and a large open-loop gain will produce a wider bandwidth. Equations (1) and (2) give the general design consideration of the low noise and broad bandwidth approach.

In this section, we report on two broadband preamplifier circuits. One is a differential transimpedance amplifier with high gain and large bandwidth,[16] and the other is an OEIC (optoelectronic integrated circuit) using a single-ended transimpedance amplifier with an long wavelength monolithic integrated MSM (metal-semiconductor-metal) photodiode.[17]

4.1. Transimpedance amplifier

4.1.1. Circuit design

The transimpedance amplifier has a differential transimpedance input stage and four differential voltage amplifier cells. The differential transimpedance amplifier, as shown in Fig. 4, was chosen because it can operate with both single-ended and differential inputs. More important, the differential transimpedance configuration exhibits good common mode noise rejection, allows self thresholding, and reduces the simultaneous switching current. The transistors Q_F and Q'_F parallel to the feedback resistor R_F (R_F, R'_F = 600 Ω) allow the external control of both the group delay and the bandwidth according to optical characteristics. The inductors L and L' (1 nH) in series with the load resistors R_1 and R'_1 reduce the noise and broaden the bandwidth.

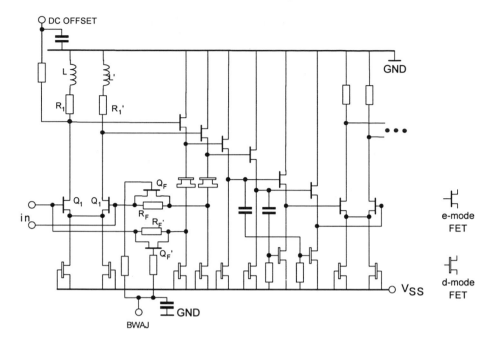

Fig. 4. Schematic of the differential transimpedance amplifier.

An application of such circuits is a photoreceiver integrating a MSM photodetector, a differential transimpedance amplifier and two limiting amplifier stages.[18] In this circuit the MSM photodiode is connected to one of the differential transimpedance inputs, for example, Q_1 (Fig. 4). Another input transistor Q_1' has a larger gate-width than that of Q_1 for the two reasons of: (1) Q_1' serves as a dummy, capacitor of the MSM photodiode, and (2) a higher current gain for Q_1' improves the symmetry of the outputs. The 1.3–1.55 μm wavelength photoreceiver OEIC has a bandwidth of 17 GHz with a high transimpedance gain of 12 kΩ. Eye diagrams were demonstrated at 20 Gb/s with an output voltage of 1 $V_{\text{p-p}}$ at an 83 μA photodiode current.

Figure 5 shows the chip micrograph of the transimpedance amplifier. The chip size is 1×1 mm^2.

4.1.2. *Circuit performance*

Figure 6 shows the transimpedance characteristics of the transimpedance amplifier at one output. The frequency domain measurements are performed using a network analyzer. The transimpedance gain is calculated from S-parameter characteristics at single-ended output using $Z_T = Z_0 \cdot S_{21}/(1 - S_{11}) = 70$ dBΩ, where Z_0 represents the 50 Ω transmission line impedance. The measured maximum 3 dB cutoff frequency is 22.2 GHz and the differential gain is 76 dBΩ. Connecting the transimpedance amplifier with a photodetector, the bandwidth of the optical receiver will be reduced

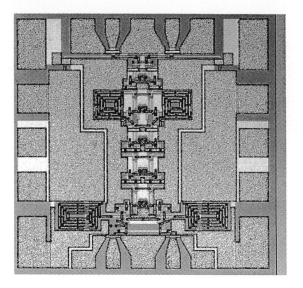

Fig. 5. Chip micrograph of the transimpedance amplifier.

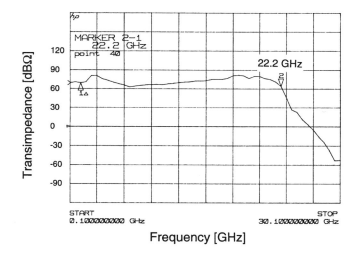

Fig. 6. Frequency characteristics of the transimpedance amplifier.

due to the dominant time constant of the photodiode capacitance combined with the input impedance of the amplifier.[18] The power dissipation of the transimpedance amplifier is 400 mW with a single power supply voltage of -4 V.

4.2. Optoelectronic receiver

4.2.1. Design and fabrication

Monolithic integrated photoreceivers have been widely investigated because of their potential for high-speed operation, increased reliability, easy packaging and reduced

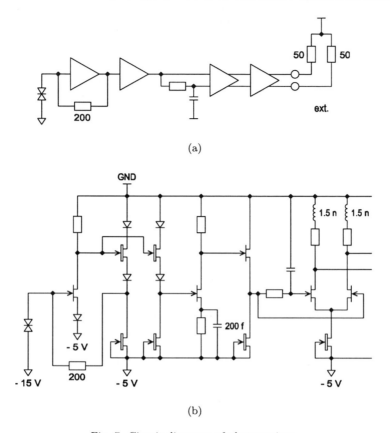

(a)

(b)

Fig. 7. Circuit diagrams of photoreceiver.

cost. Based on our advanced GaAs technology, high-speed front-end OEICs were realized for long wavelength (1.3–1.55 μm) systems.[17] Figure 7(a) shows the block diagram of the optical receiver OEIC. It consists of an MSM photodiode, a trans-impedance input stage, and two amplifier stages. Figure 7(b) depicts a section of the detailed circuit diagram. For simplicity the second differential amplifier stage has been omitted. To obtain a high bandwidth we used a peaking capacitor (200 fF) in the second amplifier stage, and airbridge inductors (1.5 nH) in the first differential amplifier stage. The MSM photodiode is formed with the absorbing InGaAs layer on the GaAs substrate, as shown in Fig. 8. Figure 9 shows a section of the photoreceiver. The integrated InGaAs MSM photodiode with a light-sensitive area of 25×25 μm^2 exhibited a front-side illumination responsivity of 0.32 A/W for 1.3 μm and 0.30 A/W for 1.55 μm incident light and 5 V bias, and a dark current of less than 9×10^{-8} A within the entire voltage range up to 10 V.

4.2.2. Circuit performance

The transimpedance at the photoreceiver outputs is 2.08 kΩ (differential). A maximum output voltage swing of 0.5 V$_{\text{p-p}}$ and a power dissipation of 290 mW were

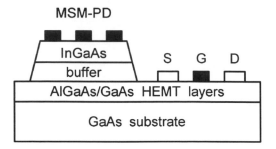

Fig. 8. Cross-section of InGaAs MSM photodiode and AlGaAs/GaAs HEMT.

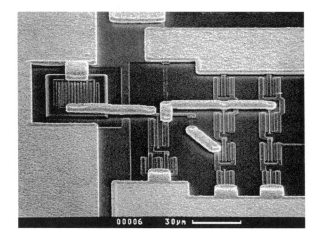

Fig. 9. Photograph showing a section of fabricated photoreceiver.

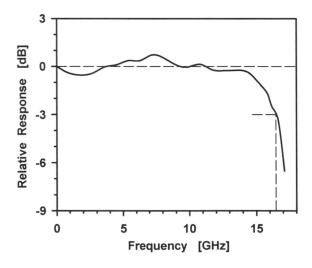

Fig. 10. Measured frequency response of photoreceiver at 1.3 μm wavelength.

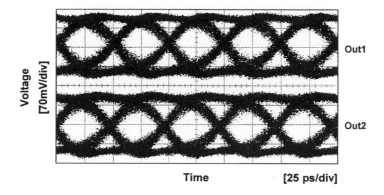

Fig. 11. Eye diagrams of the differential photoreceiver outputs at 20 Gbit/s.

measured. To determine the frequency response, the integrated photodiode was irradiated by 1.3 μm light from a calibrated laser diode via a single mode fiber. The laser diode was modulated directly.

As shown in Fig. 10 the −3 dB bandwidth is 16.5 GHz. Figure 11 depicts the eye diagrams of the two photoreceiver output voltages for a 20 Gbit/s 1.55 μm optical data stream with a PRBS pattern length of $2^{15} - 1$. The optical input power was −5 dBm.

5. AGC Amplifier and Limiting Amplifier

Because optical signal amplitude strongly depends on tolerances of fiber, connectors, splice losses, fluctuations of parameters caused by temperature and aging, for further signal processing a constant signal amplitude is desired. For this reason AGC or limiting amplifiers are used as main amplifiers. Furthermore, limiting amplifiers are also used in clock recovery circuits. Such amplifiers with a large dynamic range deliver constant amplitude to the data and clock recovery circuits even though the input signal varies constantly. Broadband amplifiers with a voltage gain above 25 dB and cutoff frequencies up to about $f_T/2$ are only possible if adequate circuit concepts are used. Such analog circuits, which include transimpedance amplifiers operating near the limit of a given technology, are more difficult to design than digital ICs, because they are more sensitive to fabrication spread, parasitic elements and circuit environment. The best published limiting amplifiers include 15 Gbit/s circuit using Si-bipolar technology,[4] and up to 15 GHz in AlGaAs/GaAs-HBT technology,[19] while the multiplexer and demultiplexer operate above 50 Gbit/s,[2–4] respectively. A 30 GHz, 7 dB distributed amplifier with source coupled FET logic (SCFL) interface for optical transmission systems has been recently reported.[20]

The AGC and limiting amplifiers presented in this section feature a differential configuration with active source followers to broaden the frequency band.[16]

5.1. *Limiting amplifier*

5.1.1. *Circuit design*

Figure 12 shows the simplified block diagram of the single-chip dc-coupled amplifier. It consists of a high impedance input stage (IB), four identical amplifier cells (A1–A4) and an output buffer (OB). An offset-control circuit with an on-chip RC-network generates a dc- voltage at the input which depends on the output offset voltage.

Each of the four amplifier cells, shown in Fig. 13, features a differential configuration combining enhancement and depletion FETs, and inductors L (L′) in series

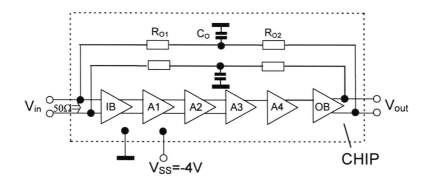

Fig. 12. Block diagram of the limiting amplifier.

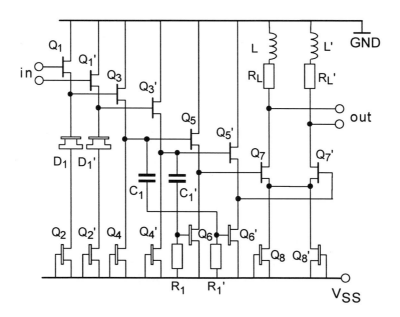

Fig. 13. Circuit diagram of one amplifier cell.

with load resistors R_L (R_L'). Differential operation was used to reduce time jitter and crosstalk and to give good common-mode suppression. The source-followers are used for level shifting and impedance transformation. Three cascaded source-followers improve the decoupling capability and increase the drain-gate voltage of the current-switch transistors (Q_7 an Q_7') allowing lower drain-gate capacitances. Q_3–Q_6 (Q_3'–Q_6') make up a differential active source follower circuit. During the signal transition, the gate voltage of Q_6 (Q_6') is temporarily modified through C_1' (C_1) and draws extra charge or discharge currents for loading, thus speeding up the transition. The time constant $\tau = R_1 C_1' = R_1' C_1$ should be smaller than half the period of the input signal at the maximum operating frequency. R_1 (R_1') determines the dynamic peak load current.

The active source followers activate the current sources of the source follower stages without an additional load at the output, resulting in an improvement of the bandwidth. During steady state operation they have the same biasing conditions and thus the same power dissipation as a conventional source follower FET logic circuit. This design is also suitable for low-power ICs due to small steady state current and large dynamic current. Simulations indicated that the differential active source-follower technique improves the bandwidth by more than 30%. Figure 14 shows a photograph of the limiting amplifier. The chip size is 1×1 mm^2.

5.1.2. *Circuit performance*

The chip was measured on-wafer using a single-ended input signal. Figure 15 gives the measured gain (S_{21}) and input return loss (S_{11}) versus frequency characteristics.

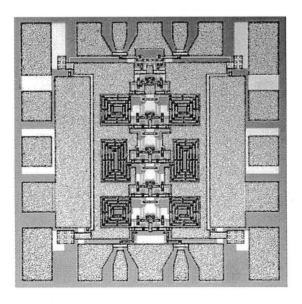

Fig. 14. Photograph of the limiting amplifier.

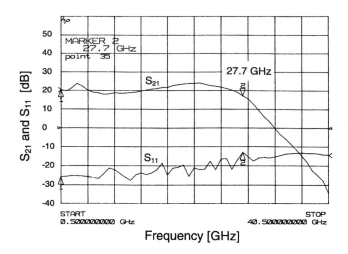

Fig. 15. S_{21} and S_{11} of the limiting amplifier for single-ended output.

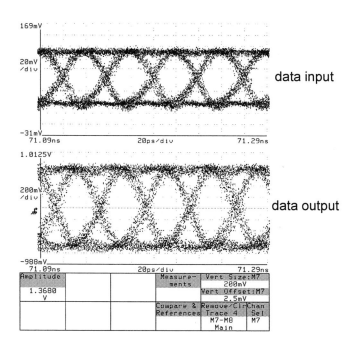

Fig. 16. Measured eye diagrams of the limiting amplifier at 25 Gb/s for single-ended input.

The amplifier has a bandwidth of 27.7 GHz, a differential gain of 26 dB and the input return loss is less than -15 dB up to 40 GHz. There is a about 3 dB gain bump at 20 GHz, but it should have no effect on the performance of the circuit if it operates in a limiting range. As shown in Fig. 16, the output eye diagrams have straight ground lines.

A large limited differential output voltage of 1.37 $V_{\text{P-P}}$ was obtained. A data bit rate of 25 Gbit/s using a single ended input of 70 $mV_{\text{p-p}}$ (PRBS of length $2^{31} - 1$) has been achieved, as shown in Fig. 16. In this case, the gain is 26 dB. The measured maximum data bit rate was limited by the measuring equipment. The circuit is capable of converting single ended data input into differential data output at high speed. The measured 3 dB bandwidth of 27.7 GHz with a gain of 26 dB represents the largest bandwidth-gain product in any technology reported so far, and the circuits are suitable for 40 Gbit/s lightwave communication systems. The power consumption is about 700 mW using a single supply voltage of -4 V.

5.2. *Automatic-gain-control (AGC) amplifier*

5.2.1. *Circuit design*

Figure 17 shows the block diagram of the AGC amplifier. The signal passes through the high-impedance input buffer (IB), the four gain controllable amplifier cells (A1–A4), and the output buffer (OB). The peak detector (PD) generates a dc voltage which depends on the amplitude of the output from the amplifier cell A3. This output is compared with that of the reference network (REF). The difference of these two output voltages is amplified by the high gain amplifier (K). The output voltage V_{AGC} of the low pass filter (LP) controls the gain of the four amplifier cells. The offset control circuit using the on-chip RC network (R_{O1}, R_{O2}, C_O) generates a dc voltage at the input depending on the output offset voltage. The AGC amplifier can also function as a limiting amplifier by switching off the automatic-gain-control.

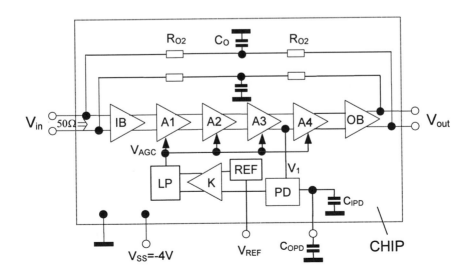

Fig. 17. Block diagram of the AGC amplifier.

Each of the four amplifier cells consists of differential amplifier with ac-coupled active source-followers, as shown in Fig. 18. The transistor Q_1 acts as a variable resistor controlled by the gate voltage V_{AGC}. The resistor R_1 (= 1 kΩ) is connected in parallel to the transistor Q_1 in order to improve the linearity of the amplifier. Using D-mode FET current sources instead of E-mode with source degeneration offers the advantage of reducing crosstalk and supply voltage independent currents due to self-biasing. Figure 19 shows the AGC amplifier chip with a chip size of 1.5×1 mm^2.

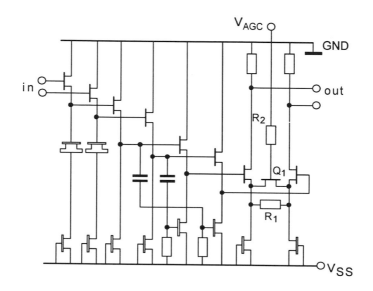

Fig. 18. Schematic diagram of one amplifier cell.

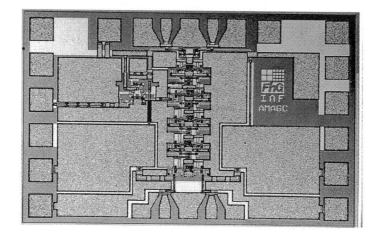

Fig. 19. Chip micrograph of the AGC amplifier.

5.2.2. Circuit performance

Figure 20 shows the measured gain versus frequency characteristics for the entire dynamic range for single-ended input and output. The AGC amplifier has a bandwidth of 18 GHz with a differential gain of 30 dB and a dynamic range of 35 dB.

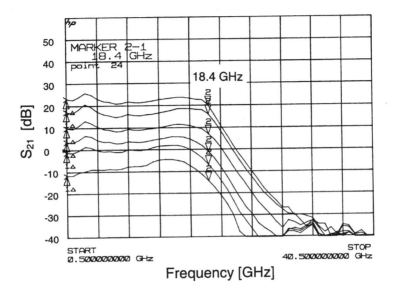

Fig. 20. Gain (S_{21}) of the AGC amplifier over the dynamic range for single-ended output.

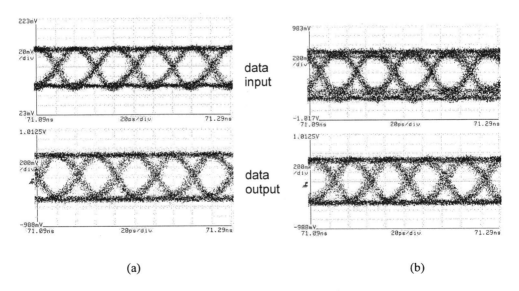

(a)   (b)

Fig. 21. Measured eye diagrams at 25 Gb/s of the AGC amplifier for single-ended input voltage swing of (a) 70 mV$_{p-p}$ and (b) 1 V$_{p-p}$.

Figure 21 displays the measured eye diagrams at 25 Gbit/s for different input amplitudes. At an input voltage swing of (a) 70 mV$_{p-p}$ and (b) 1 V$_{p-p}$ the output voltage swing, rise- and fall-times remain unchanged. The AGC amplifier with 100 Ω on-chip resistors at outputs has a differential output voltage of 900 mV$_{P-P}$. The power dissipation of the circuit is 500 mW with a single power supply voltage of -4 V.

6. Data Decision

The ability to regenerate binary data is an inherent advantage of digital transmission systems. The data decision circuit regenerates the analog input signal which includes noise and jitter. High-speed data decision circuits composed of a master-slave D-flip-flop (MS-D-FF) are frequently used as basic circuits in optical fiber communication systems and test instruments. Moreover, the MS-D-FF is the most fundamental circuit because it determines the speed performance in almost all digital circuits, e.g. 4:1 multiplexers and 1:4 demultiplexers.[23,24] In recent years, high-speed MS-D-FF circuits operating up to 19, 25 Gbit/s and 46 Gbit/s[2,4,25] have been reported. In this section we report on the design and performance of a high-speed low-power MS-D-FF circuit operating up to 20 Gbit/s with 24 mW power dissipation[29] and a parallel data decision (1:2 demultiplexer) for a bit rate of 40 Gbit/s.[30]

6.1. *Low-power MS-D-FF*

6.1.1. *Circuit design*

Lowering the power dissipation is one effective way of decreasing weight and size of equipment. In optical fiber communication systems with high-speed data rates of about 20 Gbit/s, the above mentioned 19 Gbit/s MS-D-FF IC using 0.2 μm GaAs MESFETs consumes a power of 1.5 W.[25] Therefore, low-power MS-D-FF circuits must be developed to the design of large scale ICs operating at high data bit rates.[26] In recent years, low-power MS-D-FFs operating up to 3.1 Gbit/s[27,28] have been reported. In this section we describe the design and performance of low-power high-speed data decision ICs.

Figure 22 shows the schematic of the decision circuit in source coupled FET logic. It is composed of a data input stage consisting of a source-follower pair, a clock input stage consisting of two source-follower pairs, the core of the MS-D-FF consisting of two series D-latches clocked out-of-phase, and an output buffer.

This circuit was designed to obtain low supply voltage, high speed and low power consumption, as described in the following: (a) Two-stage source followers for clock inputs instead of direct ac connection were used to simplify applications, although more current is consumed. (b) Feedback source followers of the D-latches are included to improve driving capability and provide level shifting. Omission of source followers may lower supply voltage and power dissipation, but the operating speed would decrease even more. (c) Resistors rather than HEMT active loads are

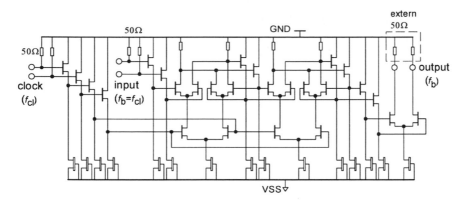

Fig. 22. Schematic of the D-FF decision circuit.

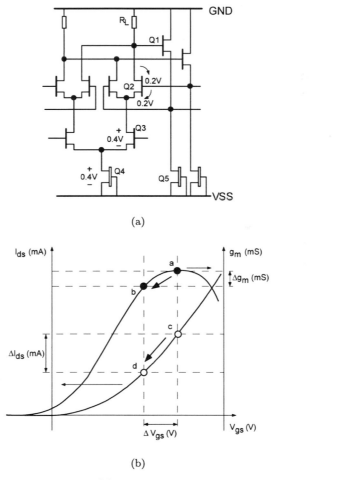

(a)

(b)

Fig. 23. (a) Schematic of D-latch and (b) variation of transconductance Δg_m and gate-source voltage ΔV_{gs} from drain current ΔI_{ds} for a given gate width.

143

used, because it often proves difficult to maintain all D-mode active load HEMTs in the saturation region with a low supply voltage. (d) D-mode HEMTs are used for current sources instead of E-mode HEMTs with source degeneration, or resistors can be used. Using only a resistor makes the current more sensitive to supply voltage variations. D-mode HEMTs as current sources need 0.4 V voltage, and do not influence each other due to self-biasing. (e) A logic swing of 0.5 V was chosen at load resistors for reliable operation. Smaller logic swing can lower supply voltage, but the noise margin will also be decreased, causing large jitter. (f) Larger gate-width for a given current is used in the upper current switches and in the latch feedback source followers to obtain a smaller gate-source voltage V_{gs}, as shown in Fig. 23 (a) and (b), although the transconductance g_m is reduced from the maximal value. Simulation indicates that V_{gs} decreases from 0.4 V to 0.2 V, while the maximum operating speed is about 8% lower. Figure 24 shows the microphotographs of the MS-D-FF. The chip size is 0.7 mm × 0.7 mm.

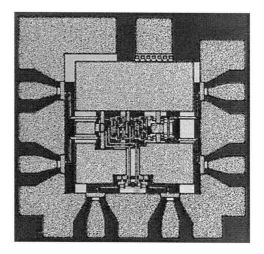

Fig. 24. Microphotographs of the MS-D-FF.

6.1.2. *Circuit performance*

The circuits have been tested on-wafer with a −1.5 V supply. Figure 25 shows the measured eye diagrams of the MS-D-FF with 20 Gbit/s single-ended data input (PRBS length $2^{31} - 1$).

   Error-free operation with widely opened eye diagrams was obtained. The clock phase margin was measured at various frequencies by adjusting the phase of the clock with respect to the data. A phase margin of 145° at 20 Gbit/s was obtained. The power dissipation of the MS-D-FF including the output buffer is 24 mW, while the output stage consumes about 10 mW. The differential output swing is 400 mV$_{P-P}$ into 50 Ω external loads. The purpose of this work is to investigate

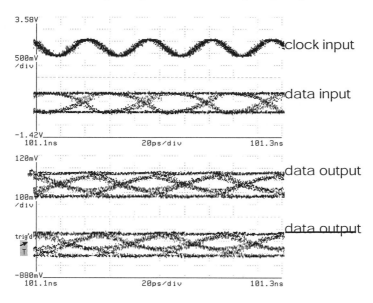

Fig. 25. Measured eye diagrams of the MS-D-FF at 20 Gbit/s.

speed performance versus power dissipation of the basic logic component. The 20 Gbit/s bit rate and 14 mW power consumption of the MS-D-FF core is suitable for building large scale ICs, for instance, the high order 1:N-demultiplexers. Higher output swing can be obtained by increasing the current of the output buffer. The measured results represent the highest data rate versus power consumption in any technologies.

6.2. *Parallel data decision*

6.2.1. *Circuit design*

Generally, for data decision there are two basic schemes: direct (using a MS-D-FF, as described in the previous section), and parallel processing.[31] For direct processing all cells have to operate at full speed, i.e., at the bit rate of the input data signal. The parallel processing is actually a 1:2 demultiplexer composed of two parallel MS-D-FF's clocked out-of-phase at half of the clock frequency.

The choice of the decision circuit also depends on clock recovery IC whether delivering half or full of the clock frequency related to the data bit rate. For these reasons we have chosen the parallel processing concept for 40 Gbit/s operation. Figure 26 shows the block diagram of the data decision circuit.

Two input buffers are used for the data and clock signal. The buffered clock signal is applied to the four D-latches in such a phase arrangement that two channels are alternatively selected. Two output buffers are used for the parallel regenerated data signals. The photograph of the chip is shown in Fig. 27.

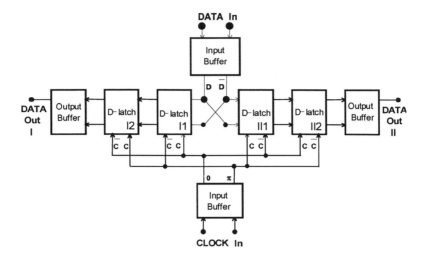

Fig. 26. Block diagram of the parallel-processing data decision.

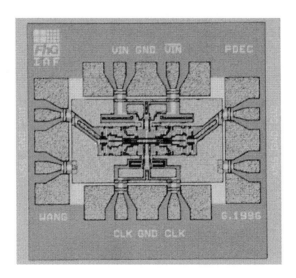

Fig. 27. Chip micrograph of the data decision.

6.2.2. *Circuit performance*

Figure 28 shows the measured eye diagrams of the input clock signal at 25 GHz, the input PRBS signal at 25 Gbit/s with a length of $2^{23} - 1$ and the two output signals at 25 Gbit/s.

In this case, the bit pattern 010 of the input data signal at 25 Gbit/s corresponds to the bit pattern 00 11 00 at 50 Gbit/s, since each pulse is sampled twice at an interval of half a clock period. The measured clock phase margin is about 170°

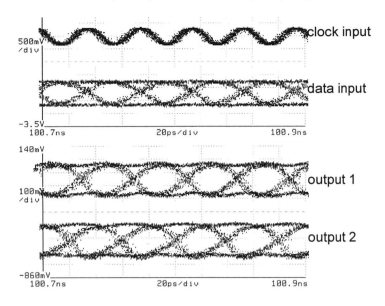

Fig. 28. Measured results at 25 Gbit/s.

in this case, which indicates the IC can reliably operate at more than 40 Gbit/s. The dc currents at −3.3 V and −5.2 V supply voltages are 70 mA and 90 mA, respectively, with power consumptions of about 230 mW and 470 mW, respectively.

7. Clock Recovery

At receiver ends of optical fiber links, clock recovery (CR) is required for data regeneration and/or demultiplexing. Clock extraction circuits for nonreturn-to-zero (NRZ) data can be grouped into two main categories: the phase-locked loops (PLL)[32–34] and the narrow bandpass filters.[35,36] Usually, NRZ signals must first undergo appropriate nonlinear preprocessing in the CR to generate the spectral energy at the clock frequency.

The basic PLL consists of a phase detector, a loop filter, and a voltage-controlled oscillator (VCO). The phase detector compares the phase relationship between the VCO output and the serial data input. A loop filter converts the phase detector output into a smooth DC voltage, and the DC voltage is input to the VCO whose frequency is varied by this voltage. The two kinds of circuits of (a) the phase- and frequency-locked loop (PFLL) increasing the lock-in range by inserting a frequency sensitive detector parallel to the phase detector[39–41]; and (b) variable delay line PLL (DLL) in which a voltage-controlled delay line can be used in a PLL instead of a VCO,[42,43] also belong to PLL circuits. PLL circuits are widely used for clock extraction due to the advantages of the monolithic integration, continually compensation for changes in the environment and the input bit rate.

Using a narrowband filter the clock signal can be obtained by passing the incoming signal through a bandpass filter. The narrowband filter methods are classified into passive filters, such as surface-acoustic wave (SAW) filters, which however are difficult to integrate monolithically and need manual adjusting, and active filters[44,45] which can be monolithic integrated and will be described below.

7.1. *Circuit design*

We have designed a monolithic IC for the clock recovery based on a regenerative frequency divider at bit rates of both 40 and 20 Gbit/s. The CR circuit consists of four sub-circuits: a preprocessor, a narrowband regenerative frequency divider (NRFD), a phase-shifting amplifier, and a limiting amplifier. A modified XOR-circuit was utilized as the preprocessor, as shown in Fig. 29.

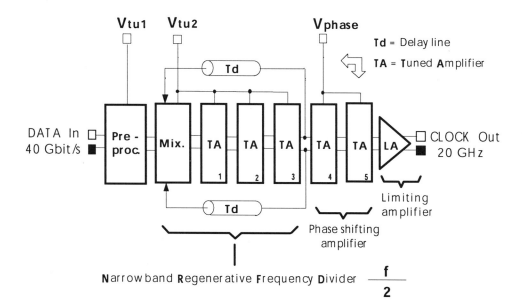

Fig. 29. Block diagram of the clock recovery circuit.

The LC-resonator filter in the loading circuit of the XOR-logic, was designed with a center frequency of f_b (= 40 GHz). The NRFD loop consists of a mixer and three stages of tuned amplifiers. The mixer has a circuit structure analog to the preprocessor. The central frequency of all resonators in the mixer and in the amplifiers is $f_b/2$ (= 20 GHz). The subcircuit of the phase shifter was designed as in Ref. 44. The limiting amplifier was optimized for application at 20 GHz. A coplanar waveguide (CPW) pair is used as transmission lines for the feedback. Figure 30 shows the circuit diagram of one tuned amplifier (TA) cell. It is a differential current amplifier (EF_1, EF_2) with a tuned resonator (L_1, L_2, and C_{dd}). EF_3 and EF_4 form

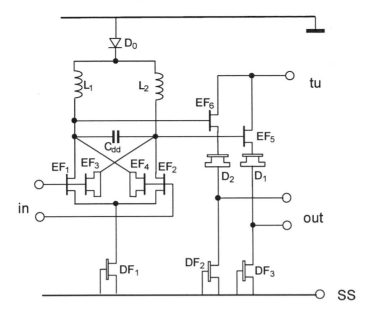

Fig. 30. Circuit diagram of a tuned amplifier cell.

the neutralizing compensation network. The level shifting D_0 and the gate-to-drain Schottky junctions of EF_5 and EF_6 along with V_{tu} are used to tune the central frequency of the resonator. There are three advantages to be gained by using the tuned resonator: (1) The gain of the TA will be increased. (2) The LC resonator shows a filtering function so that only the desired spectrum will be amplified. (3) There are no dc component in the output spectrum. This is advantageous in the CR using NRFD.

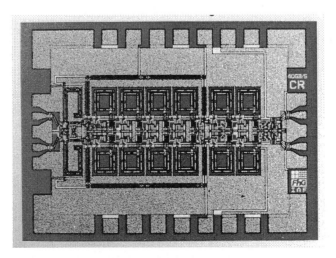

Fig. 31. Chip photograph of the clock recovery.

Following the NRFD-loop, two stages of tuned amplifiers, identical with that in the NRFD-loop, are used. Their tuning voltage of V_{phase} is used to adjust the phase of the output clock signal. Figure 31 shows the chip micrograph. On the 2×1.5 mm^2 chip 160 HEMTs and more than 100 other elements are integrated.

7.2. *Circuit performance*

Figure 32 shows the measured eye diagrams of the input data with a PRBS-length of $2^{23} - 1$ at 20 Gbit/s (upper plot) and the recovered clock signals at 20 GHz (lower plot).

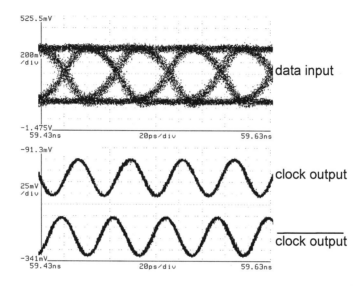

Fig. 32. Measured input data at 20 Gbit/s and the differential output clocks at 20 GHz.

The amplitude of the input pulse is 380 mV$_{pp}$ (single-ended). The amplitudes of the complementary outputs are 420 mV$_{pp}$. The 180° phase shifting function is clearly demonstrated. The measured time jitter is < 2 ps. The function at 40 Gbit/s was demonstrated by using a sinusoidal input signal at 20 GHz. The circuit can be operated in the supply voltage range from −3.3 to −5.2 V. The dc power consumption is about 500 mW at −3.3 V supply.

8. Demultiplexer

A demultiplexer is located at the receiver end to split the serial high-speed input data stream into parallel outputs. Because the principle of high-order 1:N-demultiplexer[46] is similar, we discuss here only the 1:4 demultiplexer. There are three types of demultiplexers: the parallel structure,[47] the shift register structure[48] and the tree-type structure.[5,49–51] The tree-type architecture provides high-speed performance owing to modified loads of data and clock signals in every stage of the demultiplexer.

8.1. *Circuit design*

The complete demultiplexer circuit consists of three 1:2 demultiplexers with identical circuitry, and a frequency divider by two followed by a phase shifter circuit. The first 1:2 demultiplexer divides the incoming data stream into two parallel data streams. One stream contains the even numbered data bits, the other the odd numbered ones. Between the parallel data streams, there is a time delay of half a clock period. This is compensated by an additional latch (delay latch, see Fig. 33).

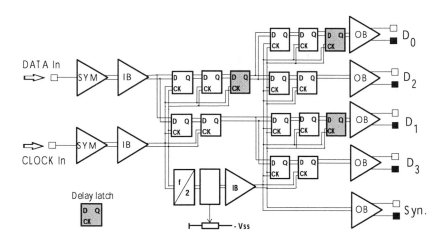

Fig. 33. Block diagram of the 1:4 demultiplexer.

Both of these data channels are used as input signals for the following two 1:2 demultiplexer stages which convert these into four parallel ones. The phase shifter circuit is able to shift the internal clock phase in a range of up to 70 degrees. With this possibility we have optimized the demultiplexer circuit for the high speed range. The differential output stages for data synchronization are able to drive 50 Ω loads. An on-chip output termination resistor of 100 Ω is provided in order to reduce the output return loss compared to open drain configurations. Measurements of the differential output signals with a 50 Ω load show a voltage swing of approximately 800 mV$_{\text{P-P}}$. In Fig. 33 the block circuit diagram of the demultiplexer is depicted. Figure 34 shows a chip photograph of the 1:4 demultiplexer circuit. The chip size is 1.5×2 mm^2 and the number of active elements about 400.

8.2. *Circuit performance*

For the measurements of the 1:4 demultiplexer a 2:1 multiplexer designed by the authors is used to generate a 40 Gbit/s data signal. Figure 35 shows the eye

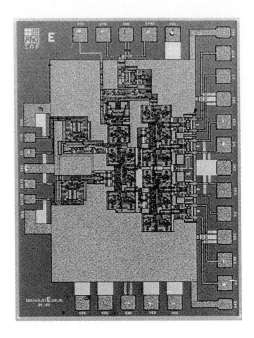

Fig. 34. Chip photograph of the 1:4 demultiplexer IC.

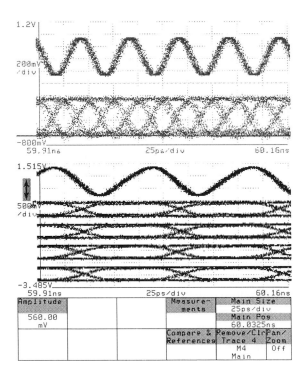

Fig. 35. Eye diagrams of a 40 Gbit/s input and four 10 Gbit/s outputs.

diagrams at 40 Gbit/s at a clock frequency of 20 GHz. Traces from top to bottom show the 20 GHz clock, the 40 Gbit/s data input, the divided by two output at 10 GHz, and four 10 Gbit/s outputs. The outputs are synchronized in time The differential output amplitude is about 1 V_{p-p}. The dc power consumption is approximately 2 W at a supply voltage of -5 V.

9. Frequency Divider

High-speed frequency dividers belong to the key components for various applications, e.g., in measurement equipment, microwave and satellite communication systems. Therefore, many different high-speed static and dynamic frequency dividers based on various kinds of devices have been developed. There are two kinds of dynamic frequency dividers: analog dividers proposed by Miller,[52] and digital dividers. The analog frequency divider is also known as the regenerative frequency divider. A 28 GHz Si bipolar analog frequency divider has been reported in Ref. 53. However, besides the steady-state problems, it suffers from substantial phase noise.[54] Two digital dynamic frequency dividers feature inverter and buffer controlled by a transfer gate, and eliminated latches compared to the static frequency divider for operating speeds up to 51 GHz and 48 GHz,[55–58] respectively. A static frequency divider consists of a MS-D-FF with the output fed back to the input. Static frequency dividers have a lower maximum frequency but a wider operating range than dynamic ones.[57–59] For this reason they are often used in broadband communication systems. We report here on a 30 GHz static frequency divider which has been used in the 1:4 demultiplexer circuit.[60]

9.1. *Circuit design*

Figure 36 shows the block circuit diagram of the complete frequency divider circuit. Between both divider stages a source follower acts as level shifter and driver. The output resistance of an FET source follower is approximately $R_{out} = 1/g_m$ for

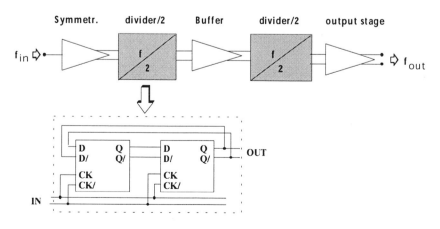

Fig. 36. Block diagram of the divide-by-four frequency divider.

low frequencies. Here g_m represents the transconductance of the FET. At higher frequencies, of course, this simple relation can no longer be applied because of the internal capacitances of the transistor. Special emphasis has been laid on the efficiency of the source follower to minimize the dynamic output resistance. The latches, applied in the frequency divider circuit, are the same as the data latches in the demultiplexer circuit. The application of double source followers speed up signal transition. Simulations show that the FET double source follower increases the gain at higher frequencies. The current sources of the divider circuit were designed using

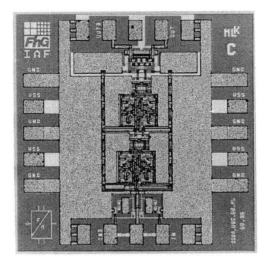

Fig. 37. Chip photograph of the frequency divider.

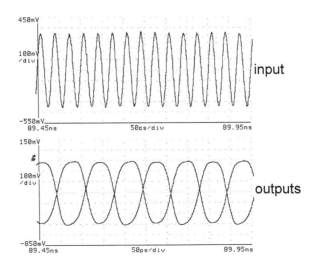

Fig. 38. Input and output signals at $f_{\text{in}} = 30$ GHz.

E-mode FETs. Figure 37 shows a photograph of the divider circuit. The chip size, including input circuit and output buffers, is 1×1 mm^2.

9.2. *Circuit performance*

Figure 38 shows the input and output signals at $f_{in} = 30$ GHz. The circuit operates for supply voltages varying from -3.3 to -5.5 V. The power consumption is about 280 mW at a supply voltage of -4 V.

10. Summary

The design and performance of numerous analog and digital ICs for an 20–40 Gbit/s optical data receiver based on 0.2 μm AlGaAs/GaAs/AlGaAs quantum well HEMT technology have been presented in this paper. For each of the circuits we have also aimed at giving readers a general view of design techniques. Stable operation for 20–40 Gbit/s and low power consumption have been demonstrated. To achieve these, suitable circuits concepts, individual optimization, and careful layout design are necessary for a given technology. Table 2 summarizes the function blocks and corresponding performances of the monolithic integrated ICs for an optical data receiver. We are continuing to push towards higher operation speed and the further integration of several functional blocks into a single chip, for example, two ICs for

Table 2. Summery of high-speed ICs for optical data receiver.

Circuits	Bandwidth/ Bit rate	Gain	Output (differential)	Power
Transimpedance amplifier [16]	22 GHz	76 dBΩ	1 V$_{p-p}$	400 mW
OEIC receiver [17]	20 Gbit/s	66 dBΩ	0.5 V$_{p-p}$	290 mW
Limiting amplifier [16]	27.7 GHz	26 dB	1.3 V$_{p-p}$	700 mW
AGC amplifier [16]	25 Gbit/s	30 dB	1 V$_{p-p}$	500 mW
Low-power MS-D-FF [29]	20 Gbit/s	-	0.4 V$_{p-p}$	24 mW
Parallel decision [30]	20–40 Gbit/s	-	0.6 V$_{p-p}$	350 mW
Clork recovery [45]	20–40 Gbit/s	-	0.2 V$_{p-p}$	500 mW
1:4 demultiplexer [51]	40 Gbit/s	-	1 V$_{p-p}$	2 W
Static frequency divider [60]	30 GHz	-	1.1 V$_{p-p}$	280 mW

an optical data receiver of one: a PIN photodiode, transimpedance and limiting amplifiers, and another one: data decision, clock recovery and demultiplexer.

Acknowledgments

The authors wish to thank G. Weimann for his encouragement and support, M. Berroth, now with University of Stuttgart, for many years of commitment, valuable advice and encouragement, and to the German Federal Ministry BMBF for financial support.

References

1. K. Hagimoto, M. Yoneyama, A. Sano, A. Hirano, T. Kataoka, T. Otsuji, K. Sano, and K. Nogushi, "Limitation and challenges of single-carrier full 40-Gbit/s repeater system based on optical equalization and new circuit design", *OFC'97*, Dig. Tech. Papers, 1997, pp. 242–243.
2. M. Yoneyyama, A. Sano, K. Hagimoto, T. Otsuji, K. Murata, Y. Imai, S. Yamagushi, T. Enoki, and K. Sano, "A 40-Gbit/s optical repeater circuits using InAlGa/InGaAs HEMT digital IC chip set", *IEEE MTT-S*, Dig. Tech. Papers, 1997, pp. 461–464.
3. A. Felder, M. Möller, J. Popp, J. Böck, and H.-M. Rein, "46 Gb/s DEMUX, 50 Gb/s MUX, and 30 GHz static frequency divider in silicon bipolar technology", *IEEE J. Solid-State Circuits* **31** (1996) 481–486.
4. H.-M. Rein, "Design considerations for very-high-speed Si-bipolar IC's operating up to 50 Gb/s", *IEEE J. Solid-State Circuits* **31** (1996) 1076–1090.
5. M. Lang, Z. Wang, Z. Lao, A. Thiede, M. Rieger-Motzer, W. Bronner, G. Kaufel, K. Köhler, A. Hülsmann, and J. Schneider, "20–40 Gbit/s 0.2 µm HEMT chip set for optical data receiver", *IEEE J. Solid-State Circuits* **32** (1997) 1384–1393.
6. A. Thiede, M. Schlechtweg, Z. Wang, M. Lang, P. Leber, Z. Lao, U. Nowotny, V. Hurm, M. Rieger-Motzer, M. Sedler, K. Köhler, W. Bronner, A. Hülsmann, G. Kaufel, B. Raynor, J. Schneider, T. Jakobus, J. Schroth, and M. Berroth, "Mixed signal circuits based on a 0.2 µm gate length AlGaAs/GaAs/AlGaAs quantum well HEMT technology", *IEEE Trans. VLSI Systems* **6** (1998) 6–17.
7. M. Schlechtweg, W. H. Haydl, A. Bangert, J. Braustein, P. J. Tasker, L. Verweyer, H. Massler, W. Bronner, A. Hülsmann, and K. Köhler, "Coplanar millimeter-wave IC's for W-band applications using 0.15 µm pseudomorphic MODFET's", *IEEE J. Solid-State Circuits* **31** (1996) 1426–1434.
8. B. G. Bosch, "Gigabit electronics — a review", *Proc. IEEE* **67** (1979) 340–377.
9. K. Runge, P. L. Pierson, P. J. Zampardi, P. B. Thomas, J. Yu, and K. C. Wang, "40 Gbit/s AlGaAs/GaAs HBT 4:1 multiplexer IC", *Electron. Lett.* **31** (1995) 876–877.
10. R. Schmid, T. F. Meister, M. Neuhäuser, A. Felder, W. Bogner, M. Rest, J. Rupeter, and H.-M. Rein, "20 Gbit/s transimpedance preamplifier and modulator driver in SiGe bipolar technology", *Electron. Lett.* **33** (1997) 1136–1137.
11. Z. Lao, U. Nowotny, A. Thiede, V. Hurm, G. Kaufel, M. Rieger-Motzer, W. Bronner, J. Seibel, and A. Hülsmann, "A 45 Gbit/s AlGaAs/GaAs HEMT multiplexer IC", *Electron. Lett.* **33** (1997) 589–590.
12. Z. Lao, A. Thiede, U. Nowotny, M. Schlechtweg, V. Hurm, W. Bronner, M. Rieger-Motzer, J. Hornung, G. Kaufel, and A. Hülsmann, "High power modulator driver ICs up to 30 Gb/s with AlGaAs/GaAs HEMTs", *IEEE GaAs IC Symp.* Tech. Dig., 1997, pp. 223–226.

13. A. Hülsmann, G. Kaufel, K. Köhler, B. Raynor, J. Schneider, and T. Jakobus, "E-beam direct-write in a dry-etched recess gate HEMT process for GaAs/AlGaAs circuits", *Jpn. J. Appl. Phys.* **29** (1990) 2317–2320.

14. Z. Lao, U. Langmann, J. N. Albers, E. Schlag, and D. Clawin, "Si bipolar 14 Gb/s 1:4-demultiplexer IC for system applications", *IEEE J. Solid-State Circuits* **31** (1996) 54–60.

15. A. Buchwald and K. Martin, *Integrated Fiber-Optic Receivers*, Kluwer Academic Publishers, Massachusetts, USA, 1995.

16. Z. Lao, M. Berroth, V. Hurm, A. Thiede, R. Bosch, P. Hofmann, A. Hülsmann, C. Moglestue, and K. Köhler, "25 Gb/s AGC amplifier, 22 GHz transimpedance amplifier and 27.7 GHz limiting amplifier ICs using AlGaAs/GaAs-HEMTs", *IEEE ISSCC,* Dig. Tech. Papers, 1997, pp. 356–357.

17. V. Hurm, W. Benz, W. Bronner, T. Fink, T. Jakobus, G. Kaufel, K. Köhler, Z. Lao, M. Ludwig, C. Moglestue, B. Raynor, J. Rosenzweig, M. Schlechtweg, and A. Thiede, "Long wavelength MSM-HEMT and PIN-HEMT photoreceivers grown on GaAs", *IEEE GaAs IC Symp.*, Tech. Dig., 1997, pp. 197–200.

18. Z. Lao, V. Hurm, M. Ludwig, A. Thiede, J. Rosenzweig, M. Schlechtweg, W. Bronner, K. Köhler, T. Fink, A. Hülsmann, and T. Jakobus, "Modulator driver and photoreceiver for 20 Gb/s optic-fiber links", *IEEE J. Lightwave Technology*, has been submitted.

19. M. Nakamura, Y. Imai, E. Sano, Y. Yamauchi, and O. Nakajima, "A 15-GHz AlGaAs/GaAs HBT limiting amplifier with low phase deviation", *IEEE GaAs IC Symp.*, Tech. Dig. Papers, 1991, pp. 45–48.

20. S. Kimura, Y. Imai, and Y. Miyamoto, "Novel distributed baseband amplifying techniques for 40-Gbit/s optical communication", *IEEE GaAs IC Symp.*, Tech. Dig. Papers, 1995, pp. 193–196.

21. Z. Lao, M. Berroth, V. Hurm, M. Ludwig, W. Bronner, and J. Schneider, "A monolithic 10-channel 10 Gbit/s amplifier array using 0.3 μm-AlGaAs/GaAs-HEMTs", *Electron. Lett.* **32** (1996) 1708–1709.

22. Z. Lao, M. Berroth, V. Hurm, M. Rieger-Motzer, A. Thiede, W. Bronner, A. Hülsmann, and B. Raynor, "A monolithic 24.9 GHz limiting amplifier using 0.2 μm AlGaAs/GaAs/AlGaAs HEMTs", *IEEE GaAs IC Symp.*, Tech. Paper Digest, 1996, pp. 211–216.

23. Z. Lao, U. Langmann, J. N. Albers, E. Schlag, and D. Clawin, "Silicon bipolar 14 Gb/s 1:4-demultiplexer IC regarding system requirements", *IEEE Bipolar/BiCMOS Circuits and Technol. M.*, Dig. Tech. Pap., 1994, pp. 103–106.

24. Z. Lao, U. Langmann, J. N. Albers, E. Schlag, and D. Clawin, "A 12 Gb/s Si bipolar 4:1-multiplexer IC for SDH systems", *IEEE J. Solid-State Circuits* **30** (1995) 129–132.

25. K. Murata *et al.*, "A novel high-speed latching operation flip-flop (HLO-FF) circuit and 1st application to a 19-Gb/s decision circuit using a 0.2-μm GaAs MESFET", *IEEE J. Solid-State Circuits* **30** (1995) 1101–1108.

26. Z. Lao and U. Langmann, "Low-power 10 Gb/s Si bipolar 1:16-demultiplexer IC", *IEEE J. Solid-State Circuits* **31** (1996) 128–131.

27. W. Wilhelm and P. Weger, "2V low-power bipolar logic", *IEEE ISSCC,* Dig. Tech. Papers, 1994, pp. 94–95.

28. K. Kishine *et al.*, "A low-power bipolar circuit for Gbit/s LSIs — current mirror control logic (CMCL)-", *IEEE Symp. VLSI Circuits*, Dig. Tech. Papers, 1995, pp. 127–128.

29. Z. Lao, M. Berroth, A. Thiede, M. Rieger-Motzer, G. Kaufel, J. Seibel, W. Bronner,

A. Hülsmann, J. Schneider, and B. Raynor, "Low power 20 Gbit/s data decision and 17 GHz static frequency divider ICs with 1.5 V supply voltage", *Electron. Lett.* **33** (1997) 289–290.

30. Z. Wang, M. Berroth, A. Thiede, M. Rieger-Motzer, P. Hofmann, A. Hülsmann, G. Kaufel, K. Köhler, B. Raynor, and J. Schneider, "Low power data decision IC for 20–40 Gb/s data links using 0.2 mm AlGaAs/GaAs HEMTs", *Electron. Lett.* **32** (1996) 1855–1856.

31. D. Clawin and U. Langmann, "Multigigabit/second silicon decision circuit", *IEEE ISSCC*, Dig. Tech. Papers, 1985, pp. 222–223.

32. B. Razavi and J. Sung, "A 6 GHz 60 mW BiCMOS phase-locked loop with 2V supply", *IEEE ISSCC*, Dig. Tech. Papers, 1994, pp. 114–115.

33. J. Hauenschild, C. Dorschky, T. W. Mohrenfels, and R. Seitz, "A 10 Gb/s BiCMOS clock and data recovering 1:4-demultiplexer in a standard plastic package with external VCO", *IEEE ISSCC*, Dig. Tech. Papers, 1996, pp. 202–203.

34. A. Pottbäcker and U. Langmann, "An 8 GHz silicon bipolar clock-recovery and data-regenerator IC", *IEEE ISSCC*, Dig. Tech. Papers, 1994, pp. 116–117.

35. H. Ichino, M. Togashi, M. Ohhata, Y. Imai, N. Ishihara, and E. Sano, "Over-10-Gb/s IC's for future lightwave communications", *IEEE J. Lightwave Technology* **12** (1994) 308–319.

36. D. Briggmann, G. Hanke, U. Langmann, and A. Pottbäcker, "Clock recovery circuits up to 20 Gbit/s for optical transmission systems", *IEEE MTT-S*, Dig. Tech. Papers, 1994, pp. 1093–1096.

37. Z. Lao, A. Pottbäcker, J. N. Albers, and U. Langmann, "Taktrückgewinnung und 1:4-Demultiplexer als empfangsseitige Funktionseinheit bei 10 Gbit/s für ein optisches Übertragungssystem", *6. ITG-Fachkonferenz*, ITG-Fachbericht 127, VDE-Verlag, Berlin, Germany, 1994, pp. 201–204.

38. Z. Wang, V. Hurm, M. Lang, M. Berroth, M. Ludwig, T. Fink, K. Köhler, and B. Raynor, "10 Gbit/s monolithic optoelectronic integrated receiver with clock recovery, data decision, and 1:4 demultiplexer", *ESSCIRC*, Dig. Tech. Papers, 1995, pp. 354–357.

39. A. Pottbäcker, U. Langmann, and H. U. Schneider, "A 8 Gb/s Si bipolar phase and frequency detector IC for clock recovery", *IEEE ISSCC*, Dig. Tech. Papers, 1992, pp. 162–163.

40. B. Razavi, "A 2.5-Gb/s 15-mW clock recovery circuit", *IEEE J. Solid-State Circuits* **31** (1996) 472–480.

41. T. W. Yoo and M. S. Park, "10 Gb/s clock extraction and data regeneration circuit implemented with phase-locked loop", *IEEE MTT-S*, Dig. Tech. Papers, 1997, pp. 1713–1716.

42. M. G. Johnson and E. L. Hudson, "A variable delay line PLL for CPU-coprocessor synchronization", *IEEE J. Solid-State Circuits* **23** (1988) 1218–1223.

43. T. H. Lee and J. F. Bulzacchelli, "A 155 MHz clock recovery delay- and phase-locked loop", *IEEE ISSCC*, Dig. Tech. Papers, 1992, pp. 160–161.

44. Z.-G. Wang, M. Berroth, A. Thiede, M. Rieger-Motzer, P. Hoffmann, A. Hülsmann, K. Köhler, B. Raynor, and J. Schneider, "10 and 20 Gb/s clock recovery GaAs IC with a 288° phase shifting function", *Electron. Lett.* **32** (1996) 1498–1499.

45. Z.-G. Wang, M. Berroth, A. Thiede, M. Rieger Motzer, P. Hoffmann, A. Hülsmann, K. Köhler, B. Raynor, and J. Schneider, "Circuit techniques for 10 and 20 Gbit/s clock recovery using a fully-balanced narrowband regenerative frequency divider with 0.3 μm HEMTs", *IEEE ISSCC*, Dig. Tech. Papers, 1996, pp. 204–205.

46. Z. Lao, U. Langmann, J. N. Albers, E. Schlag, and D. Clawin, "Si bipolar 1:16-demultiplexer for a 10 Gbit/s fiber optic communication system", *Electron. Lett.* **30** (1994) 1214–1216.

47. K. J. Negus, "Multi-Gbit/s silicon bipolar multiplexer and demultiplexer with interleaved architectures", *IEEE Bipolar Circuits and Technology Meeting*, Dig. Tech. Papers, 1991, pp. 35–38.

48. M. Ohuchi, T. Okamura, A. Sawairi, F. Kuniba, K. Matsumoto, T. Tashiro, S. Hatakeyama, and K. Okuyama, "A Si bipolar 5-Gb/s 8:1 multiplexer and 4.2-Gb/s 1:8 demultiplexer", *IEEE J. Solid-State Circuits* **27** (1992) 664–667.

49. M. Bagheri, K. C. Wang, M. C. Chang, R. B. Nubling, P. M. Asbeck, and A. Chen, "11.6-GHz 1:4 regenerating demultiplexer with bit-rotation control and 6.1-GHz auto-latching phase-aligner IC's using AlGaAs/GaAs HBT technology", *IEEE J. Solid-State Circuits* **27** (1992) 1787–1793.

50. Z. Lao, J. N. Albers, U. Langmann, and E. Schlag, "A 20 Gb/s silicon bipolar 1:4-demultiplexer IC", *IEEE J. Lightwave Technology* **12** (1994) 320–324.

51. M. Lang, U. Nowotny, Z. Wang, Z. Lao, A. Thiede, M. Rieger-Motzer, W. Bronner, G. Kaufel, K. Köhler, B. Raynor, and J. Schneider, "GaAs HEMT ICs for 40 Gbit/s data transmission systems", *European Microelectronics Application Conf.*, Dig. Tech. Papers, 1997, pp. 85–88.

52. R. L. Miller, "Fractional frequency generators utilizing regenerative modulation", *IRE*, Tech. Dig., **27**, 1939, pp. 446–457.

53. M. Kurisu, G. Uemura, M. Ohuchi, C. Ogawa, H. Takemura, T. Morikawa, and T. Tashiro, "A Si bipolar 28 GHz dynamic frequency divider", *IEEE ISSCC*, Dig. Tech. Papers, 1992, pp. 92–93.

54. B. Razavi, "Challenges in the design of frequency synthesizers for wireless applications", *IEEE CICC*, Dig. Tech. Papers, 1997, pp. 395–402.

55. A. Thiede, M. Berroth, U. Nowotny, J. Seibel, R. Bosch, K. Köhler, B. Raynor, and J. Schneider, "An 18-34 GHz dynamic frequency divider based on 0.2 μm AlGaAs/GaAs/AlGaAs quantum-well transistors", *IEEE ISSCC*, Dig. Tech. Papers, 1993, pp. 176–177.

56. A. Thiede, P. Tasker, A. Hülsmann, K. Köhler, W. Bronner, M. Schlechtweg, M. Berroth, J. Braunstein, and U. Nowotny, "28-51 GHz dynamic frequency divider based on 0.15 μm T-gate $Al_{0.2}Ga_{0.8}As/In_{0.25}Ga_{0.75}As$ MODFETs", *Electron. Lett.* **29** (1993) 933–934.

57. Z. Lao, M. Berroth, M. Rieger-Motzer, A. Thiede, V. Hurm, M. Sedler, W. Bronner, A. Hülsmann, and B. Raynor, "31 GHz static and 39 GHz dynamic frequency divider ICs using 0.2 μm-AlGaAs/GaAs-HEMTs", *European Solid-State Circuits Conf.*, Dig. Tech. Papers, 1996, pp. 424–427.

58. Z. Lao, W. Bronner, A. Thiede, M. Schlechtweg, A. Hülsmann, M. Rieger-Motzer, G. Kaufel, B. Raynor, and M. Sedler, "35 GHz static and 48 GHz dynamic frequency divider ICs using 0.2 μm-AlGaAs/GaAs-HEMTs", *IEEE J. Solid-State Circuits* **32** (1997) 1556–1562.

59. M. Wurzer, T. F. Meister, H. Schäfer, H. Knapp, J. Böck, R. Stengl, K. Aufinger, M. Franosch, M. Rest, M. Möller, H.-M. Rein, and A. Felder, "42 GHz static frequency divider in a Si/SiGe bipolar technology", *IEEE ISSCC*, Dig. Tech. Papers, 1997, pp. 122–123.

60. M. Lang, M. Berroth, M. Rieger-Motzer, A. Hülsmann, G. Kaufel, K. Köhler, and B. Raynor, "30 GHz static frequency divider using a 0.2 μm AlGaAs/GaAs/AlGaAs HEMT technology", *Electron. Lett.* **24** (1996) 2111–2112.

61. M. Berroth, M. Lang, Z. Wang, Z. Lao, A. Thiede, M. Rieger-Motzer, W. Bronner, G. Kaufel, K. Köhler, A. Hülsmann, and J. Schneider, "20–40 Gbit/s 0.2 μm HEMT chip set for optical data receiver", *IEEE GaAs IC Symp.*, Dig. Tech. Papers, 1996, pp. 133–136.

62. Z. Lao, V. Hurm, W. Bronner, A. Hülsmann, T. Jakobus, K. Köhler, M. Ludwig, B. Raynor, J. Rosenzweig, M. Schlechtweg, and A. Thiede, "20 Gb/s, 14 kΩ trans-impedance long wavelength MSM-HEMT photoreceiver OEIC", *IEEE Photonics Technology Letters*, in press.

63. U. Nowotny, Z. Lao, A. Thiede, H. Lienhart, J. Hornung, G. Kaufel, K. Köhler, and K. Glorer, "44 Gbit/s 4:1 multiplexer and 50 Gbit/s 2:1 multiplexer in pseudomorphic AlGaAs/GaAs-HEMT technology", *IEEE Int. Symp. Circuits and Syst.*, Monteray, CA, May 31–June 3, 1998.

64. A. Felder, R. Stengl, J. Hausenschild, H.-M. Rein, and T. F. Meister, "25 to 40 Gb/s Si ICs in selective epitaxial bipolar technology", *IEEE ISSCC*, Dig. Tech. Papers, 1997, pp. 122–123.

International Journal of High Speed Electronics and Systems, Vol. 9, No. 2 (1998) 473–503

AlGaAs/GaAs HBT CIRCUITS
FOR OPTICAL TDM COMMUNICATIONS

K. RUNGE, P. J. ZAMPARDI, R. L. PIERSON, R. YU,
P. B. THOMAS, S. M. BECCUE, and K. C. WANG

Rockwell International Science Center, 1049 Camino Dos Rios
Thousand Oaks, California, 91360, USA

We describe experimental ultra-high-speed HBT circuits for lightwave communications applications. High speed circuits such as multiplexer/demultiplexers, variable gain amplifiers, (VGAs), and transimpedance amplifiers operating at high bit rates (> 30 Gb/s) are required for the realization of high-performance lightwave systems using TDM or WDM. We have demonstrated 40 Gb/s 4:1 multiplexers, > 30 Gb/s 1:4 demultiplexers, DC-26 GHz VGAs, DC-25 GHz transimpedance amplifiers, 30 Gb/s data and clock regenerators, 40 Gb/s differentiate-and-rectify timing recovery circuits, and 40 Gb/s delay-and-multiply timing recovery circuits, for use in such systems using a manufacturable hybrid digital/microwave HBT process.

1. Introduction

In this paper we discuss the application of the AlGaAs/GaAs HBT (heterojunction bipolar transistor) to high speed fiber optic systems operating at 30 Gb/s and beyond. In recent years the emphasis in both TDM (time division multiplexed) and WDM (wavelength division multiplexed) lightwave systems research has been on increasing the overall capacity of the system. In TDM systems, the emphasis has been on increasing the raw speed per channel from 2.5 Gb/s to 10 Gb/s with current laboratory work moving towards data rates of 40 Gb/s, and beyond. WDM system capacity has also been growing dramatically with an increase in the number of wavelengths transmitted, and also with the per channel transmission rate. The WDM per channel transmission rate has traditionally lagged that of the TDM system. Fiber dispersion management may limit both the TDM and WDM systems capacity in the future. But current research is pointing the way for 40 Gb/s and beyond per channel transmission rates. The electronic circuits for ultra high-speed fiber optic transmission have always been seen as a bottleneck to systems performance. The perceived severity of this "bottleneck" has varied from year to year with the advancement of both optical technology (ex. fiber amplifiers, low drive voltage external modulators), coupled with improvements in circuit design techniques, and a great increase in the speed (f_T/f_{MAX}) of IC technologies. The rapid development of optical and electrical technologies and design techniques shows no sign of slowing down.

The design of high-speed circuits for lightwave systems requires not only advanced device technology, but also clever circuit design. In this paper we will talk about the AlGaAs/GaAs HBT transistor and its application to lightwave circuits. We will also show examples of key circuits which demonstrate circuit principles.

The AlGaAs/GaAs heterojunction bipolar transistor (HBT) is ideal for the design of lightwave circuits. The AlGaAs/GaAs transistor features high f_T, f_{MAX} combined with high breakdown voltage. High f_T is needed for circuits in the analog path, while high f_{MAX} and f_T, are needed for the digital circuits. High breakdown allows for high drive voltage for the modulator driver, although recent progress in the development of modulators has relaxed this requirement. The semi-insulating substrate also aids in the design of high-speed circuits by having a lower C_{cs} capacitance in comparison to silicon bipolar technology. It also allows for high-Q passive components, unlike conductive or highly resistive substrates in other technologies. The AlGaAs/GaAs technology may also exhibit comparable power dissipation when compared to InP technology in many circuit applications. This is attributable to the fact that a given f_T/f_{MAX} often is reached in both technologies at a comparable collector emitter voltage (V_{CE}) and at practical collector currents (I_C). A GaAs based substrate compares favorably to Si based substrates in that the GaAs is semi-insulating (S. I.). This reduces collector to substrate capacitance, which may play a key role in high-speed and high integration level silicon circuits.

2. Processing Description

Circuits and devices were fabricated on 3" commercially available, carbon-doped, MOCVD-grown AlGaAs/GaAs HBT wafers. A schematic cross section of this process is shown in Fig. 1. This HBT process relies on self-alignment of the emitter-base structure and extrinsic C_{bc} reduction via implant. The fabrication process includes the use of a mesa for base isolation, which is then planarized using Rockwells' SADAP process.[1] The monolithic integration of Schottky diodes is critical for the design and fabrication of A/D converters and shock-line structures with these ultra high-speed HBTs.

This process results in excellent device uniformity. The DC current gain (β) is typically in the range of 20–30. The RF characteristics from this process are also

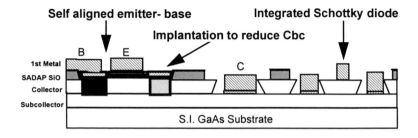

Fig. 1. Schematic cross-section of an HBT monolithically integrated with a Schottky diode.

excellent. For CML logic, the transistors typically experience V_{ce} of 1–2 volts. A typical design current density used is less than or equal to 5×10^4 A/cm^2. Three transistor structures were used to fabricate the circuits in this paper,[2] with high f_T transistors used in the analog (AGC and preamplifier) circuits. f_T/f_{MAX} values quoted for each circuit include the parasitic capacitance from the test pad structure which is about 15 FF for both the input and output pads.

3. Device Characteristics

This process results in excellent device uniformity, as shown in the Gummel plots of Fig. 2 for 1.4×3 μm^2 HBTs across a 3″ wafer. The current gain is typically in the range of 20–30. The RF characteristics from this process are also excellent. For CML logic, the transistors typically experience V_{ce} of 1–2 volts. Therefore, RF measurements were made at $J_c = 5 \times 10^4$ A/cm^2 and $V_{ce} = 1.5$ volts. Three typical structures for circuit fabrication are shown in Table 1. Structure A has an f_T of

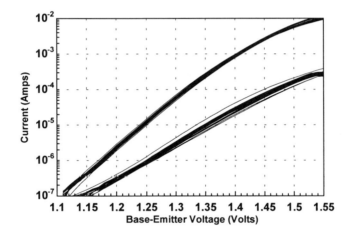

Fig. 2. DC device uniformity across a 3″ wafer ($23 - 1.4 \times 3$ μm^2 devices, including edge fields).

Table 1. Simplified MOCVD Epi-structures.

Layer	Structure A	Structure B	Structure C
Cap I	n-InGaAs 800 A	n-InGaAs 800 A	n-InGaAs 800 A
Cap II	n-GaAs 5e18, 1200 A	n-GaAs 5e18, 1200 A	n-GaAs 5e18, 1200 A
Emitter	n-Al$_x$Ga$_{1-x}$As (x = 0.25) 5e17 800 A	n-Al$_x$Ga$_{1-x}$As (x = 0.25) 5e17 800 A	n-Al$_x$Ga$_{1-x}$As (x = 0.25) 5e17 800 A
Base	p-GaAs, 7e19, 350 A	p-GaAs, 7e19, 500 A	p-GaAs, 7e19, 700 A
Collector	n-GaAs, 2e16 $\Rightarrow$ 3e17, 7000 A	n-GaAs, 3e16, 3000 A	n-GaAs, 3.5e16 $\Rightarrow$ 3e17, 7000 A
Sub-Collector	n+-GaAs, 5e18, 8000 A	n+-GaAs, 5e18, 8000 A	n+-GaAs, 5e18, 8000 A

60 GHz and f_{max} of 111 GHz biased as above and was used to design the AGC circuit. As shown in Fig. 3, the RF uniformity of this structure was good. The peak f_T is 95 GHz at 1.2×10^5 A/cm^2 with f_{max} of 130 GHz. Structure B has an f_T of 53 GHz and f_{max} of 120 GHz (at $J_c = 5 \times 10^4$ A/cm^2) with the peak f_T, f_{max} of 96 GHz, 160 GHz at 1.67×10^5 A/cm^2. This wafer had the best pre-amplifier performance. Finally, structure C had an f_T of 40 GHz and f_{max} of 105 GHz; models based on this structure were used in the design of the 4:1 MUX. The peak f_T was 60 GHz, with f_{max} 125 GHz at 1.2×10^5 A/cm^2. It is important to note that these values include the parasitic capacitance from the test pad structure which is about 15 FF for both the input and output pads.

The f_T/f_{MAX} curves (at $V_{CE} = 1.5$ V) versus current density, for structure C, are plotted in Fig. 4. The common emitter characteristics of a device of area

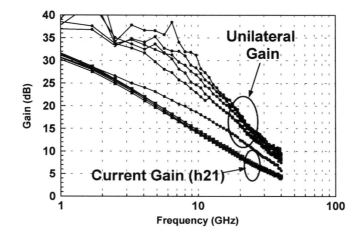

Fig. 3. RF 1.4×3 μm^2 device uniformity across a 3″ wafer.

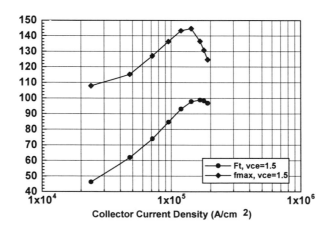

Fig. 4. f_T, f_{MAX} for a 1.4×6 μm^2 device {Structure C} at $V_{CE} = 1.5$ V.

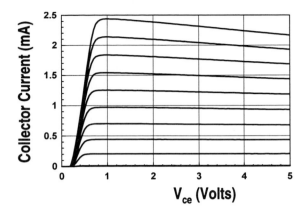

Fig. 5. I_C/V_{CE} device characteristics for 1.4×3 μm^2 device.

1.4×3 μm^2 are plotted in Fig. 5. The AlGaAs/GaAs HBT exhibits a large Beta-Early voltage (βV_A) product, which is important in data conversion applications. The negative slope of the collector current I_C versus V_{CE}, at high V_{CE} and high I_C, is due to self-heating effects.

A SEM cross-section of a two-finger 1.4×3 μm^2 device is shown in Fig. 6. The device is cut to reveal the material layers of the HBT.

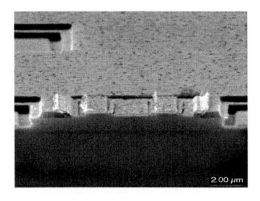

Fig. 6. Cross-section of a 1.4×3 μm^2 two-finger HBT.

Using this process, we have demonstrated many key circuits for lightwave communications above 30 Gb/s (Table 2). Circuits presented here are a 40 Gb/s multiplexer (MUX), a > 30 Gb/s 1:4 demultiplexer (DMUX), a DC-26 GHz VGA, and a 25 GHz bandwidth pre-amplifier, a 30 Gb/s data and clock regenerator, a 40 Gb/s differentiate-and-rectify circuit, and a 40 Gb/s delay-and-multiply circuit. We will discuss the application of these circuits in a typical lightwave transmission system, as well as the circuits themselves.

Table 2. Summary of high speed circuits.

Circuit	Speed	Power
4:1 MUX	40 Gb/s	2.5
1:4 DEMUX	> 30 Gb/s	2.7
Variable Gain Amplifier	DC-26 GHz BW	0.9
Transimpedance Preamplifier	25 GHz	0.3
Data and clock Regenerator	30 Gb/s	2.1
Differentiate and rectify	40 Gb/s	0.19
Delay and multiply	40 Gb/s	0.15

4. Lightwave System Description and Component Examples

4.1. *Lightwave system block diagram*

A block diagram of a typical lightwave transmission system is shown in Fig. 7. The approximate transistor count for each circuit block is shown at the top of the figure. The multiplexer and demultiplexer drawn here are shown as 2:1, where 4:1 or higher order multiplexing/demultiplexing is desirable for system applications. After the multiplexer on the transmitter side, an external modulator is used to modulate the optical signal. At bit rates below 20 Gb/s, direct modulation of high speed lasers may be used. What limits the bit rate achievable with a laser is the laser bandwidth, and the need for a large AC modulation current into an impedance close to that of a forward biased diode. Impedance matching the laser to 50 Ω may help. The external modulator, on the other hand, has an input impedance which is close to 50 Ω, and may currently require a drive voltage as low 0.9 $V_{p\text{-}p}$.[3] After transmission over the fiber, a photodiode converts the optical signal into an equivalent electrical current. The receiver converts the electrical current into a voltage. This is typically implemented with a transimpedance amplifier. Two or more stages of AGC or limiting amplification are used to amplify the received signal to and amplitude sufficient for the decision circuit or regenerating demultiplexer. These circuits regenerate and retime the received data. The data retiming (NRZ data), is performed by a clock recovery circuit. This is necessary, because the spectrum of NRZ data contains no energy at the baud rate. By using nonlinear processing in the form of

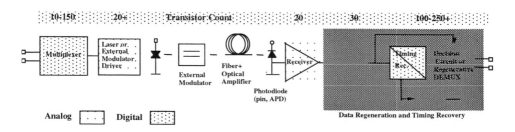

Fig. 7. Block diagram of a generic lightwave transmission system.

differentiation and rectification or delay-and-multiplication a spectral component is generated at the baud rate. The recovered spectral component is then filtered by a high-Q filter. At the high data rates described in this article, high-Q filtering is achieved with either a dielectric resonator or a phase locked loop. In either case, some post amplification is required to raise the recovered signal to a level required for the decision circuit or regenerative demultiplexer. In the case of the dielectric resonator approach, the post amplification needs to be in the form of a limiting or AGC amplifier. This is because the output level of the filter will decay exponentially during periods of non-transition in NRZ data.

We will discuss the functionality of each component in more detail, and show an example of an AlGaAs/GaAs HBT circuit that implements each function.

4.2. *Data multiplexing*

High-speed time division multiplexers are key elements in digital communications, signal processing systems, and test equipment. The multiplexer combines the incoming parallel data streams into a serial output data stream for the laser or external modulator driver. The multiplexer or preceding ICs should feature internal retiming (alignment of input data streams in time), to ensure timing margin. At present most published designs feature 2:1 multiplexing because of its relative simplicity and high speed potential. One of the data channels needs to be delayed by half a bit period relative to the other, because this circuit is a simple interleaver. The input clock rate is equal to half the output bit rate. However, 4:1 or higher multiplexing schemes are required for most systems applications. This can be accomplished by combining (packaged or in a hybrid) two 2:1 multiplexers, two half bit period delay lines (assuming differential input signals), and a frequency divider. It can also be accomplished by integrating all these functions on a single IC. The required delays can be implemented with flip-flops.

4.2.1. *4:1 multiplexer*

The experimental 40 Gb/s MUX[4] presented here features 4:1 multiplexing, including data acquisition and delay, which are performed by a multifunctional circuit (thought to be most efficient with respect to low power consumption and device count). The circuit topology was described by Daniel,[5] and is architecturally similar to a 27 Gb/s 4:1 MUX.[6]

Figure 8 illustrates the circuit principle, which features a two-stage architecture. Only the final 2:1 MUX stage operates at the maximum bit rate; all the other circuitry operates at half of the bit rate. In addition, because the front-end MUXes operate at lower speed, the circuit features non-differential data inputs without sacrificing performance.

First, compared to a straightforward design with four master-slave D-type flip-flops (MS-DFFs), the power consumption and device count are reduced considerably. Second, there is no need to generate a 90 degree shifted clock of CK2, which allows the use of a low power regenerative divider instead of a MS-DFF divider.

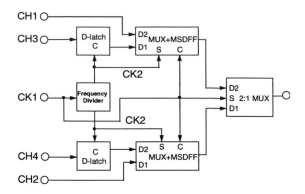

Fig. 8. Block diagram of 4:1 multiplexer circuit.

However, in this implementation, a MS-DFF divider was used to generate CK2, to maximize the operating frequency range of the circuit. All circuits were clocked with a 50% duty cycle differential clock.

4.2.2. *Circuit design*

For the various circuit building blocks, the current-mode logic (CML) series gating technique has been applied. The latches, flip-flops, and MUXes used are standard ECL type circuits. Only the muxlatches[7] require an additional transistor level to perform the multiplexing and latching. Differential signals are used throughout the circuit.

Single-ended input data is amplified and converted to a differential signal by a cascade input buffer. The frequency divider was carefully designed for high-speed operation, as well as broad frequency of operation. Careful design of the internal clock driver, as well as layout of the clock path, was required for high-speed operation. The output buffer (open-collector) of the final MUX generated a differential output signal of $0.8V_{p\text{-}p}$ into a 50 Ω load, and was back-terminated with 100 Ω to reduce potential reflections.

Optimization of internal circuit waveforms as well as device operating points was used to insure maximum speed, and to minimize any relaxation oscillation. SPICE simulations for the MUX were used to verify the design, as well as the optimization of the circuit. Layout parasitics were carefully considered in critical paths.

The circuit was tested, on-wafer, using RF probes. Input test data signals were generated by a 10 Gb/s bit-error-rate tester. The 4:1 multiplexer operated up to a maximum bit rate of 40 Gb/s, with data input levels of 125 $mV_{p\text{-}p}$ per channel, and a single-ended clock input level of 500 $mV_{p\text{-}p}$. The output eye diagrams at 40 Gb/s (both data and inverted data) are shown in Fig. 9(a). Proper multiplexer operation was verified by interleaving four 10 Gb/s input data channels, each four bits in length, and observing the resulting output bit pattern. (Bit-error rates were taken up to 30 Gb/s, the speed limit of a 1:4 demultiplexer IC, with an error rate of less

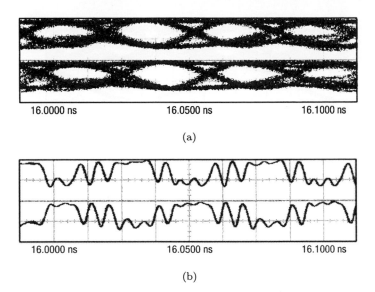

16.0000 ns   16.0500 ns   16.1000 ns

(a)

16.0000 ns   16.0500 ns   16.1000 ns

(b)

Fig. 9. Output diagrams (data and inverted data) at 40 Gb/s. Vertical scale: 400 mV/div.

than 1×10^{-15} being measured). The 40 Gb/s output bit pattern for pseudorandom input data is shown in Fig. 9(b). The 4:1 multiplexer circuit occupies an area of 2 mm^2, with a single power supply voltage of -7.5 V. The power is dissipated mainly in the clock driver and emitter follower stages. A microphotograph of the 4:1 MUX is shown in Fig. 10.

Transistors used in the 4:1 MUX had an f_T of 40 GHz and f_{MAX} of 105 GHz at the gate bias point. The peak f_T for these devices was 60 GHz, with f_{max} 125 GHz at 1.2×10^5 A/cm^2.

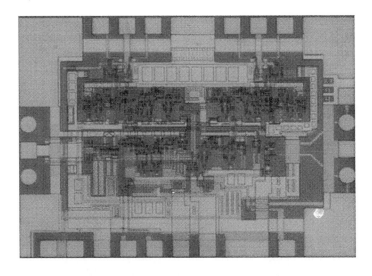

Fig. 10. Microphotograph of 4:1 MUX chip.

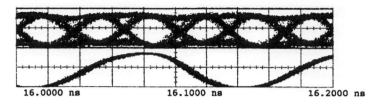

16.0000 ns   16.1000 ns   16.2000 ns

Fig. 11.   Output data eye diagram, and divided clock output at 30 Gb/s.   Vertical scale: 400 mV/div.

The 4:1 MUX design was also fabricated at Rockwell Semiconductor Systems GaAs production facility in Newbury Park, CA.[8] Operation beyond 30 Gb/s was measured, with the resulting data eye diagrams and divided clock output show in Fig. 11.

4.3. *Photodiode*

The two candidates for operation in the 1.3 and 1.55 μm wavelength range are the P-I-N photodiode and the avalanche photodiode. PIN photodiodes operate at frequencies in excess of 40 GHz, with a typical responsivity of 0.7–0.95 A/W. Research devices include a new class of photodetectors, which are based on transmission line structures. These traveling wave photodetectors have been demonstrated with a bandwidth of 370 GHz,[9] and a responsivity of approximately 20% at 800 nm. Commercially available photodetectors are typically sold as a packaged unit comprising of a butt coupled single mode fiber and high-speed electrical contacts. Commercial photodetectors are sold with bandwidths in excess of 40 GHz. The reliability of these detectors is very high.

The avalanche photodiode (APD) is a less well developed photodetector for high bit rate operation, although some commercially available APDs operate with a gain bandwidth product of greater than 25 GHz. Devices with a gain-bandwidth product in excess of 500 GHz[10] have been demonstrated in the laboratory. A typical bias level for these devices is in the range of 40–80 V_{DC}, and is a function of the device composition.

Fiber amplifiers have been utilized to realize more sensitive receivers. This approach will be discussed following the pre-amplifier section.

4.4. *Preamplifier*

The high impedance amplifier and the transimpedance amplifier (Tz amp.) have been researched for a number of years. Low impedance amplifiers are generally not used due to poor sensitivity. The high impedance amplifier features high sensitivity, but acts as an integrator. This requires subsequent equalization, and also tends to limit the dynamic range and the low frequency content possible in the input signal. The transimpedance amplifier achieves comparable sensitivity to the high impedance amplifier, but with improved dynamic range, low frequency response, and without subsequent equalization. Transimpedance amplifier ICs are commercially available for a few hundred dollars at data rates up to 10 Gb/s, using a variety of IC processes ranging from Si bipolar to GaAs FET.

To achieve higher performance than an OEIC (optoelectronic integrated circuit), hybrid preamplifiers can been realized (using discrete HEMT or FET devices) and are expected to be manufactured. Integrated OEIC preamplifiers, having the benefits of lower cost, higher reliability, etc., will require an advanced HEMT, FET, submicron silicon bipolar, or heterojunction bipolar process (HBT). Most reported transimpedance amplifiers (which are single-ended) show considerable gain peaking and bandwidth limitation caused by feedback/input capacitance time constants. Only with careful design can such peaking be eliminated.

4.4.1. *Addition of erbium doped fiber to preamplifier*

The advent of erbium doped fiber amplifiers has led to their use as an initial amplifier to amplify the light input to the photodetector. To illustrate this, the following examples (at 10 Gb/s) will be discussed. In the case of an externally modulated P-I-N front-end, the 10 Gb/s sensitivity at 1.5 μm was increased from approximately -20.4 dBm to -33.8 dBm (8 GHz 10 pA/Hz PIN-FET receiver, 1.48 μm pump, 5.6 dB NF amplifier, and a 1 nm optical filter).[11] Using a 0.98 μm pump, receiver sensitivities of -37.2 dBm using a 3 nm optical filter and a PIN-FET receiver have been achieved.[12] This compares quite favorably to the -22.5 dBm sensitivity achieved without fiber amplifiers and an APD. For the case of APDs and fiber amplifiers, work at 10 Gb/s has not been widely reported. The sensitivity of a standard 2.3 Gb/s Alcatel APD system operating with a sensitivity of -33.3 dBm was increased to -43.3 dBm.[13] It was found that an APD did not greatly improve receiver sensitivity over a PIN unless a narrow bandwidth optical filter was used.[14] However, the multiplication factor of the APD allowed the fiber amplifier to be operated at lower gain (i.e., lower pump power) to achieve comparable sensitivity (both systems were beat noise limited).

Note: In order to enhance the sensitivity of an existing receiver (APD or PIN), a narrow (on the order of 1 nm) passband filter must be inserted between the fiber preamp and the photodetector. In order to keep the laser wavelength within the filter passband, a high degree of laser wavelength stability is required (typical wavelength variation is on the order of 1 Å/°C). WDM systems can further compound this issue.

4.4.2. *Preamplifier*

The preamplifier was a transimpedance amplifier designed to have maximum flatness (± 0.5 dB) from 0 to 30 GHz. A second version consisted of a flip-chip compatible T_z amp and an integrated T_z amp + Schottky detector (for $\lambda = 0.86$ μm) circuit. A microphotograph of the flip-chip compatible version is shown in Fig. 12 and the schematic in Fig. 13. The chip was measured using an optical spectrum analyzer.

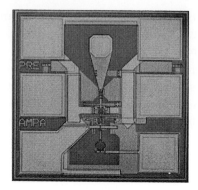

Fig. 12. Microphotograph of pre-amplifier circuit.

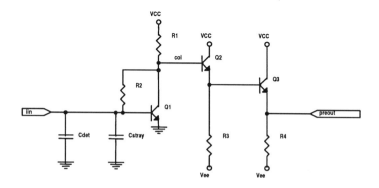

Fig. 13. Schematic diagram of pre-amplifier circuit.

The results are shown in Fig. 14. The 3 dB bandwidth of this chip was 25 GHz. Note the excellent gain flatness and good phase response. Several different chips on each wafer were tested. Preamplifiers on the same wafer showed very similar responses, again indicating the excellent process uniformity. The preamplifier was fabricated with a transistor structure which had an f_T of 53 GHz and f_{max} of 120 GHz (at $J_c = 5 \times 10^4$ A/cm^2) with the peak f_T, f_{max} of 96 GHz, 160 GHz at 1.67×10^5 A/cm^2.

4.5. *Gain control amplifier*

After the preamplifier, amplification is required with a large gain and input dynamic range to compensate for signal level fluctuations. Broadband variable gain amplifiers are also key elements for optical communication systems. Receivers for such systems require linear channel response with little magnitude variation, constant group delay, and frequency response down to DC to achieve good bit error rates. The two most common gain blocks are the automatic gain control (AGC) amplifier, and the limiting amplifier. Table 3 compares the two alternatives.

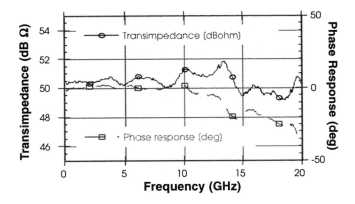

Fig. 14. Transimpedance and phase for transimpedance amplifier.

Table 3. Comparison between limiting and AGC amplifiers.

Limiting	AGC
Higher Speed	Slightly slower
Lower power consumption (P_d)	Higher P_d due to added transistor stacking, linearity requirements
Easier design	
Smaller chip area	
Simpler application	
Lower noise figure (NF)	NF is a function of noise in amplifier plus control loop
No control loop	Control loop determines transient response, recovery time
Limiting may be incompatible with some modulation formats	Linear amplifier
	Improper design may result in output DC level shifts with input level
Limiting may hamper sensitivity of APD based systems which benefit from threshold optimization	Linear amplifier

4.5.1. *Variable-gain limiting amplifier*

We have demonstrated a packaged amplifier with these properties from DC to 26 GHz.[15] A microphotograph of this chip is shown in Fig. 15(a). The circuit consists of an input buffer, two amplifier stages connected by emitter follower buffers, and an output buffer. The input and output buffers provide good terminations for 50 Ω source and load. Figure 15(b) shows a simplified circuit diagram for one of the amplifier stages. This is a modified transadmittance/transimpedance (TA/TI) amplifier pair.[16] Q1 and Q2 form the TA portion, while Q3–Q6 and the R_f's form

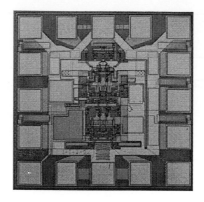

(a)

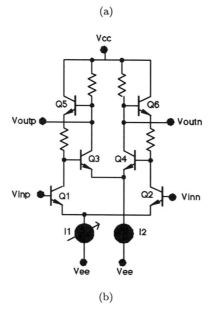

(b)

Fig. 15. Microphotograph of VGA and simplified circuit diagram of TA/TI stage.

the TI portion. Variation of I_1 by an external gain controlled voltage changes the transimpedance (g_m) of the TA amplifier which in turn varies the overall gain of the amplifier. Q5 and Q6 in the parallel feedback path of the TI amplifier buffer the output from the feedback network which results in higher gain, wider bandwidth, and constant output DC voltage independent of the gain-controlled voltages.[17] The emitter followers between the stages provide level shifting and improve the input/output impedance matching between the output TI portion of the first amplifier stage and the input TA portion of the second amplifier stage. The measured performance of this VGA is shown in Fig. 16. Note that this circuit has a gain variation of only ±1 dB, constant group delay within the passband, and a gain controlled range of 10–16 dB. This performance is suitable for use in > 30 Gb/s fiber-optic TDM transmission systems.[18] The AGC amplifier was fabricated with a transistor structure

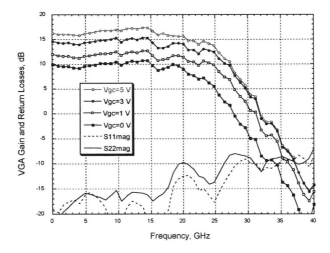

Fig. 16. VGA output after packaging.

which exhibited a f_T of 60 GHz and f_{max} of 111 GHz at the amplifier bias point. The peak f_T was 95 GHz at 1.2×10^5 A/cm^2 with f_{max} of 130 GHz.

4.6. *Data demultiplexing*

After amplification by the circuits in the linear channel (preamplifier, gain control amplifiers), the received data needs to be digitally regenerated, retimed and de-multiplexed. The regeneration and retiming are typically performed by a MS-DFF which performs the functions of data regeneration and retiming (decision circuit). The subsequent demultiplexing can be combined with the decision circuit to form a regenerative demultiplexer (DEMUX). With a half-rate clock input and using two MS-DFFs clocked anti-phase, a 1:2 demultiplexer IC can be formed. However, as with the case of the multiplexer, 1:4 demultiplexing is more desirable for systems applications. Again, two 1:2 demultiplexers, two delay lines, and a frequency divider can be combined (packed or in a hybrid) for this function. It can also be accomplished by a single IC.

4.6.1. *1:4 demultiplexer*

The 1:4 DEMUX was designed with similar multifunctional circuits as the 4:1 MUX.[6,7] Figure 17 shows the circuit principal, which features a novel two-stage architecture.[19]

The master-slave (MS) flip-flops in the initial 1:2 demultiplexing are used to regenerate and stabilize the incoming data. A slave {S} D-type latch is added to the lower master-slave D-type flip-flop (MS-DFF) for bit alignment. The final stage of demultiplexing is accomplished with six freeze-type {F} latches.[7] The input clock CK1 (with a frequency half that of the input data rate) is used to time all latches (D and F type). CK2 is used as a select signal by the F latches, and is delayed

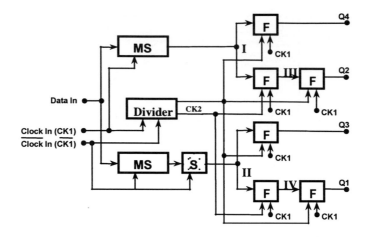

Fig. 17. 1:4 DMUX schematic block diagram.

relative to CK1 by the gate delay of the frequency divider. Because only one phase of the divide clock (CK2) is used, a dynamic frequency divider may be used in place of a conventional MS static frequency divider with reduced power dissipation and potentially increased circuit operating speed. However, in this implementation a MS-DFF divider was used to generate CK2, to maximize the operating frequency range of the circuit. The simulated output eye diagrams (15 Gb/s), with 60 Gb/s input data, are shown in Fig. 18.

The circuit was tested on-wafer using a 30 Gb/s test signal generated by connecting the on-wafer 4:1 MUX and 1:4 DMUX with 0.8 m of semi-rigid cable.[20] The input test signal is shown in Fig. 19(a). The resulting demultiplexed output eye diagrams (CH1-3) and divided clock outputs are shown in Fig. 19(b). The divided clock output was designed with a falling edge, which is independent of data rate, in its relative position to the output data channels. This clock is used to time subsequent circuits in a system.

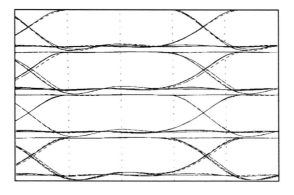

Fig. 18. Simulated output eye diagram for 1:4 DMUX.

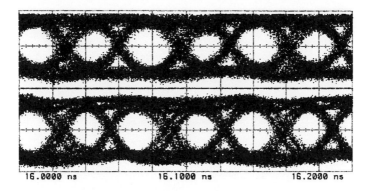

Fig. 19(a). Input test signal to 1:4 DMUX at 30 Gb/s. Vertical scale: 160 mV/div.

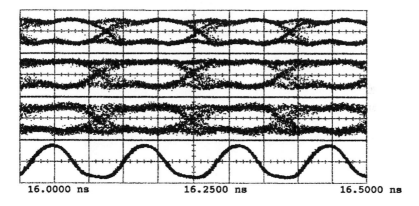

Fig. 19(b). Output demultiplexed eye diagrams (CH1-3) and output divided clock. Vertical scale: 600 mV/div.

The MUX/DMUX pair were measured back-to-back with a BER test set to have error-free performance at 30 Gb/s. The transistors in the 1:4 DMUX were identical to those in the 4:1 MUX. A microphotograph of the DMUX is shown in Fig. 20.

We also packaged and tested the 1:4 demultiplexer. Error free operation was measured up to a maximum bit rate of 30 Gb/s. With a 30 Gb/s test signal generated by the 4:1 MUX, the resulting demultiplexed output data channel is shown in Fig. 21. The differential data input sensitivity of the packaged IC was 200 m$V_{\text{p-p}}$ at 2^{31}-1 PRBS, with a differential clock amplitude of 0.9 $V_{\text{p-p}}$ and a phase margin of 22^0 in the 7.5 GHz clock (approximately 1/4 of the 30 Gb/s input data period). The differential data input sensitivity was approximately 15 m$V_{\text{p-p}}$ at 10 Gb/s, and the phase margin was 310^0 (under full-rate clock input).

4.7. *Decision and timing recovery*

The data regeneration function has long been recognized as the most critical building block in a direct detection lightwave system, primarily because it requires

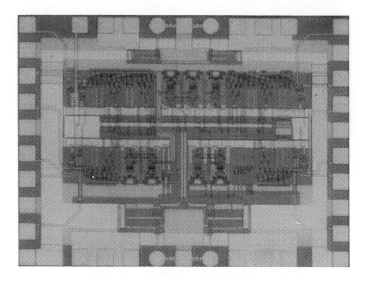

Fig. 20. Die microphotograph of 1:4 DMUX.

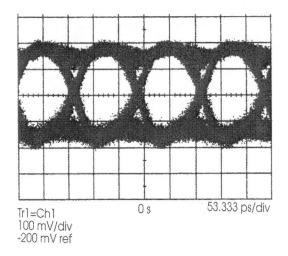

Trl=Chl
100 mV/div
-200 mV ref

0 s   53.333 ps/div

Fig. 21. Output 7.5 Gb/s eye diagram with a 30 Gb/s input from a packaged 1:4 demultiplexer.

nonlinear timing signal generation (for NRZ data format) and subsequent high-Q filtering. Monolithic integration of data regeneration function (timing recovery circuit along with decision circuit) represents many advantages from the viewpoint of cost, size, reliability, and performance. This leads to the option of using phase-locked loops (PLL) to implement the high-Q bandpass filter, which usually is a dielectric resonator filter at multi-gigabit rates, and a discrete surface acoustic wave (SAW) device at low bit rates. The SAW device is not physically realizable at 10 Gb/s (due to the small dimensions required for the electrical to acoustical transducers).

The 10 Gb/s SONET hierarchy (OC-192) was the first SONET rate at which the effect of unscrambled A1, A2, and C1 bytes on timing recovery had come into question. This leads to the issue of whether a bit or byte multiplexing architecture is required. Sufficient transition density in these bytes guaranteed clock recovery was possible. Long periods of non-transition in the data tax the clock recovery.

The data regeneration function constitutes two major building blocks: the decision circuit and the timing recovery circuit. The decision circuit function is accomplished through a purely digital D-type flip-flop. The timing recovery circuit, however, poses the most challenging circuit design issues since it requires various very-high-speed signal (linear, nonlinear, analog, digital) processing techniques. A spectral null normally exists at the baud rate for NRZ data. For this reason it cannot simply be filtered from the incoming data stream, as is the case for return-to-zero (RZ) data. Both analog and digital methods may be used in the timing recovery circuit. These two methods are briefly discussed.

Analog

An analog differentiator and/or rectifier may be used in the front end of the timing recovering circuit to generate a pulse stream with very strong spectral component at the clock frequency, prior to the clock extraction by the high-Q filter. No physical half-bit delay lines are required, but the single-ended output of the rectifier may require a band-limiting conversion to a differential signal for monolithic integration. Differentiate-and-rectify is well suited for monolithic integration. The differentiate and rectify method has typically higher loss than that of the digital methods. This loss in the circuit will degrade the signal-to-noise ratio (SNR) of the recovered clock spectral component. But the differentiate and rectify circuit has the advantage of effectively high-pass filtering the spectrum (the differentiation function) and also low-pass filtering due to the inherent bandwidth of the circuit. It has been shown that bandpass filtering NRZ data around the half baud prior to clock recovery, the signal-to-noise ratio of the resulting recovered clock component will be greater than that of the non-filtered spectrum. Some researchers have proposed double differentiation to further enhance this signal-to-noise ratio.

Digital

Digital approaches (for example, a delay-and-multiply circuit utilizing an exclusive OR or AND gate) have an inherent speed advantage over analog methods, and provide differential output signals. However, physical half-bit delays must be generated; which are difficult to implement at high (>= 10 Gb/s) data rates and require precise tuning. The circuit must be readjusted for different data rates. The digital circuits typically exhibit gain, whereas analog methods exhibit loss. By implementing an on-chip high pass filter, at the appropriate frequency, it is possible to perform similar prefiltering of the input spectrum as was described in the analog case. The high-pass filter would typically be implemented as a buffered RC network.

179

4.8. High-Q filter

Dielectric resonator versus phase locked loop

Tables 4 and 5 detail the technological trade-off between the two methods of high-Q filtering.

Phase locked loop

The reliability and robustness (practical yield achieved with large-scale circuits operating at close to the maximum speed for a given process) are still in question for most IC technologies at data rates > 10 Gb/s, and may limit the availability of integrated PLL and data regeneration circuits.

To increase the loop's capture range while maintaining the same noise bandwidth, the timing recovery circuit will adopt the dual-loop scheme, phase- and frequency-locked loop (PFLL), to help the frequency acquisition. One architecture suitable for high-speed all-analog PFLL approach requires a voltage-controlled oscillator (VCO) with quadrature outputs, two analog phase comparators, and loop filters along with other low-frequency control circuits. The VCO will undoubtedly

Table 4. Advantages of PLL over Dielectric Resonators.

Center frequency is broadly adjustable, tracks clock automatically	Center frequency is fixed, may shift over time
Bandwidth independently controllable	Bandwidth is fixed, and must be carefully chosen Q too low: High jitter Q too high: Static phase offsets
Active device, low loss	Added amplification required
May be implemented as an IC	Physical size a function of frequency
Tolerates large frequency offsets without static phase errors	Frequency offsets introduce large static phase errors
Long strings of ones or zeros are tolerated with almost no effect on timing phase or amplitude	System must be designed with a sufficiently large Q*D product. Q = filter Q, D = post limiting amplifier dynamic range

Table 5. Advantages of Dielectric Resonators over PLL.

Simple to implement	IC design difficult at multi-gigabit rates
Simple circuitry	Complicated circuitry, may be difficult to adjust at multi-gigabit rates
Reliability analysis possible	Reliability analysis more complicated
Potential Problems: Aging and temperature changes	Potential Problems: Same as dielectric resonator, plus design errors such as false lock, injection locking, jitter peaking, and RF interference

be the most important component for such an approach, considering requirements on phase jitter, tuning range, quadrature phase, and high-frequency capability. The possible candidates are relaxation oscillators and ring oscillators; both are suitable for monolithic integration. The analog phase comparators can be implemented in a wideband Gilbert-type multiplier, which can operate near the transistor's unity-gain frequency. The loop filters, which control the noise bandwidth and Q factor, can be realized on or off the chip. Since it only involves low-frequency circuit, speed will not be the issue in the filter design.

Dielectric resonator filter

As Tables 3 and 4 suggest, passive filtering for clock recovery circuits is practical at 10 GHz using dielectric resonator materials. Filter loaded Q's ranging from about 300 to 1000 are achievable from materials yielding unloaded Q's of 10000. At 10 GHz, resonator sizes are on the order of 0.2 inches in diameter by 0.1 inches in thickness, resulting in simple to implement, relatively compact packages. While temperature characteristics are of concern, the thermal properties of the materials can be tailored to the housing parameters so that resonant frequency stability over temperature is well controlled. It is unlikely that these ceramic devices, which have been fired at extremely high temperatures under high pressure, will show serious aging problems. Like the SAW device, the physical size of the filter decreases with frequency, making devices for data rates of 20 or 30 Gb/s very small.

4.8.1. *Clock and data regenerator*

We have designed a multifunctional circuit, fabricated in a baseline HBT techno-logy, for use in laboratory systems experiments.[21] The IC, when combined with an external transimpedance amplifier, and a high-Q filter (PLL or dielectric resonator) can be used as a regenerator for nonreturn-to-zero (NRZ) lightwave systems. By performing the nonlinear operation of differentiation followed by rectification, a strong spectral component is generated at the baud rate.

Data entering the the input port of the circuit (Fig. 22) was on-chip terminated into 50 Ω, and amplified by a two-stage (37 dB gain) limiting amplifier. Each stage was implemented with localized series feedback followed by localized shunt feedback,[16] as shown in Fig 23. The series feedback utilizes transistor emitter resistance (r_E), while the shunt feedback had emitter follower feedback for improved bandwidth.[17] Amplified data was then applied to the on-chip MS-DFF, and to the timing recovery portion of the IC.

The decision circuit (data regeneration and retiming) was implemented in a master-slave configuration with static D-type latches, with two emitter followers in the feedback path for maximum speed. The decision circuit operated up to 22 Gb/s, under full rate clock and data input.

The timing recovery portion of the IC used on-chip nonlinear processing (differentiate-and-rectify circuit), to generate a clock spectral component at the

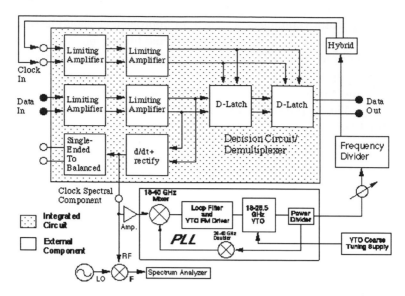

Fig. 22. Block diagram of the data regeneration/clock extraction chip, including external PLL as narrow band filter.

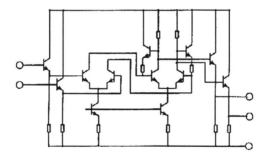

Fig. 23. Limiting amplifier detail.

bit rate (Fig. 24). The differentiator was implemented with a differential pair with its emitters connected by BE and BC junction capacitors.[22] The capacitance was adjustable, with an externally applied reverse bias. Its operation was measured to be insensitive to the applied voltage. Rectification was implemented with two transistors connected at both the collectors and the emitters. A series 39 Ω resistor matched this output (single-ended) to 50 Ω. An optional differential output was generated with an on-chip single-ended to differential converter with open collector output drivers. The converter acted as an amplifier, and bandlimited the differential output to about 15 GHz.

The clock input signal is also on-chip terminated to 50 Ω, and is amplified by two stages of limiting amplification (similar to the data path), before being applied to the input of the on-chip MSDFF. If the recovered clock, after external high-Q

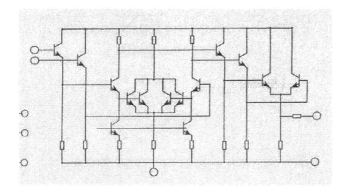

Fig. 24. Timing recovery circuit detail.

filtering, is placed through a 1:N frequency divider, then the output of the MSDFF will be one of the N constituent data streams multiplexed together in the input data stream. In our case N = 4 (the PLL provided an internal divide-by-two, together with an external 1:2 divider). For use in systems experiments, the IC was mounted in a research package,[23] similar to a hybrid circuit module.[24] Measured data was derived from packaged parts.

The output of the on-chip differentiate-and-rectify circuit at 24 Gb/s is shown in Fig. 25, with a 2^{10}-1 pseudo-random bit sequence (PRBS) at the input to the MUX. The recovered spectral component at 24 Gb/s is shown in Fig. 26, with a 2^{15}-1 PRBS input to the MUX. The 24 Gb/s spectrum was measured, with a down conversion mixer (20 dB conversion loss, 10 dBm LO at 19.63 GHz), to have a power level of −32.27 dBm (a front-end mixer was used due to spectrum analyzer limitations). The spurious clock spectral component generated from non-data sources (primarily MUX clock feedthrough) was measured to be −47.35 dBm at 24 Gb/s. At 30 Gb/s, the signal and spurious levels were measured to be −47.33 and −53.76 dBm, respectively. No measures were taken to reduce the magnitude of the spurious component.

The on-chip generated recovered clock spectral component was high-Q filtered with an off-chip PLL constructed from discrete microwave components. The PLL

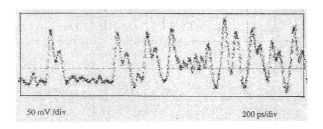

50 mV /div   200 ps/div

Fig. 25. Output waveform of on-chip d/dt+rectify.

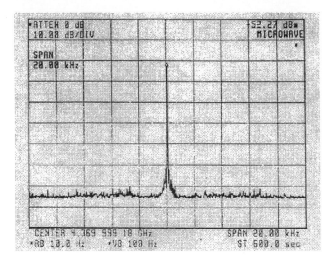

Fig. 26. Output spectrum of d/dt+rectify.

consisted of a third order loop with a YIG oscillator VCO, a microwave mixer, power divider, and a loop filter using a commercial operational amplifier. The loop bandwidth was 300 kHz. Because the PLL required an input signal level of between −10 dBm to +10 dBm, the differentiate-and-rectify output of the IC was amplified by external microwave amplifiers. To place this amplification on-chip would have risked chip-wide oscillation. The output power of the VCO was +13 dBm. A frequency doubler within the PLL loop creates an inherent division by two within the loop. The PLL readily acquired and maintained phase lock, demonstrating the suitability for system applications of this component.

4.8.2. *Delay-and-multiply timing recovery*

In addition to the differentiate-and-rectify method of timing recovery, the required nonlinear processing for timing recovery may also be done by performing a delay-and-multiply operation on the incoming data. This operation is implemented by splitting the incoming data stream in two, delaying one path by half a bit period, and then performing an analog multiplication (Fig. 27). Multiplication is obtained with a high-speed Gilbert multiplier. A variable delay line is incorporated to allow optimal operation for data bit-rates from 30 Gb/s to 40 Gb/s. The variable delay line is implemented with a high-impedance transmission line periodically loaded with reverse-biased Base-emitter junctions of HBTs. The bias voltages of the BE junctions are varied to vary the loading capacitance of the transmission line, hence the delay. The strongest spectral component is generated when the delay branch provides exactly a half bit period delay. The IC features fully differential 50 Ω inputs (125 m$V_{\text{p-p}}$) and outputs frequency multiplication. Output power is greater

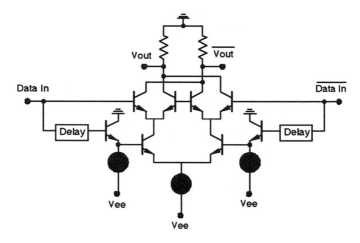

Fig. 27. Schematic of delay-and-multiply circuit.

than -10 dBm into a 50 Ω load. The die size is 1.62 mm by 1.12 mm. The delay line is biased with a voltage from -5 V to 0 V.

The delay-and-multiply circuit was tested with a 20 GHz sinusoidal input test signal 125 m$V_{\text{p-p}}$ in amplitude. The resulting sinusoidal output is shown in Fig. 28. The resulting multiplied output waveform was at twice the frequency of the input signal (40 GHz). The delay-and-multiply circuit was designed to be used with a PLL similar to that shown in Fig. 29. It was designed to be a direct alternative to the differentiate-and-rectify circuit used. Both circuits feature the same input impedance, and hence the same loading on the previous stage, but the delay-and-multiply circuit is less lossy than the differentiate-and-rectify. It has an output signal which is typically larger in amplitude, due to the gain of the Gilbert mixer cell.

4.9. *High speed package*

In order to exploit the ICs similar to those described in this paper for systems use, a method of packaging them must be developed. Due to the high number of I/O

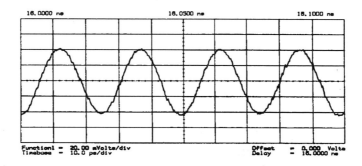

Fig. 28. Output waveform (40 GHz) of delay-and-multiply circuit, with 20 GHz sinusoidal input.

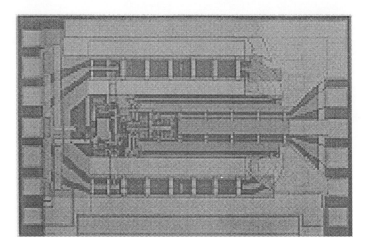

Fig. 29. Die microphotograph of delay-and-multiply circuit.

pins required, the connector size becomes a dominant factor in the package size and performance. For circuits such as the 4:1 multiplexer, a ten pin package was required (differential clock in, four single-ended data inputs, differential data out, two divided clock output signals). This made the package large, and as a result, made the transmission line attenuation an important consideration. With the development of new smaller high speed connectors, the package size can be dramatically decreased, thereby reducing package losses even further. Two approaches to packaging were considered. Microstrip transmission lines, and coplanar waveguide transmission lines. Microstrip lines are slightly less lossy than coplanar waveguides for this application, but require a backside ground plane. Coplanar waveguide offers a top side ground plane. At frequencies approaching 40 GHz, microstrip applications will require backside vias to reduce ground inductance. Therefore, we chose to utilize a coplanar waveguide based package for simplicity.

4.9.1. *Research prototype package*

The coplanar fixture, of dimensions 2.5 cm × 2.5 cm × 1 cm, features 8 high-speed input/output signal lines, and 4-power supply connections (Fig. 30). High-speed input/output signals were brought into the package with Wiltron K connectors having a 1.25 cm center-to-center pitch. Mounted inside the fixture is a 0.64 mm thick quartz substrate with coplanar waveguides etched in the 5 um thick gold plated top surface (there is no metalization on the lower surface). Thin-film resistors were created with a 50 Ω per square TaN layer, located below the metalization layer. The substrate edges were soldered or conductive epoxied to the walls of the fixture, creating a uniform and low inductance ground. To create a uniform ground, and suppressing moding, ground sections were interconnected with wire bonds. Bypass chip capacitors were placed as close as possible to the power supply bond pads

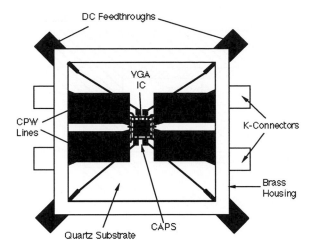

Fig. 30. Top view of package used for VGA.

of the IC to suppress undesirable noise. Larger, surface-mount ceramic capacitors were also connected to internal power supply lines, for further noise suppression.

Coplanar waveguide offers several advantages over microstrip lines for this application. Coplanar waveguides are planar, having both signal lines and ground on the top surface of the substrate, resulting in a low inductance ground. In addition, crosstalk between adjacent coplanar waveguides is suppressed. Finally, coplanar waveguides may be tapered from waveguides with thick center conductors at the K connector launchers to waveguides with thin center conductors at the IC bond pads (with an appropriate reduction in gap width), while maintaining a constant impedance. This allows multiple signal and power lines to converge on an IC (typically 1 mm square), with low ohmic losses.

A cross-section of the package is shown in Fig. 31. The quartz substrate is suspended in the center of the package by the edge of the package and in the center of the package where it has a hole cut into it to allow the IC to be mounted flush with the substrate. This allows for the ribbon bonds between the IC and the transmission lines to be as short as possible.

4.9.2. *Package model*

To determine the effect of the package on the eye diagram of the circuits described in this paper, a model of the package was developed, suitable for use in SPICE. The equivalent circuit for the package is shown in Fig. 32. The input K connectors are assumed to be on the left of the figure, and the IC die is at the end of the network on the right of the figure. The input connector is modeled by a series inductance L_C and a stray shunt capacitance C_C. Next a lossy 50 Ω transmission line model is used to represent the attenuation of the line. This is followed by an ideal transmission line to simulate the signal delay, and finally the bond wire to the IC is modeled

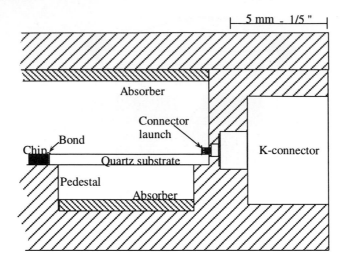

Fig. 31. Cross-section of package.

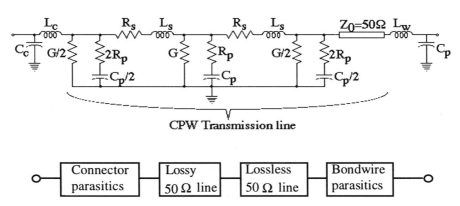

Fig. 32. SPICE model of package, connector parasitics, and interconnecting bond wires.

with a series inductance L_W and a stray shunt capacitance C_P. Figure 33 shows the measured package frequency response, the predicted performance based on calculations, and the response given by the SPICE model. All are in close agreement. Shown at the bottom of the figure is the measured and predicted (by the SPICE model) return loss for the package. Both are in close agreement. The return loss is an important measure of a system. Return loss (db) = 10 log P_{REFL}/P_{IN}, where P_{IN} is the incident power and P_{REFL} is the reflected power. A low value of return loss is important in a systems application, because reflected waves between two components will cause degradation of the desired transmitted signals between those components. Reflections can cause a serious degradation in system performance, and may contribute to a significant system power penalty (amount of received optical power at the photodetector to achieve a given bit error ratio; usually 1×10^{-9}). A photograph of the package used for the VGA is shown in Fig. 34.

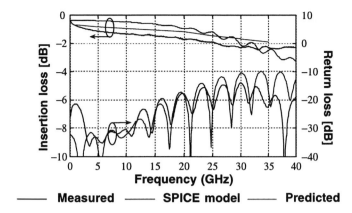

Fig. 33. Left: Measured package frequency response, SPICE model, and calculated. Right: Return loss measured and SPICE model.

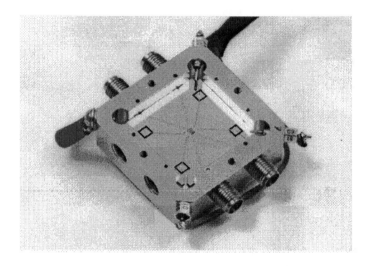

Fig. 34. Photograph of packaged VGA amplifier.

5. Summary

In this paper we have presented many ultra-high speed lightwave circuits, fabricated in a 1.4 μm AlGaAs/GaAs HBT technology. These circuits illustrate that circuit performance up to 40 Gb/s can be achieved with a rather conservative HBT technology. It is expected with aggressive device scaling and operation of AlGaAs/GaAs HBTs at higher current densities, an area of operation not yet investigated, the AlGaAs/GaAs HBT will realize extremely high f_T/f_{MAX}; and as a result will be utilized in future lightwave systems operating well beyond 40 Gb/s. In addition, the semi-insulating GaAs substrate will allow for the design of ever increasingly complex circuits which operate at ultra-high bit-rates, not only at the output of the IC, but within the IC itself.

Acknowledgments

We would like to thank the processing staff at Rockwell Science Center for their efforts and acknowledge the support and guidance of Jon Rode, John Bowers, and Michael Melliar-Smith. This work is supported by ARPA, under contract DABT63-93-C0039, Thunder and Lightning Program.

References

1. M. F. Chang *et al.*, "Self-aligned dielectric assisted planarization process (SADAP)", US Patent No. 4,996,165.
2. P. J. Zampardi, R. L. Pierson, K. Runge, R. Yu, S. M. Beccue, J. Yu, and K. C. Wang, "Hybrid digital/microwave HBTs for > 30 Gb/s optical communication", *IEDM Technical Digest*, 1995, pp. 803–806.
3. K. Yoshino, K. Wakita, I. Kota, S. Kondo, Y. Noguchi, S. Kuwano, N. Takachio, T. Otsuji, Y. Imai, and T. Enoki, "40-Gbit/s MWQ EA modulator module with very low driving-voltage", *NTT 1997 Annual Report*, p. 18.
4. K. Runge, R. L. Pierson, P. J. Zampardi, P. B. Thomas, J. Yu, and K. C. Wang, "40 Gbit/s AlGaAs/GaAs HBT 4:1 multiplexer IC", *Electron. Lett.* **31**, 11 (1995) 876–877.
5. D. Daniel, "Novel 4:1 multiplexer circuit for Gbit/s data rates", *Electron. Lett.* **26**, (1990).
6. K. Runge, D. Daniel, and J. L. Gimlett, "A 27 Gb/s AlGaAs/GaAs HBT 4:1 multiplexer IC", *Proc. 1991 GaAs IC Symp.*, pp. 233–236.
7. R. G. Swartz, "Ultra-high speed multiplexer/demultiplexer architectures", *Int. J. High Speed Circuits* **1**, 1 (1991).
8. Fabricated under DARPA/NraD ADC program. Contract#N66001-94-C-6000.
9. Y.-J. Chiu, S. B. Fleischer, D. Lasaosa, and J. E. Bowers, "Ultrafast (370 GHz bandwidth) p-i-n travelling wave photodetector using low-temperature-grown GaAs", *Appl. Phys. Lett.* **71**, 17 (1997).
10. W. Wu, A. Hawkins, and J. E. Bowers, "Silicon based telecommunication avalanche photodetectors", Invited paper, *OECC'97*, Seoul, Korea, July 1997, pp. 446–447.
11. B. L. Batel, "High performance optical transmission system using an Erbium-doped fiber preamplifier pumped at 1480 nm", *OFC 92 Technical Digest*, 1992, p. 246.
12. T. Saito, Presentation at Bellcore, February 1992.
13. P. M. Gabla *et al.*, *OFC 1992 Conf. Proc.*, p. 245.
14. P. M. Gabla *et al.*, *ECOC'91*, Paris, France, Sept. 1991, paper WeC9.3.
15. R. Yu, S. Beccue, P. Zampardi, R. Pierson, A. Petersen, K. C. Wang, and J. Bower, "A packaged broadband monolithic variable gain amplifier implemented in AlGaAs/GaAs HBT technology", *Proc. 1995 GaAs IC Sym.*
16. B. M. Cherry and D. E. Hooper, "The design of wide-band transistor feedback amplifiers", *Proc. IEE*, 1963, pp. 375–389.
17. H. Ichino, N. Ishihara, M. Suzuki, and S. Konaka, "18 GHz 1/8 dynamic frequency divider using Si bipolar technologies", *IEEE J. Solid-State Circuits* **24**, 6 (1989).
18. A. K. Petersen *et al.*, "3 MHz-30 GHz traveling-wave optical front-end receiver", *Proc. OFC'95*, pp. 157–158, San Diego, CA, Feb. 1995.
19. K. Runge, "1:4 demultiplexer architecture for Gbit/s lightwave systems", *Electron. Lett.* **27**, 9 (1991).
20. K. Runge, R. L. Pierson, P. J. Zampardi, P. B. Thomas, J. Yu, and K. C. Wang, "30 Gbit/s 1:4 demultiplexer IC using AlGaAs/GaAs HBTs", *Electron. Lett.* **33**, 9 (1997).

21. K. Runge, R. Y. Yu, P. J. Zampardi, R. L. Pierson, and K. C. Wang, "Packaged 30 Gbit/s data demultiplexing and clock extraction IC fabricated in a AlGaAs/GaAs HBT technology", *Electron. Lett.* **32**, 6 (1996).
22. Z. G. Wang, U. Langmann, and B. G. Bosch, "Multi-Gb/s Silicon bipolar clock recovery IC", *IEEE J. Selected Areas in Communications* **9**, 5 (1991).
23. A. K. Petersen, R. Y. Yu, K. Runge, J. E. Bowers, and K. C. Wang, "Microwave packages for 30 Gbit/s analog and digital circuits", *Proc. Electrical Performance of Electronic Packaging*, Portland Oregon, October 1995.
24. K. Runge, M. Bagheri, and J. Young, "High performance hybrid circuit modules for lightwave systems operating at data rates of 10 Gb/s and higher", *Electron. Lett.* **27**, 3 (1991).
25. N. Ishihara, O. Nakajima, H. Ichino, and Y. Yamauchi, "9 GHz bandwidth, 8-20 dB controllable-gain monolithic amplifier using AlGaAs/GaAs HBT technology", *Electron. Lett.* **25**, 19 (1989) 1317–1318.

International Journal of High Speed Electronics and Systems, Vol. 9, No. 2 (1998) 505–548

HIGH SPEED CROSSPOINT SWITCHES

CHARLES E. CHANG,* K. C. WANG and ARLENDA D. CAMPAÑA*

*Rockwell Semiconductor Systems, 2101B Corporate Center Drive,
Newbury Park, CA 91320, USA*

ANDRE G. METZGER and PETER M. ASBECK

*University of California, San Diego,
Department of Electrical and Computer Engineering,
9500 Gilman Drive, La Jolla, CA 92093, USA*

STEVE M. BECCUE

*Rockwell Science Center, 1049 Camino Dos Rios,
Thousand Oaks, CA 91360, USA*

Technology and performance of electronic crosspoint switches with data rates above 1 Gb/s/channel are reviewed. Switch applications and architectures are described, as well as the principal problems in achieving low output jitter. Recent results for different IC technologies are summarized. As a particular example, a 10 Gb/s crosspoint switch implemented in GaAlAs/GaAs-HBT technology is described, including details of design, packaging, testing methodology, and performance results.

1. Introduction

With the current explosion of data transmissions in our present information age, data communication and telecommunication networks are rapidly growing in numbers and data rates. The key data routing function in such data networks is frequently performed by crosspoint switches, which allow incoming data streams to be directed to specified output channels. The function of a crosspoint switch is schematically shown in Fig. 1. Here data inputs enter the switch from the left, and data outputs exit from the bottom. In analogy to the earliest electromechanical telephone switching apparatus, a cross-bar or crosspoint connection is made between each output and one particular input.

The vector of output indices is a permutation of the vector of input indices. The particular output selection corresponds to one of N! possibilities, which is typically specified by a binary input control word containing at least $Nlog_2(N)$ bits. The input control word governs the switching elements within the crosspoint (and is typically stored in memory within the crosspoint).

---

*Charles Chang, K. C. Wang, and Arlenda Campaña were formerly with the Rockwell Science Center and are now with Rockwell Semiconductor Systems, Newbury Park.

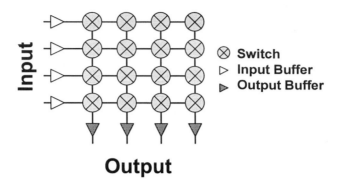

Output

Fig. 1. Schematic representations of a crosspoint switch. Inputs are from the left, output are at the bottom. A switch resides at each crossover point.

Although crosspoint switches for analog signals are widely useful, the focus of this paper is on digital data streams, typically in non-return-to-zero format. In principle, different input data streams can share a common clock and have a particular synchronized timing relationship with one another. However, the more frequently encountered situation is that of input data streams originating from different, poorly synchronized clocks (which makes retiming extremely difficult with a single clock).

Crosspoint switches are widely used in local, metropolitan and wide area network (LAN, MAN and WAN) applications. They form the switching core for connecting processors in a multiprocessor system; they are important components of the packet switches for data networks; and they are employed within telecommunication central offices to reconfigure networks to respond to changing load conditions or to emergency situations.

As data throughput requirements in data networks increase, increasing demands are placed on the crosspoint switches within them. Currently there is a need to develop crosspoint switch ICs with ultrahigh data rates, and as large a value of N (the number of inputs and outputs channels), as possible. Recently reported crosspoint switch IC size and data rates are shown in Fig. 2. There is an inherent tradeoff possible between N and maximum data rate, R. The product NR is the aggregate data throughput of the switch, and represents an important figure of merit. At present, crosspoint switches have the highest aggregate data throughput of any IC ever demonstrated. As such, they represent important demonstration vehicles for IC technology development, and in the design of high speed ICs, most, if not all, of the critical problems for ultrahigh speed digital ICs are faced.

This article reviews the present status of technology for crosspoint switches, emphasizing high data rate switches (corresponding to data rates above 1 Gb/s/channel). Initially, the applications of such switches are reviewed. Their key specifications (principally jitter) are then discussed. Architectures for switch implementation are described, followed by a review of integrated circuit technologies that have been used for switches and a summary of the resulting performance. The

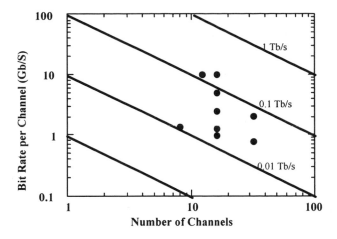

Fig. 2. A plot of the bit rate per switch channel versus the number of inputs channels for a variety of commercial and research crosspoint switches.

problems of packaging high speed switch ICs is subsequently discussed, together with approaches for combining multiple ICs to form a composite unit with greater capacity. The design and technology issues are then described in detail for a particular example that has been recently reported, a 12×12 switch for 10 Gb/s/channel data streams implemented using GaAlAs/GaAs HBT IC technology. The chapter ends with a discussion of future technology possibilities and a summary.

2. Applications of Crosspoint Switches

The crosspoint switch is an important building block for digital communication systems that have to share expensive resources. These switches have been used in a wide gamut of systems ranging from workstations to computer networks. In recent times, the shift towards graphical and video information display coupled with the wide spread popularity of the Internet and Java has significantly increased the demand for bandwidth and connectivity. Although this demand can be met by increasing the number of parallel connections, economic factors have traditionally favored increasing the bandwidth in each connection, especially when it comes to expensive installed fiber optic lines. As a result, there is significant demand in current and near-future network applications for multi-gigahertz per channel crosspoint switches. This section will focus on the potential application of high-speed crosspoint switches for telecommunication and computer networks.

Although the telephone network was designed for voice, the telcom network has been evolving into a digital network. During the mid 80's, new international broadband data communication standards for voice and data were formed. The resulting Synchronous Optical Network (SONET) (North American) and Synchronous Digital Hierarchy (SDH) (European/Asia) standards were defined to accommodate increasing bit-rates.[1] The baseline rates for SONET (OC-1) is 51.84 Mbps and the baseline bit rate for SDH (STM-1) is three times OC-1 or 155.52 Mbps. Currently telecom networks operating up to OC-12 (OC-1 bit − rate × 12 = 622.08 Mbps) are widely deployed. The crosspoint switches for OC-12 and lower bit rates can be readily realized with silicon based BJT or CMOS VLSI technology.

The explosive demand for Internet access is fueling the need to increase the bandwidth of networks. At present, laser/modulator drivers, preamplifiers, limiting/AGC amplifiers, and clock/data regenerators (CDR) fabricated with production GaAs HBT and/or GaAs MESFET technology combined with semiconductor based optical components can comfortably address bit rates at OC-48 (2.488 Gb/s) and OC-192 (9.953 Gb/s). Advanced III-V technologies are addressing OC-768 (39.813 Gb/s) bit rates and higher in the research labs. At present OC-48/OC-192 WDM links with 8 to 80 wavelengths are being commercially deployed using electronic technology; however, a switching element is needed transform a link into a network in order to meet the connectivity requirement. This is especially true with the increased popularity of switched Ethernet or Gigabit Ethernet using the SONET backbone. In the near future, WDM links with up to 128 channels or more may become reality. The crosspoint switch for these networks would require a larger number of channels (8 to 128), high-isolation between channels, and the ability to translate wavelengths. One approach to realize this network switch is to combine the commercially available optical/electrical and electrical/optical link technology with an electronic crosspoint switch.[2] On the other hand, the research community is actively developing the technology for an all-optical switch that would be naturally transparent to the bit rate and somewhat transparent to the data type (digital or analog). Current research topics for an all-optical switch are focused on improving switch isolation, reducing insertion loss, and wavelength translation. Although the optical switch has much future potential, the network switch formed with an electronic crosspoint switch (based on mature electronic technology) which leverages production link electronics can be readily demonstrated for system use.

For example, a 12 × 12 or 16 × 16 crosspoint network switch can have several applications in a 4-channel, 10 Gb/s/channel WDM SONET network. In SONET networks, the switching time is much shorter than the typical transmitted data and the switch setting only needs to be changed to optimize the long-term load or in emergencies. The SONET reconfiguration time (approx. 60 ms) can easily be meet with either PLL or SAW filter based clock/data regeneration (CDR). The electronic switching time typically does not limit the switch rate. An application of crosspoint switches is shown in Fig. 3, switch S6 can also be used as an add-drop node where the network crosspoint can be used to drop a specific wavelength from the input node, add a new channel to the output node, and translate wavelengths that pass through. With the addition of electronic MUX/DEMUXs to the network switch, slower bit rates can be added/dropped from each of the OC-192 channels. This allows slower equipment to access the high-speed network. A 16 × 16 crosspoint network switch or two back-to-back 12 × 12 crosspoint network switches, such as switch S3 in Fig. 3, can be used to interconnect two SONET rings. Other applications of

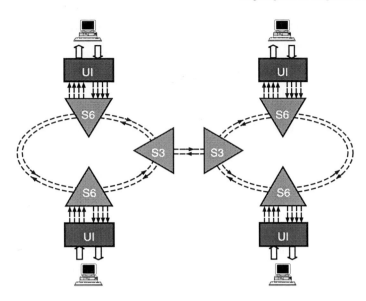

Fig. 3. An application of a crosspoint switch in a 4-channel WDM SONET ring network. The 12 × 12 switch can be used as both an add-drop node (S6) and to interconnect two rings (S3).

these electronic crosspoint switches include either local or custom high-speed data communication networks operating in the electrical or optical domain which use much of the same electronic technology in telecommunications. Such applications of these networks may include distributed supercomputing where vast amounts of data needs to be processed in real-time. To address future WDM systems with large numbers of channels (32-512), electronic switches can be cascaded to form larger network elements.

In ATM networks where short packets are switched at high-speeds, the multi-gigabit electronic crosspoint switches can also find application. Recent estimates show that large ATM networks elements need a throughput from 100 to 1000 Gb/s to meet expected demand.[3,4] Although, low speed packet sorting and switching can be achieved with standard CMOS VLSI technology, the amount of data flowing between two points (such as major cities) can be quite large, which necessitates the use of high-speed electronics. Although electronic crosspoints can have potentially fast switch times, at high data rates, it is not economical to reconfigure the switch for each packet. Instead, ATM packet concentrators can be used to collect packets to form high-speed data with the length of several packets for more efficient transmission.[4] Even with concentrators, there may be a need to improve the switching time of crosspoint switches. To minimize power and die size, typical crosspoint switches are programmed with a serial interface. For example, a 16 × 16 switch can be programmed with 64-bit pattern at 200 MHz for a program time of 0.32 μs. Once programmed for the next setting, the switch can be set in less than 1 ns. For faster switching and programming times, parallel interfaces with gigahertz clocks can be designed with an associated increase in power dissipation and die size. Although the synergetic combination of several state of the art electronic technologies would be required to synthesize such a complex ATM switch, the heart of the switch would be a multi-gigabit crosspoint switch with a large number of channels used to route signals across high-speed pipelines.

3. Crosspoint Switch Performance Criteria

Key considerations for a crosspoint switch, in addition to bit rate, are jitter introduction and switch reconfiguration time. In the following, design considerations to meet these objectives are described.

3.1. Jitter

A fundamental consideration for switch operation is that the outputs be faithful replicas of the input data streams. Because suitable amplification and limiting can recover amplitude or waveform degradation, the key issue is to preserve the zero-crossing times of the data streams going through the switch. Unfortunately, typical switches introduce randomness known as jitter into these zero-crossing times. This noise is analogous to the phase noise of narrow-band analog signals. If the overall system is synchronous (having a well-established system clock) then the jitter of the crosspoint switch can also be mitigated. However, the typical situation is that of an asynchronous system so that jitter introduced by the switch is hard to correct. To do so might involve clock recovery for each channel, for example.

The jitter of the output is specified as the RMS time deviation from the average clock period or peak-to-peak spread, over a finite measurement duration. In practical situations, the jitter is determined from histograms over a period of 20–30 sec. As a rule of thumb, peak-to-peak jitter is typically 6 to 10 times the RMS jitter.

Jitter is exhibited in the eye diagram, and leads to eye closure. The presence of jitter will add to the system error rate, in a way which is exacerbated when (a) many switches are concatenated (increasing jitter progressively); or (b) the crosspoint switch feeds a system which has very substantial amplitude and timing randomness (such as a laser/fiber/detector channel). The contribution of the switch jitter to the overall error budget is dependent on the jitter statistics, particularly the probability of large jitter (at the extremes of the distribution).

When a system is composed of a chain of different elements with jitter, the net jitter (t_net) of the overall system is related to the jitter contribution t_j of the elements in a way that depends on the associated physical mechanisms. For pattern-dependent jitter, different jitter contributions are correlated, and tend to add directly:

$$t_\text{net} = \sum a_j t_j, \tag{1}$$

a_j is $+/-1$ depending on the nature of the jitter source. Most frequently, all the a_j's have the same sign. For jitter introduction from random sources, the jitter is uncorrelated and the jitter variances are additive:

$$t_{\text{net}} = \sqrt{\sum t_j^2}.$$ (2)

Sources of jitter in a circuit are numerous, and include the following:

Random noise

Noise from transistors and resistors can be regarded as an additive noise voltage source, v_n, that appears together with the input signal voltage, v_s. For noise levels substantially smaller than the input levels, the associated shift in output waveform timing is given by

$$t_n = v_n/dV_{\text{in}}/dt,$$ (3)

where dV_{in}/dt is the slew rate of the input voltage; for a digital signal with amplitude V_{logic} and clock period T, dV_{in}/dt is about 3 V_{logic}/T. V_n for thermal noise of a resistor R is given by

$$(v_{n,\text{rms}})^2 = 4\,KTBR,$$ (4)

where B is the bandwidth, of order $0.7/T$. The resulting noise contribution is appreciable at high data rates. For example, assuming $V_{\text{logic}} = 400$ mV, and $R_{\text{load}} = 400$ ohms for a 10 Gb/s digital signal, t_n is calculated to be 0.018 ps. Transistor noise is greater than this, to an extent given by the effective noise figure of the device (averaged over frequency and bias conditions). The jitter contributions from each gate add according to Eq. (2); for large number of gates within each signal path, the contributions can be significant. For example, if an effective noise figure of 4 dB is assumed, and 8 stages are taken into account, then the RMS jitter for 10 Gb/s is estimated to be 0.7 ps.

$1/f$ noise is an important factor in addition to white noise in most transistor circuits. A corresponding jitter component is introduced, which alters timing timescales of the order of milliseconds or longer. This component is not eliminated with PLL clock data recovery circuits since the jitter component due to $1/f$ noise falls within the loop bandwidth of the PLL.

Finite circuit frequency response

Jitter is introduced by circuit elements for which the input to output delay is dependent on the pattern of the input. Typically for fast systems, a 10101010 digital signal will have a different delay through digital logic gates than a 11110000 pattern. Linear circuits experience the same phenomenon, which corresponds to a variation in the group delay $d\Phi/d\omega$ with angular frequency ω (where Φ is the phase of the output). The group delay variation of an RC network becomes significant for frequencies above $1/2\pi RC$. A chain of digital gates will display pattern dependent jitter, which can be understood as the phase delay of the associated input RC network, compounded with a nonlinear amplifier of the logic gate structure. This logic gate pattern-dependent jitter accumulates linearly with the number of gates in the signal path. It must be minimized by speeding up the performance of each gate. For large crosspoint structures that arise in switch layouts, the capacitance of the interconnects can become large, and avoidance of the RC induced phase delays is a significant problem.

A related source of jitter is associated with reflections on high-speed transmission lines used on-chip or at inputs and outputs. If there are reflections associated with terminations, or impedance mismatches along the lines, then jitter can be introduced if the reflected signals overlap the risetime or falltime of the proper signal. The time variation t_r due to the reflection is given by

$$t_r = v_r/dV_{\text{in}}/dt \sim 3G_1G_2T \,, \tag{5}$$

where G is the round trip voltage reflection coefficient for the line. To avoid this jitter source, proper terminations of transmission lines is of paramount importance.

Cross-talk between signals

The presence of multiple signals within the crosspoint switch gives rise to cross-talk. The interfering signals within the logic structure produce voltage variations during the logic transitions, and cause jitter. To avoid problems, the level of cross-talk voltages must be limited to values on the order of 0.01 V_{logic}. As the frequency increases, it is increasingly difficult to guarantee such low levels of cross-talk. Sources of cross-talk include:

(a) cross-capacitances between metal signal lines, particularly at crossovers.
(b) parasitic capacitances within the elementary transistor circuits, such as multiplexers. Output to input feedback capacitance (C_{bc} in bipolar transistors & C_{gd} in FETs) is a particularly problematic parasitic;
(c) cross-talk associated with common ground, common power supply, or common voltage reference lines. At high frequency, the inductance of the common lines is a more dramatic problem than resistance, and must be carefully minimized.

A basic strategy for minimizing cross-talk is to use differential signals throughout the IC, preserving parallel routing of the signal and its complement. The value of reducing cross-talk far outweighs the additional capacitance from the greater number of metal traces. With a differential structure, the high frequency current on the ground and power supply lines is also reduced dramatically, reducing cross-talk.

In the design of a high-speed crosspoint switch, a jitter budget, which allocates allowable jitter to the different elements of the circuit and to the different sources, is an important element. It is noteworthy that some of the jitter sources mentioned above produce "recoverable jitter", that is, timing variations that can be reduced by an equalizer following the switch. In principle, all of the pattern-dependent jitter is recoverable (up to the point where the accumulated jitter produces output errors within the switch itself). For example, a chain of RC or other low pass filters, even if separated by nonlinear amplifiers, can be compensated by a chain of RL or other high pass filters (although there will be random noise jitter unavoidably introduced in doing so).

3.2. *Switch reconfiguration time*

In various applications, it is desirable to change the connection pattern of the switch in a very short period of time. An example is packet switching, where each packet destination is contained in the packet header. It is not possible, without considerable difficulty, to switch a crosspoint without losing a few bits of the data streams, since in general the data streams do not share a common clock, and it is not simple to find a switching instant that is guaranteed not to upset either data stream. It is often possible to minimize the number of upset bits, however, to only a handful. In practice, with conventional electronic crosspoint switches, the limiting factor in switch reconfiguration is the time it takes to read in the data specifying the switch configuration. To save space and power, the read-in can be done serially, at relatively low data rates (compatible with control computers). The control data is stored, and at a strobe command, the data streams are switched. To speed up switching time for packet applications, considerable attention to the memory buffer and interconnect to the mux control is required.

4. Architectures

The ideal digital crosspoint switch would have several desired features and performance requirements. Features of an ideal switch would include: (1) the ability to switch a large number of channels; (2) the ability to provide non-blocking switching (sending input n to output m would not prevent other inputs from using any of the remaining outputs); (3) the ability to broadcast (sending an input to more than one output); and (4) the ability to form scaleable switches (able to synthesize larger switches with the same building blocks). The ideal switch would also perform with low inter-channel cross-talk, minimize pattern dependent jitter, pass signals at high bit-rates, and operate with low-power dissipation while being realized with a low transistor count using simple interconnect routing. In practice, the ideal crosspoint switch requirements are very difficult to realize, especially at high-bit rates; thus, several different architectures, each having its advantages and design trade-off have been used. In the following, various switch architectures used in high-speed crosspoint switches are summarized.

4.1. *Matrix architecture crosspoint switches*

The most natural realization of a crosspoint switch is a matrix of n inputs by m outputs interconnected with switches at each intersection as shown in Fig. 1. Due to the large number of independent switches, this design is inherently non-blocking and has the ability to broadcast one input to as many outputs as desired. The matrix architecture can be best realized with a FET or CMOS technology where pass gates are used as the switching elements. The primary advantage of this type of switch is its potential to pass both small-signal analog and digital signals since the "on" FET can be thought of as a somewhat linear low-value resistor connecting an input to an output. This architecture is extensively employed in commercial low-speed analog/digital switches.

At high-speeds, the matrix architecture has several key limitations for digital signals. Since the input and output lines needs to be connected to a larger number of nodes, the interconnect lines tend to be long which results in a large capacitance that is difficult to drive at high-speed with low jitter. In addition, both the input and output interconnects are attached to a large number of "off" devices which further increases the load capacitance of the driving stage. In the pass-gate configuration, the active device does not have much gain or drive capability; thus, the input drive stage needs to drive all of the parasitic capacitance from the input to the output. The situation can be significantly improved by placing buffers between each crosspoint; however, this further increased the size, transistor count, and complexity of the circuit. Due to these limitations, high-speed crosspoint switches do not typically use this design.

4.2. *Single multiplexer architecture crosspoint switches*

To adapt the matrix architecture to high-speed crosspoint switches, it is desirable to combine the inter-stage buffering with the switching function into one stage. This is naturally done with digital multiplexers (MUX). Figure 4 illustrates the topology for a 4 × 4 crosspoint switch using four 4:1 MUX in parallel. Each output is driven by one multiplexer that has access to the four inputs. In comparison with

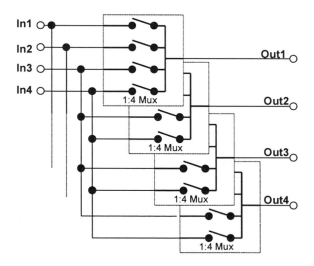

Fig. 4. Schematic representation of a 4-channel single multiplexer crosspoint switch architecture implemented with four 4:1 MUXes.

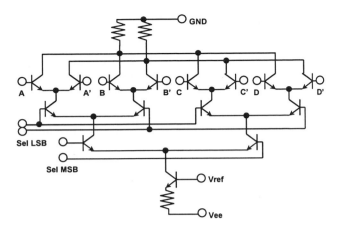

Fig. 5. Typical 4:1 MUX implemented with bipolar transistors using current mode logic (CML).

the matrix architecture, the MUX architecture is virtually identical to the matrix architecture with the MUX replacing one of output legs in the matrix. Since each output has its own multiplexer, the switch is both non-blocking and has the ability to broadcast. Furthermore, unlike the matrix switch where each pass-transistor needs to be switched, each multiplexer is naturally digitally programmable which simplifies the control logic.

Figure 5 shows a typical three-level 4:1 multiplexer implemented with bipolar current mode logic. In general, this topology can be implemented with bipolar or FET technology. The switching function is contained in the bottom two levels of differential pairs which steers current to the selected top stage. These transistors function as four differential pairs that share the same output resistors. The top stage functions as both the switch and gain stage. In the bipolar case, when current is steered to the desired stage, the voltage gain (A_v) is $g_m^* R_{\text{load}}$ where g_m is given by qI_c/nkT. Without input/output loading, the upper frequency limit of the switch is approximately the f_t/A_v where f_t is the short circuit current gain cutoff-frequency. In practice, both the loading of the next stage as well as the significant parasitic interconnect capacitance will lower the maximum speed. For the non-selected inputs, both transistors are off which isolates the input from the common output. Typical high-speed multiplexer sizes are 2:1, 4:1, or 8:1 and voltage headroom considerations and parasitic capacitance limit the size. As a result, larger multiplexers are typically formed by cascading smaller MUXs. These cascaded approaches typically have improved performance over, for example, a one-stage 16:1 multiplexer at the expense of increased power dissipation. Many commercially available crosspoint switches that have channels operating in the low gigabit region employ this topology.[5–7]

Although this is a proven and effective approach to realize high-speed crosspoint switches, for very high bit rates (> 5 Gb/s), the single multiplexer approach has limitations in terms of isolation. Using HBT technology as an example and assuming the common emitter point of the "off" differential pair is common ground, the isolation is provided by a cascade of two voltage dividers, R_b & C_{be} followed by R_{load} & C_{bc} (where the latter dominates). For typical production HBT technology operating at 10 Gb/s, the isolation from input to output is about -18 dB assuming frequency $= 5$ GHz, $R_{\text{load}} = 150\ \Omega$, $C_{bc,\text{off}} = 25$ fF, $C_{be,\text{off}} = 10$ fF, $R_b = 75\ \Omega$, and $R_c = 12\ \Omega$. Since the signals are differential, the differential isolation is 6 dB worse. With a 4:1 multiplexer, the three "off" channels couple to the "on" channel leading to poor isolation since the "off" data is not typically correlated. Traditionally, digital logic is known for its high isolation especially if the coupled "noise" does not exceed the noise margin of the gate. This is only true if the output is in either a logic high or logic low state where the multiplexer's isolation is near infinite. However, when the output is switching, the coupled "noise" from the other three channels will modulate the differential cross-over point causing jitter. At high bit rates (> 5 GHz), the jitter caused by the low-isolation of the multiplexer can be a primary contributor to the overall jitter; however, at lower bit rates (below 2 Gb/s), the magnitude of this jitter mechanism is reduced assuming the same technology is used.

4.3. *DEMUX/MUX architecture crosspoint switches*

At high-bit rates where the isolation of the multiplexer becomes an issue, a DEMUX/MUX architecture can be used to increase the isolation. It this case, a DEMUX is used to preselect the channel to the MUX, which minimizes the number of active channels presented to the MUX for reduced cross-talk. Details of this architecture are discussed in Sec. 8. From a design point-of-view, the DEMUX/MUX approach has many advantages; however, the ability to broadcast is sacrificed.

4.4. *High-isolation multiplexer architecture crosspoint switches*

One method to maintain the broadcast ability while maintaining low cross-talk is to employ a high-isolation multiplexer. Figure 6 illustrates one possible high-isolation two stage multiplexer. The first multiplexer stage consists of four individual differential buffers riding on a current steering select tree. The second multiplexer stage is a standard multiplexer. When selected, the first stage behaves as a standard differential buffer. Due to the differential nature of the signal, the multiplexer current stage does not impact the speed performance of the buffer. With the buffer off, each transistor provides an additional -18 dB of voltage isolation resulting in a net voltage isolation of nearly -40 dB at the critical differential crossover point. Furthermore, since the output load resistors are not shared between the different inputs of the first multiplexer stage, the isolation is further improved. Figures 7 and 8 show the simulated results comparing the standard multiplexer to the high-isolation two-stage multiplexer. The results show that the high-isolation buffer can significantly reduce jitter due to cross-talk.

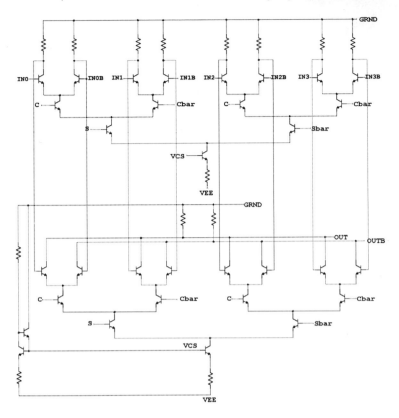

Fig. 6. The high-isolation MUX architecture schematic representation where the selectable buffer before the MUX provides additional isolation and low power buffering.

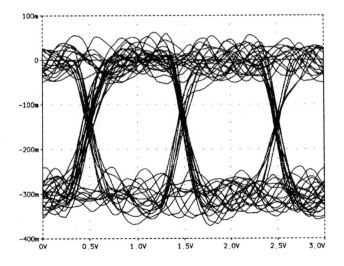

Fig. 7. Simulated output of the standard 4:1 MUX (Fig. 5) at 10 Gb/s with three asynchronous 10 Gb/s interfering signals presented to the "off" channels showing high jitter due to cross-talk.

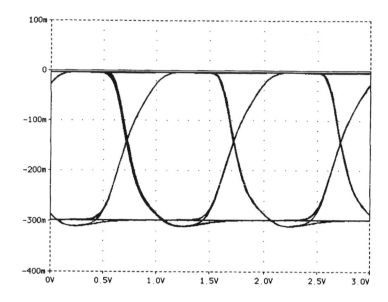

Fig. 8. Simulated output of the high-isolation 4:1 MUX (Fig. 6) at 10 Gb/s with three asynchronous 10 Gb/s interfering signals presented to the "off" channels. The results show the low jitter due to the improved isolation of the MUX.

In high-speed switches, inter-stage buffers are typically used to drive the interconnect capacitance between the multiplexers for added performance. Thus the added gain of the first stage of the high-isolation multiplexer is needed anyway. For a power point of view, this high-isolation buffer offers a reduction in power since only one of the four needed buffers are switched on at a time. To form larger multiplexers, the entire buffer stage of the unused multiplexers can be shut down for a further reduction in power dissipation.

For high-speed operation, emitter followers typically serve as a buffer between the two stages. For reduced power dissipation, the emitter followers can also use a multiplexer current steering stage (with some modification); however, the added capacitance of the current steering stage will reduce the bandwidth and add to the pattern dependant jitter. Simulations show that this unique high-isolation multiplexer has promise and 12 × 12 and 16 × 16 switches with this new two-stage high-isolation multiplexer have been designed, fabricated, and the switches are currently under test.

4.5. Clocked structure architecture crosspoint switches

One method to minimize the added jitter (due to either thermal noise, low-isolation, or limited bandwidth) in a crosspoint switch is to retime the data with a low-jitter clock and wide-phase margin high-speed flip-flop. If the internal jitter of the different stages of the crosspoint switch is less than the phase-margin of the flip-flop, then the data only needs to be retimed at the output. If not, then intermediate retiming stages needs to be employed.

In some very specialized cases, all of the input signals can be referenced to a single clock. If so, the complexity and power dissipation of the switch can be greatly reduced since only one low-jitter clock needs to be distributed to all of the retiming flip-flops; however, the skew between the channels needs to be minimized which is extremely difficult at very high bit rates. In commercial crosspoint switches[7,8] operating below 2 GHz, retiming flip-flops can be found at either the input, output, or both. At high-speeds, the input data phase with respect to clock phase is limited by the phase-margin of the input flip-flop. With a large number of inputs, aligning all of the input data to the same clock phase is quite difficult so such retiming schemes are limited to either slower or smaller sized gigabit switches.

For most applications, the different input signals may be referenced to clocks with slightly different frequencies originating from different sources; thus, it is not possible to distribute a single clock to all of the re-timing flip-flops. Instead, it is necessary to switch both the clock and data to the proper output switch. This greatly contributes to the complexity of the circuit. Although the frequency of the clock is twice the maximum frequency of the data, the clock does not suffer from jitter due to inter-symbol interference. This implies that the clock switch bandwidth overhead can be less than the data switch bandwidth overhead. Although the need for high-isolation between data channels is reduced with re-timing, the high-isolation requirement is passed to the clock switch where isolation issues are more complex. Any jitter in the clock will be passed to the data after retiming. To maximize the advantage of having to switch and distribute both the clock and data, the jitter can be minimized if the data is re-timed after each stage in a switch. For a technology where the bit rate is approaching the maximum operational frequency, this can be a powerful technique for developing a low jitter switch with its additional hit in power dissipation and complexity.

In general, re-timing is a powerful tool for reducing jitter due to inter-symbol interference provided a high-isolation method is used to distribute the different low-jitter clocks to the retiming flip-flop. The re-timing method does require (1) more I/O pads which can significantly limit the size of the switch, (2) higher power dissipation, and (3) more complex interconnect routing.

4.6. *Application of clock/data recovery to retiming crosspoint switches*

The difficulties in distributing numerous phase aligned clock/data pairs to the input of the switch and switching high-speed clocks with high-isolation can be remedied with the introduction of a phase-lock loop based (PLL) clock-data recovery (CDR) unit after and/or before the switch. A CDR unit would recover the clock from the data in order to re-time the data to remove the jitter. Within the loop bandwidth of a PLL, the input jitter is passed. Above this frequency, the input jitter is removed from the signal and the remaining output jitter is from the clock used for retiming (typically due to phase noise of the VCO which is potentially low due the high-Q).

Although PLLs are complex circuits, they are typically employed in fiber-optics links to remove jitter added by other components and regenerate the data. As a result, the addition of several PLLs to the switch outputs to form a network switch system is a very practical approach to alleviate the needed to distribute, switch, and phase align the data and the clock to the retiming flip-flop while offering the jitter reduction and data regeneration.

The CDR also significantly increases the complexity of the crosspoint switch. At present, a fully integrated single chip CDR unit operating at either 2.5 or 10 Gb/s is under development by several companies. Due to the large number of CDRs required, the integration of numerous independent CDRs on one chip is complicated by the tendency of different VCO's to phase-lock to each other rather than to the different inputs, especially, with the limited isolation due to close proximity and lack of shielding in most semiconductor technologies.

5. IC Technologies

High-speed crosspoint switches have been demonstrated in a variety of IC technologies. This section discusses the unique characteristics of each technology for switch implementation, and reviews reported switch characteristics.

5.1. *CMOS*

While potentially the least costly circuit technology, CMOS conventionally lacks the speed needed for the fastest applications. CMOS speed is rapidly increasing, however, with scaling of gate lengths. A 16×16 switch was demonstrated in 1.25 μm technology that operated up to 250 Mb/s, with jitter of 160 ps. Power dissipation was limited to 0.9 W. Compact structures for the crosspoint memory circuits and switches were achieved, as shown in Fig. 4. Switch memory was stored in the same planar array as the switches, addressed with row/column addressing as in a RAM. In order to minimize cross-talk, the voltage swing on the output sense lines was limited to only 0.5 V (with a 5 V supply), again in a fashion reminiscent of RAM structure. Since the time of this demonstration, considerable strides have been made in CMOS device performance with progressive decrease of channel length, such that considerably improved performance can be expected.

5.2. *Si bipolar and BiCMOS*

Si bipolar switching devices traditionally have allowed faster operation than CMOS. Crosspoint switch integrated circuits with clock frequency up to 1.4 Gb/s are being offered in commercial products. A 5 Gb/s 16×16 crosspoint switch has been reported.[13] Implementation of multiplexers in bipolar technology is most often done with ECL, using multiple levels of gating, such as in the 2:1, 4:1 and 8:1 multiplexers (4:1 MUX is shown in Fig. 5). Use of 4-level trees as in the 8:1 circuit is problematic because of the power supply voltage needed, which tends to increase power dissipation for the entire circuit, and leads to concerns about possible breakdown voltage problems. The regular array of gates makes implementation with gate arrays relatively straightforward. Routing on-chip is generally carried out differentially, reducing cross-talk from cross-capacitance as well as reducing variations in current drawn from the supplies, reducing common inductance pickup. Memory for switch settings can be based on ECL latches, or with BiCMOS technology, and lower power dissipation CMOS latches.

5.3. *GaAs FET*

GaAs MESFET technology provides high f_t and rapid digital switching capability, together with levels of integration up to VLSI. A number of efforts have been made to demonstrate and commercialize high speed crosspoint switches with GaAs MESFET, using Enhancement/Depletion (E/D) technology.[9-11] The best performance in terms of jitter and data rata achieved is based on Source-Coupled FET Logic (SCFL), a direct analog of bipolar ECL. Differential signal routing was done to reduce cross-talk. A 16×16 crosspoint switch has been reported, with operation to above 2.5 Gb/s, and jitter (peak-to-peak) less than 70 ps.[10] Power dissipation was 8 W (a significant fraction of which is needed to drive 16 high speed outputs). Die size was 5.2 mm $\times$ 5.2 mm, when implemented with cells from a standard cell library. A related IC consisted of a 64×33 switch tested up to 2.125 Gb/s per channel (as required for double speed Fiber Channel[11]). The structure has 64 inputs to allow easy paralleling of two chips to form a 64×64 composite switch. The technology used had a gate length of 0.6 μm, and four layers of metal. Power dissipation was 13.5 W. Packaging is a significant issue for such large switches and a custom 304 BGA was designed for this chip.

An alternative circuit approach was demonstrated using superbuffer FET logic, a version of Enhancement/Depletion FET logic that provides greater output drive capability by the inclusion of a totem pole driver.[9] A 16×16 crosspoint was implemented, with an architecture employing a 16:1 multiplexer for each output, and distributing a replica of the input to each of them. Operation at a bit rate of 1.7 Gb/s was demonstrated, with a power consumption of 0.8 W. The figure of merit throughput/dc power attained a value of 34 Gb/s/W, the highest of any technology reported to date.

With the use of heterojunctions (such as a MODFET or HEMT), FET technology can achieve higher performance levels, including higher g_m, higher f_t, lower noise, etc. It can be expected that superior crosspoint switches can be produced. However, within the heterojunction research area there has been less emphasis on large scale integration, as needed for crosspoint switches. Consequently little work has been reported. Heterojunctions have provided vastly improved p-channel devices compared with homostructures. Complementary HEMT technology has emerged, using n-channel and n-channel FETs fabricated in a merged process that features low power (near zero static power) along with high speed. A 16×16

crosspoint switch has been reported with this technology, with operation to above 1 Gb/s, and at a power consumption of 510 mW.[12]

5.4. *GaAs HBT*

Heterojunction bipolar transistors provide the advantages of high drive capability, small size and excellent threshold voltage control characteristic of bipolar transistors, with higher f_t and f_{max} values, as needed to reach 10 Gb/s operation. Technologies based on GaAs substrates, and epitaxial layers of AlGaAs/GaAs, or GaInP/GaAs, have been extensively developed, and are capable of providing device integration levels above 10 000 transistors. The semi-insulating GaAs substrates decrease capacitances associated with long lines and bond pads. As a result of these advantages, GaAs based HBTs have been used to implement the 10 Gb/s crosspoint switches reported to date.[14,15]

Kerry Lowe at Nortel reported a 16 × 16 switch for operation at 10 Gb/s.[14] The architecture was based on the use of 16:1 multiplexers for each output, with each input distributed to all of the multiplexers. To reduce the capacitive loading of the input drivers by the 16 successive multiplexers, a novel distributed structure was utilized. In a manner similar to that of a distributed amplifier, distributed inductance was employed to tune out the capacitance of each input in a broad-band fashion. The input lines were thus made into the equivalent of 50 Ω transmission lines, which were terminated on the chip (on the side opposite the input pads). Power dissipation was 10.5 W, of which approximately 50% was associated with the input and output drivers. The circuit attained an rms jitter of 7.7 ps.

A 12 × 12 crosspoint switch has been implemented by Metzger *et al.* to meet the low jitter requirements of OC-192 systems. This circuit, which achieves RMS jitter below 4 ps for operation at 10 Gb/s, is detailed extensively in Sec. 8.

Heterojunction bipolar transistors based on SiGe have been researched extensively. It has been shown that they provide a speed advantage over conventional (homojunction) Si bipolar technology of the order of 20–30%. Digital circuits with the complexity required for a 16 × 16 switch should be possible.

6. Packaging Issues

The design of packages for high-speed crosspoint switches with channels operating at 10 Gb/s or higher are complicated by the large number of channels. The objective of the high-speed transmission lines within the package is to maintain high-isolation between channels while providing a impedance matched low reflection transition between the smaller die I/O bonding pads and the larger package I/O. Besides signal I/O, the package should also provide low inductance power path, low impedance path to the ground plane, and high thermal conductivity for thermal management. At high-bit rates (10 Gb/s or greater), the radiated power from the transmission lines can result in significant cross-talk, noise, and interference with other portions of the system; thus, the various high-speed sub-systems are typically contained in shielded modules interconnected with coaxial cable and by-passed power feedthroughs. The effects of radiated power is further complicated by the broad band nature of high-speed data when compared to narrow band analog RF/microwave. Due to the shield requirement, the "package domain" may be defined from die I/O pad to the coaxial connector. The package can be designed to go from directly from the die to the connectors or through a conventional package to the PCB then to the connector. In either case, the fundamental issues are similar. Since differential signals are difficult to phase match, single ended outputs are also used at high speeds.

At high bit rates (> 5 GHz) and with a large number of high-speed channels (> 24), most commercially available packages will not provide the required performance level. Although commercial high-speed packages have been developed to operate at frequencies up to 20 GHz such as in Ref. 16, they are typically limited to small dice with a few I/O pins (typically less than 10). In general these well-designed packages are too small for moderate-sized crosspoint switches (8×8 and greater). There are also several packages available commercially that have the required I/O (around 100 signal, ground, and power pins) and are rated for operation below 3 GHz.[17] In other cases, some plastic and ceramic packages not rated by the manufacturer for high speed, have worked at high bit rates (< 10 Gb/s); however, the high-speed I/Os are typically limited to the few center straight pins.[18] In either case, these packages have two main drawbacks. The first is a die cavity that is too large (> 1×1 cm^2) which would necessitate long bondwires resulting in a large performance degradation preventing use at 10 Gb/s or higher. Second, the packages are not designed for low cross-talk resulting in potential isolation problems at high-speed. As a result of these limitations, typical switch packages need to be custom designed.

One of the key issues in high-speed packages operating at 10 Gb/s or greater is the connection from the die ground pad to the ground plane of the package. At high speeds, the current of the desired signal flows through the conductor of the transmission line from the package input to the die. An equal current is returned through the ground path from the die to the package. When the ground path between the die and ground plane has non-zero impedance (particularly associated with inductance at these high-bit rates), the return current causes a voltage drop that modulates the ground potential at the chip. In Fig. 9, a standard differential pair output driver is shown with parasitic inductance from the ground plane to chip ground and from the chip output to package output. The ac current that modulates the chip ground will effect the differential cross-over point at the package output (measured with respect to the ground plane). Although the chip ground is designed for low inductance ground distribution, the parasitic capacitance to the ground plane and parasitic resistance is not controlled; thus, there is no guarantee that the ground bounce differential cross-over modulation will be the same for all transitions, leading to pattern dependent ISI jitter.

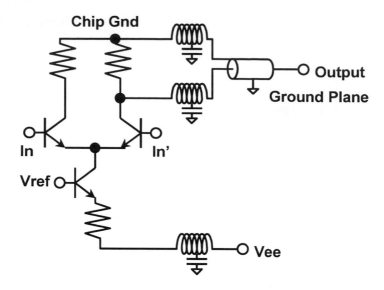

Fig. 9. Ground inductance between output and 50 ohm driver.

The problem is made worse if there are two signals sharing the same ground, the voltage drop of the first signal causes voltage modulation in the second channel and vice-versa leading to ground coupled cross-talk. The cross-talk is caused by the transition noise of the first channel modulating the differential cross-over point of the output second channel. This ground coupled cross-talk level is increased if there are many channels sharing the same ground. In a typical module, the ground plane is the metal housing surrounding the PCB or substrate. The ground plane can be successfully extended to the package ground plane with little difficulty due to the low inductance of multiple ground contacts with large package pins and connectors. (Packages without a ground plane result in increased ground complexities). The point where most of the ground issues arise is in the package to the die transition. The inductance for 1 mm bondwire is approximately 0.7 nH/mm. Typical short bondwires are on the order of 1 to 1.5 mm. In the case where the substrate height and die height match, the bondwire distance between the die and package is seldom shorter than 0.5 mm. Assuming a 5 GHz signal for 10 Gb/s data, the magnitude of impedance is around 22 Ω for 1 mm of bondwire. Since the bondwire has a serious impact on performance, many pad I/O patterns attempt to minimize the length of the bondwires.

The choice of transmission line design also impacts the design of the package and die I/O pattern. Microstrip transmission lines have been proven for operating at frequencies much higher than 20 GHz in MMIC; thus, it seems like it would be a good choice for the interconnect on the package. In an environment with dense high-speed interconnects, the coupling (both odd and even modes) between adjacent and alternate microstrip lines is quite high, especially near the die, even with moderate separation. With many signals changing at the same time, the actual impedance will become a function of the data pattern in the adjacent lines. Furthermore, the substrate height, dielectric constant, and the microstrip width set the impedance of the microstrip line. As a result, the ideal width requirement is different at the small IC pad and large connector/package pin. In order to maintain a quality transition from the die, the microstrip line has to be very narrow resulting in high-loss and poor connector transition. The situation could be slightly improved with differential microstrip transmission lines since the differential signals would be closer together and adjacent signals would be further apart for higher isolation. Also, the differential microstrip would allow some flexibility between microstrip width and spacing to handle the transition between die to pin/connector. Differential outputs at high speeds are quite difficult to use due to the need to phase match the lines. In general, microstrips offer high-performance interconnects when the number of lines are small and excellent ground can be provided.

Instead of microstrips, co-planar wave guides (CPW) offer several advantages over the microstrip waveguide. First, the isolation is improved due to the presence of a shielding ground between adjacent lines. This also helps to maintain a stable characteristic impedance in spite of transitions on adjacent lines. Second, the 50 Ω impedance can be maintained for a wide variety of center conductor widths to meet the large dynamic range between die I/O width and package pin width for low loss. This substrate thickness required to support the width changes needs to be carefully selected to minimize parasitic modes in the substrate.[19] Finally, most 50 Ω test probes for high-frequency on-wafer test have a GSG pattern so CPWs minimize the effects of mode transition when tested.

With these considerations in mind, the die I/O pattern should be selected with the highest density capable of providing the required performance while minimizing the complexity of the package and die size. The most dense die I/O pattern would be consecutive high-speed signals ($GS_1S_2S_3S_4\cdots S_nG$) with ground pads provided where convenient. Microstrip lines would be used to connect the die I/O to the package I/O. Although the die size and package pin count would be the smallest, the isolation would be worst due to (1) high coupling from the microstrip transmission lines, and (2) excessive ground bounce jitter and cross-talk due to the large number of I/O sharing a common ground bondwire to the ground plane.

Another solution for reduced ground bounce and to offer higher isolation is to use the $GS_1GS_2GS_3G$ pad I/O pattern. The transmission line isolation is improved due to the use of co-planar lines. The number of required I/O pads have doubled which increases the size of the die. The ground-bounce isolation is improved since only two channels share the same ground but the cross-talk issue can still be a problem. One could use this pattern with microstrip lines and directly connect the ground pin to the ground plane to reduce ground bounce jitter. In practice, this is very difficult to achieve. Bonding the ground pads directly to the package ground results in long bond wires for the signals resulting in poor performance. Another approach is to put ground vias into the substrate. However, the typical design rules for the minimum sized via would preclude placing them near the pads. For example, for a 25 mil Alumina substrate, the via diameter is typically $0.6\times$ the substrate thickness which is about 400 μm. The via-to-via minimum spacing is about $2.5\times$ the substrate height which is about 1600 μm. The chip pad-to-pad spacing is 150 μm leaving about 300 μm between the vias. This shows that the vias would be placed at a significant distance from the ground pad leading to increased inductance.

To minimize the ground coupling through the bondwire, it is possible to use a $G_1S_1G_1G_2S_2G_2G_3S_3G_3$ I/O pattern with CPW waveguides. In this case, each output buffer has its own ground pads so there is no coupling through the bond-wire. Due to the increased spacing between high-speed signals, the transmission line coupling is also reduced. The remaining ground coupling is due to the shared ground in the package substrate where the trace is particularly narrow at the die end. The minimum via size would prohibit placement near the die so the via would be placed where the ground plane is wide and the inductance is low. This tends to suggest that substrate vias will not significantly improve the performance of the package. In general, this pad I/O requirement has the same pad density as a differential output with a $GS_{1a}S_{1b}GS_{2a}S_{2b}G$; however, this allows for low jitter single-ended output which are easier to use. Due to the complex nature of the cross-talk coupling mechanisms and the large number of high-speed channels, the switch and package typically needs to be co-designed for optimal performance.

7. Scaling of Crosspoint Switches

With the explosive demand for bandwidth, WDM links are operating with an increasing number of channels at high bit rates. In order to meet the network requirements for these WDM systems, large high-speed crosspoint switches need to be formed with 64 to 512 channels operating at 2.5 Gb/s, 10 Gb/s, or higher bit rates. The aggregate throughput bandwidth of these switches can be phenomenally high (a 128×128 crosspoint switch operating at 10 Gb/s/channel has a throughput bitrate of 1.28 Tbps). Furthermore, these terabit crosspoint switches tax the practical limit of a semiconductor and package technology in terms of speed, integration level, yield, and size.

At present, the largest monolithic crosspoint switch realized in typical high-speed semiconductor technology (capable of operating at bit rates of > 10 Gb/s) is approximately a 16×16 switch.[14] At lower speeds (< 3 Gb/s) when ground bounce, bondwire inductance, and package cross-talk requirements are reduced, larger switches can be formed (32×32 at 1.5 Gb/s[20]). For 12×12 and 16×16 switches operating at 10 Gb/s, the chip size is typically greater than 5×5 mm^2 and in some cases, the pad I/O start to dominate the chip size. For low jitter, switches typically use either single row of $GS_1GS_2GS_3G\ldots$ or $GS_1GGS_2GGS_3G\ldots$ I/O pad patterns. For high-speed probe and package considerations, typical pad sizes are 100 μm with about 50 μm spacing between pads. As the switch size increases, the die size will be significantly larger requiring a very high-yield high-speed process. In production III-V technologies for switches operating above 10 Gb/s, the yield and integration level limits can probably yield switch cores on the order of 32×32 in a very large die. This tends to suggest that it may be economical to form larger terabit switches through the cascade of smaller building blocks. This section will focus on key issues and techniques for forming larger switches.

The scaling methodology to form a large system crosspoint switch for network systems need to properly exploit advantages of the electronic and package technologies. With one scheme, each switch die can be individually packaged and interconnected on a multi-layer PCB. With another scheme, all of the dies can be packaged and interconnected with a multi-layer multi-chip module (MCM). In both cases, a desired switch scaling scheme would incorporate (1) a compact design with a small number of ICs, (2) simplified interconnect routing to minimize package interconnect layers, (3) low signal degradation through minimize signal path, (4) low power dissipation for simplified heat management. There are numerous methods for scaling switches; however, in this work, only methodologies resulting in a non-blocking design are considered. The next sections focus on the tiled and three-stage crosspoint switches.

The most straightforward switch scaling methodology is to continue the crosspoint switch matrix through several ICs. This approach will be referred to as tile switch. Figure 10 shows an example of a 4×4 crosspoint tile switch. With the tile switch, the crosspoint matrix is continued since the left side inputs of a tile is passed to the right outputs and the top inputs are passed to the bottom outputs. To form a larger switch, individual tiles can be interconnected with a very direct

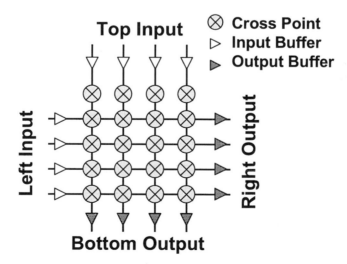

Fig. 10. Tile switch schematic showing input buffers on the top and left and output buffers on the right and bottom. When combined with the additional row of switches, the tile switch can be used to continue the crosspoint design beyond one IC.

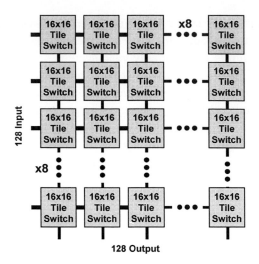

Fig. 11. A 128 × 128 crosspoint switch realized with 64 16 × 16 tiles.

interconnect scheme. Figure 11 shows a 128 × 128 crosspoint switch realized with 64 16 × 16 tile switches. Although the number of interconnections are quite large, the number of high-speed interconnect crossovers is next to none which simplifies the package design. To form a $N \times N$ switch with $n \times n$ tiles, a total of $(N/n)^2$ dies are needed. From a signal point of view, the input signal may have to pass through a maximum of $2N/n - 1$ switches. In this example, this overall switch jitter will increase the worst case accumulated thermal jitter by 3.9 times and ISI (inter-symbol interference) jitter by 16 times over the tile switch element.

To adopt the conventional crosspoint to the tile scheme, several key modifications are needed. Since the outputs can be selected for either the n right inputs or from the top previous stage inputs, an $n \times n$ switch needs $n \times (n+1)$ switches. In comparison between an $N \times N$ tile switch and conventional switch (if such large switches were realizable with current semiconductor technology), the tile switch requires N^2/n additional crosspoints (6.25% more in this example). For fair sized tiles, this should not result in a significant increase in power dissipation. The primary increase in power dissipation is in the $2n$ output drivers in each tile switch, which results in a total of $2N^2/n$ output drivers for the tile switch. In this example, the tile switch would require 1920 additional output drivers over a conventional 128 × 128 switch. If each output were to drive 400 mV into a 50 Ω load using a −3 V power supply, the additional power dissipation is about 92 Watts! This suggests that n (limited by technology) should be as large as possible. One of the main problems with cascaded switches is the unavoidable increase in power dissipation due to the increase in output drivers. If the tile switch is housed in a MCM environment, the distance between tiles can be quite short. To reduce the power dissipation, it would be possible to design two different types of tiles. The first n tiles would have 50 Ω output buffers with low S_{22} to drive the off-package transmission lines (standard buffers). The remaining $n(n-1)$ tiles can have relaxed S_{22} specification, interconnected with high impedance transmission lines (or just bondwires), and lower voltage swing. For example, with a 500 Ω back-match into a 100 Ω load with 300 mV swing, the additional power dissipation due to the output drives can be reduced to a manageable 21 W. Furthermore, with a standard 16×16 switch, the pad I/O periphery starts to dominate the die size and the problem is made worst when the number of I/Os double in the tile switch. Due to the close proximity of switch dies in the tile approach, differential output with a tight $S_{1a}S_{1b}S_{2a}S_{2b}$ may be used. Smaller pads may also be used up to the minimum feature limit of the package technology.

In a generalized crosspoint switch, there are more switching points then necessary to fabricated a non-blocking switch. One way to reduce the number of crosspoints is to utilize a three stage switch as shown in Fig. 12 where each switch is a conventional type. In this three stage switch, the first stage switch has a dimension of $n \times k$ (8 input $\times$ 15 output in the example of Fig. 12). The second stage has dimension of $N/n \times N/n$ (16×16 in the example). The output stage is $k \times n$ (15×8). In this switch design, each of the k outputs of the N/n first stage switches are fanned to the N/n inputs of k second stage switches. Since $k > n$, there are Nk/n inputs to the second stage which is more than N inputs to the first stage to prevent blocking. The N/n outputs of the k second stage switches feed the k inputs of N/n third stage switches. To minimize the number of overall crosspoints in a $N \times N$ switch while maintaining the non-blocking feature, C. Clos has shown that $k >= 2n - 1$. An excellent review of the Clos three-stage switch can be found

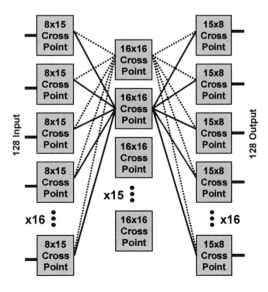

Fig. 12. 128 $\times$ 128 crosspoint switch implemented with a Clos three stage design.

in Ref. 21. In the example, the 128×128 Clos switch is formed by three different building blocks for each of the three stages using a total of 47 dies $(2N/n + k)$. One advantage of the Clos design is the signal passes through only three stages which will increase the thermal jitter by 1.73× and ISI jitter by 3×. In some very-high bandwidth applications, this approach may not require inter-stage re-timing (by switched clock or CDR). As seen in Fig. 12, the interconnect requirement for the Clos switch is quite complicated with many crossovers.

In the Clos three stage switch design, the number of crosspoints is $2kN + k(N/n)^2$ which is typically less then a standard $N \times N$ cross point switch. In this example, the number of crosspoints are reduced by 53%, which can lead to a sizable reduction in the power dissipation. Since the Clos switches are formed with smaller standard crosspoint switches, the I/O and added output buffer problem in the tile approach is reduced. The Clos switch requires $2Nk/n + N$ output drivers. With the standard 50 Ω output buffers, the tile switch would require 480 additional buffers which would increase the power dissipation by up to a reasonable 23 W. By moving to a 100 Ω environment and operating with a lower voltage swing of 300 mV, the output buffer power dissipation would only be increased by 8.64 W. This increase is quite acceptable since this output buffer power increase is offset by the significant reduction in the number of crosspoints.

Since a large monolithic switch at 10 Gb/s or greater is not feasible given the limits of current semiconductor processes, either the tile or Clos approach allows the formation of larger switches with the largest switch yieldable by a given technology. The Clos approach has the advantage of low die count, low power, and low added jitter; however, the interconnection between the switch dies can be quite complex. On the other hand, the tile approach offers simplified interconnection at the expense of increased power dissipation and die count. In the best case where the added jitter per crosspoint IC is small, the overall jitter of the cascaded switch may meet system requirements. However, in cases where the technology speed headroom is small, it is possible that the accumulated jitter may need to be removed through PLL based CDRs. When the excessive jitter is not too bad, then only N CDRs need to be added at the output. If the number of stages is high or if there is jitter that is not correctable at the output, then data regeneration would be required between switch stages which would significantly complicate the design. In situations where jitter may be a potential problem, the Clos switch with only three stages may have a significant advantage over the tile approach. The key merits such as number of crosspoints, output driver count, and IC count for the tile and Clos switches are compared in Fig. 13 as a function of switch size. All comparisons are made where the tile size equals the second stage Clos switch size. The figure shows that the advantage goes to the tile switch for small order switches and the Clos switch has a significant advantage for very large switches. The breakeven point for the number of ICs occurs at approximately for a 50×50 crosspoint. The breakeven point, in terms of the number of crosspoints, occur at a lower point (around an 18×18).

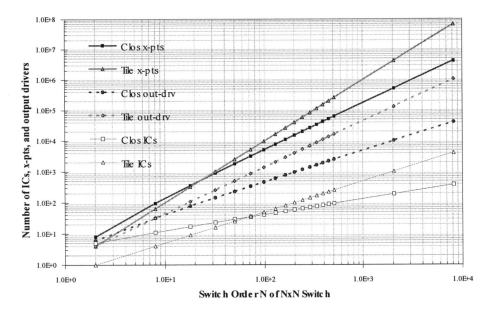

Fig. 13. Comparison of the key parameters in scaled switches by both tile and Clos switch versus the number of channels. For comparison purposes, the tile switch element is the same size as the optimal switch size of the middle Clos building block.

In terms of the number of output drivers, the cross-over point occurs for a 8 × 8 switch; however, since a tile switch output driver can consume lower power than the Clos switch, the actual output buffer power dissipation cross-over point is nearer to a 20 × 20. Considering the increase in power dissipation, die count balanced by the difficulty in packaging, the subjective Tile to Clos cross-over point is likely near a 64 × 64 switch. At 10 Gb/s/channel, this translates to about 1/2 terabit throughput.

Although the practical limit of a monolithic crosspoint switch is near 16 × 16, larger switches approaching the terabit level of throughput can be synthesized by cascading smaller monolithic through a symbiotic combination of packaging and semiconductor technology.

8. A 10 Gb/s 12 × 12 Crosspoint Switch Implemented with AlGaAs/GaAs HBTs

This section describes the design and performance of a high-speed 12 × 12 crosspoint switch IC and package with data channels operating at 10 Gb/s for use in a OC-192 compliant switch as a detailed example of high-speed crosspoint switch characteristics. This section will focus on the key aspects of high-speed circuit design and layout to optimize inter-stage drive for low jitter high-speed switching. Other trade-offs to minimize both the power dissipation and die area are considered. Critical issues in on-wafer testing and packaging are also summarized.

8.1. *High isolation DEMUX/MUX architecture*

One of the primary considerations in the choice of switch architecture is the reduction of added jitter through high-speed operation with high-isolation. For a crosspoint switching system for use in a SONET environment, the tolerable added jitter for the crosspoint switch is near 2 ps RMS for 10 Gb/s operation without additional CDRs on the switch output. Other key concerns for determining the architecture choice are power density, interconnect length, die size, and thermal issues. As shown in Sec. 4.2, the miller feedback capacitance in a standard single multiplexer approach can result in low isolation on the order of 20 dB which is too low to suppress inter-channel cross-talk.

One method to improve the isolation is to shield the inputs of the multiplexers from the interfering channels through the use of demultiplexers (DEMUX). In this approach, only one input to the MUX is active at any time which prevents coupling through C_{bc}. This should provide a similar degree of isolation (near 40 dB) as the double MUX approach proposed in Sec. 4.4. From an interconnect point of view, the DEMUX/MUX architecture encourages a natural fan-out of the input signals to the outputs which helps to shorten interconnect lengths. Although the number of interconnect crossovers are high, the DEMUX minimizes the number of active lines which helps the isolation. Compared to a single MUX design, the DEMUX/MUX architecture will require additional area and double the number of programming bits. From a power point of view, the reduced fan-out and shorter interconnect may simplify the design of inter-channel buffers which may compensate the added power of the DEMUX. The DEMUX/MUX design does, however, sacrifice the ability to broadcast. The initial design specifications of the 10 Gb/s 12×12 Crosspoint switch is shown in Table 1.

Table 1. 10 Gb/s 12×12 Crosspoint Switch Design Specifications.

Data Rate:	0 to 10 Gb/s
Added Jitter:	< 2 ps
Bit Error Rate:	$< 10^{-10}$ with a 2^{23} bit pattern
Output Rise/Fall Time (20–80%):	< 40 ps
I/O Logic Levels:	$0/-400$ mV $(+/-50$ mV$)$
Output return loss (100 kHz–5 GHz):	> 15 dB
Output return loss (5–10 GHz):	> 7 dB
Switching Time	< 1500 ps
Switch Program Time	< 10 μs
Isolation	> 40 dB
Power Dissipation	< 10 W

8.2. *Topology implementation*

Due to the high-speed operation (10 Gb/s) and large size (12 × 12 switch), the crosspoint switch requires a high-speed process ($f_t > 50$ GHz) capable of providing high-yield and high integration level; thus, a production AlGaAs/GaAs HBT technology was used. High-speed operation with minimized power dissipation is achieved through the mixed use of current mode logic (CML) and emitter coupled logic (ECL) depending on the fan-out and drive requirement.

To form the 12:1 MUX within the head-room limitation while maintaining high speed, two stages of multiplexers are needed. Here, the outputs from the first four 4:1 multiplexers are fed into a second stage 3:1 multiplexer (4:1 MUX with one input grounded). As shown in Fig. 14, twelve 1:3 DEMUX pre-switch the 12 inputs to the 12 blocks of 4 × 4 switch blocks. Although the interconnect crossover count is high, only 12 lines are active to minimize the cross-talk. The 4 × 4 blocks are formed with four 1:4 DEMUX back-to-back with four 1:4 MUX. High-isolation is maintained due to (1) MUX input isolation due to the DEMUX, and (2) minimized interconnect crossover coupling (only four active lines out of 16). The standard MUX approach would not have the DEMUX shielding and all 16 lines would be active. The 4 × 4 crosspoint switch array is passed to the output with a 3:1 MUX. The 4 × 4 switch array insures that only 12 of the many interconnects are active. This design has the advantage of a low-number of active lines, where the interconnect cross-over density is high for low cross-talk, and the building blocks for the switch is highly repetitive to simplify the layout.

To further minimize cross-talk, high speed signals are run differentially throughout the circuit. Differential lines crossing over one another have a canceling effect

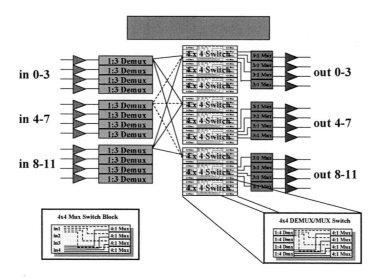

Fig. 14. 12 × 12 DEMUX/MUX switch architecture block diagram.

on the induced cross-talk[22] which justifies the increased die area. Due to the relaxed reconfiguration time and to minimize the power dissipation, long single-ended DC control signals are used where the large capacitance on the lines would actually aid in maintaining a switch setting in the presence of noise.

To further reduce the high impedance interconnect line lengths, 50 Ω transmission lines were extended all the way from the IC wire bonding pad to the input/output buffer. This aided in signal distribution while avoiding the cost in rise time associated with long high impedance on-chip interconnect. Both the package and on-chip transmission lines are CPW to minimize interface issues (such as mode conversion).

Design simulation considerations

Preserving signal quality and data transfer through the various stages of the switch involves countless compromises between speed (design margin), power, and interconnect length. Operating all the stages of the switch at high power levels provides the greatest drive with the shortest rise and fall times. However, there are also constraints from thermal effects, power line bounces, and ringing on signal lines. The key is to (1) keep the power dissipation density as low as possible to reduce thermal effects, (2) minimize large switching currents to avoid crosstalk through parasitics while maintaining the speed and design margin needed for OC-192 operation. Critical parasitic impedance must be included in the simulation. Figure 15 shows the model that was used to characterize long interconnects and power supply connections where long lines are modeled as an RLC network (best if simulated as a *distributed* RLC network[10]).

Switch core circuit implementation

Circuit design was based primarily on lower power Current Mode Logic (CML), which employs differential pair gates and typically makes use of a small 400 mV

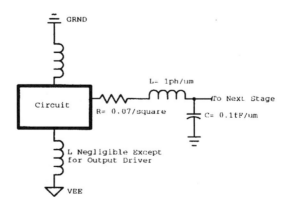

Fig. 15. RLC model used to represent the parasitics of a long interconnect.

(differential) signal swing. CML gates present a large input capacitance due to miller capacitance; thus they are used where the fan-out or interconnect length is small to minimize the power dissipation. The output speed of the differential pair is governed by the RC time constant presented by the resistive load, C_{bc} of the driving stage, the next stage capacitance, and parasitic interconnect. For increased performance, CML bias current can be increased or output logic voltage swing can be reduced for decreased load resistance. For on-chip differential signals, the minimum logic swing used in this design is 280 mV.

The emitter follower (EF) buffer (CML + EF = ECL) circuit provides a high impedance input, a low impedance output, and a voltage gain of approximately unity[23] which is ideal for driving long lines and high capacitance loads at high speeds. Due to a V_{be} drop of 1.4 V for a GaAs HBT, emitter follower circuits are seldom used to drive other emitter followers as commonly used in FETs. The design of the EF is complicated by the tendency for uneven rise and fall times (due to a changes in input impedance) and ringing (due to a possible negative input impedance for a large capacitive loads). The output of the EF can be inductive in nature, and the value of this inductance increases with bias current[23] which may lead to LC resonance or even oscillation if care is not taken in the EF design. As a result, EF are used as interstage buffers between CML gates for high-speed drive where the fan-out is large or the interconnect length is long. For example, the 1.2 mm interconnect lines between the 1:3 DEMUX to the 4 × 4 switch array use emitter followers. By adjusting the bias in the differential pair and emitter follower, the power dissipation for each stage can be tailored for the drive requirement to minimize the overall power dissipation.

Switch control circuitry

A Master-Dual Slave flip-flop architecture was used to implement the 96-bit serial flip flop memory bank which holds the switch configuration. Level shifting and the flip-flop output allowed for the direct drive of the selector tree in the DEMUXs and MUXs. The 96 master-slave-slave flip-flops were run at greatly reduced power levels which was still sufficient for fast data read in (< 200 MHz with commercial ICs).

Control lines were run in parallel along long lines intersecting the high speed signal paths to all of the multiplexers and demultiplexers from the flip-flop memory. Single ended logic was implemented to reduce the wiring requirements. Noise induced on the control lines could be potentially problematic. However, because the control lines were run over large distances in very close vicinity, their large distributed parasitic capacitance can stabilize the control voltage level and suppress high frequency noise.

I/O driver circuits and transmission line design

Output drivers with a separate power line and coplanar transmission lines were implemented to improve isolation at the output stage. To reduce power consumption, a separate -5.2 V (standard ECL) supply was used for the output stage only. Transmission line geometry was optimized for a 50 Ω characteristic impedance using a 3D electromagnetic simulator to take into account the multiple thin-film dielectrics. GSGGSG pad configuration was selected for high-isolation and simplified testing. A 50 Ω match with low S_{11}/S_{22} was obtained by nulling bondwire inductance by tailoring the transmission line capacitance for 10 Gb/s operation.

Due to system requirements for single-ended operation, only one signal was brought to the wire bonding pad from the input/output buffer. In this manner, difficulties of matching delays between two high-speed lines are avoided. However, common mode cancellation is no longer available which may result in increased jitter. This is especially true for the output drivers where the common current-source reference coupling is significant. To avoid this, each output driver was designed with an individual isolated voltage reference generator. A similar strategy with ground lines and generators was incorporated for the high-speed differential input circuitry. For flexibility, 12 individual input reference levels are supplied from off-chip which minimizes the jitter.

The traditional method bringing the low current drive level of the switch core to the relatively high-current output driver (with an associated large input capacitance) with low jitter is to provide many different intermediate stages with increasing drive and input load capacitance.[24,25] In this work, a novel low-power three stage output is used. The power dissipation of the output driver is important due to the large number and close proximity of the output drivers in the switch.

Figure 16 shows a representative block diagram of the multi-stage output buffer that was implemented for the design of the 12 $\times$ 12 switch. Because the output driver was required to drive 400 mV into a 50 Ω termination while maintaining low S_{22} with a 50 Ω back-match resistor, the output stage minimum bias is 16 mA. To

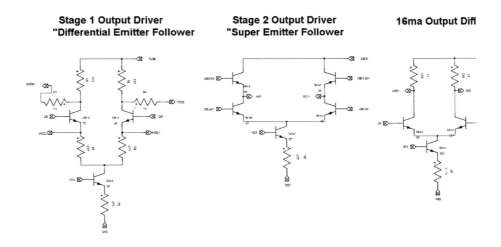

Fig. 16. A three-stage output buffer with a novel second stage push-pull driver.

keep the output buffer balanced to minimize common-mode jitter, the non-utilized output of the differential pair is loaded with a 25 Ω resistor.

The second stage of this multi-stage output driver is a push/pull circuit that could be viewed as two emitter follower circuits sharing a current steered bias source. When the high level inputs (to the emitter followers) are driven with the complements of the low level inputs (to the current steering differential pair) the circuit will function as a push/pull amplifier. If the inputs to the circuit are perfectly synchronized (no time delay between the high inputs and the low inputs) this circuit will be able to provide a large power gain and drive a large capacitive load with short rise and fall times. However, if the inputs were to arrive with a time delay, large amounts of jitter would be introduced through this stage. The first stage buffer was implemented to ensure that the push/pull circuit would be driven with a negligible time delay between high and low input levels. The first stage of the output driver is a differential pair where the output is taken not only at the collector but at the emitter as well, by using emitter degeneration resistors.

SPICE simulation results

To complete the verification of 10 Gb/s signal transfer, a full worst case single channel SPICE model was simulated. Parasitic RLCs from long interconnects and cross-overs were manually extracted with extreme care. Various pseudo random signal patterns were input to the circuit, and all simulations showed that the

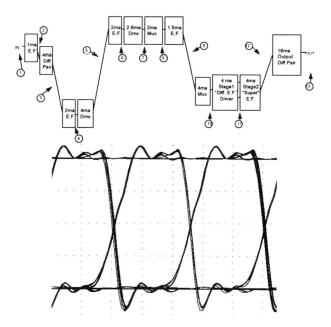

Fig. 17. SPICE simulation of 12 × 12 switch signal path. Eye diagram is at the output (node 13). The vertical scale is 100 mV/division and horizontal scale is 41.66 ps/division.

maximum output jitter was no greater than 6 ps at 10 Gb/s. The critical nodes of the simulation that were analyzed and the output eye diagram are shown in Fig. 17.

The simulation takes into account a variety of jitter sources including the finite time constant of the logic gate response, feedthrough from one signal to another through nominally non-conducting transistors, and feedthrough from one gate to another via common ground inductance. Other sources of jitter, including random noise contributions, are not taken into account.

IC layout

Figure 18 illustrates a photograph of the 12×12 switch die. The high speed switch core (DEMUX/MUX) is located near the center of the IC and the flip-flop memory is located in the left and right blocks. Control lines are run from the memory vertically between the DEMUXs and MUXs. The layout uses repetitive blocks as much as possible. Placement of the output drivers is staggered to lower the power density. This made global circuit modifications much simpler. Power lines are routed mostly

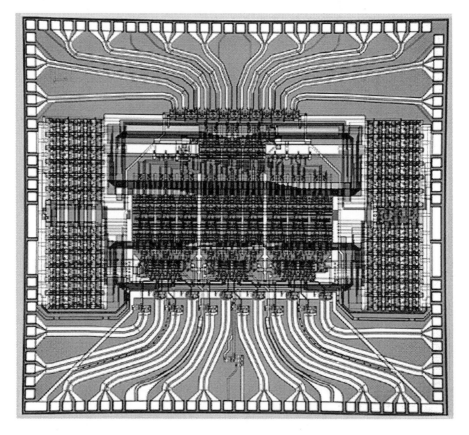

Fig. 18. Chip photograph of a 12×12 switch IC showing the chip core (center), switch state memory (left/right), and staggered output buffers (bottom center).

from the top and bottom, but additional grounds are also fed in on the output side of the IC. A low inductive power grid on metal 3 was used to supplement an ample power grid that ensured low IR drops on all of the power lines and low inductance on the ground lines. VCCS connections (the grounds to the current sources) were run with thick and highly capacitive metal lines. The transistor counts and power budgets for the 12 × 12 switch are shown in Table 2. The die dimensions are 4.8 × 5.1 mm².

Table 2. 12 × 12 Switch Total Power and Transistor Count.

	Power (mW)	Transistor Count
Switch Circuits	3816	1823
Output Buffers	1825	211
Input Buffers	235	36
Control Input	112	31
Control Memory	1122	2279
Auxiliary Circuits	278	192
Total	7388	4572

Thermal modeling

Excessive heating from the active devices can cause performance reductions in HBTs; thus, high power density was considered to be dangerous for switch performance. It was suspected that the large power consumption in the flip-flop control circuitry could sufficiently raise the background temperature of the IC at steady state to cause overheating in the high-speed stages. Much time and consideration was put into conducting a temperature simulation of the switch. A spreadsheet based modeling tool was used to model the thermal effects and to plot the spatial thermal distribution. After extracting the emitter locations from the layout, the temperature distribution near the critical output driver stages was calculated. The spatial simulation (Fig. 19) shows that the temperature rise due to the output drivers was not high enough to endanger high-speed results. This result illustrates one of the advantages of the staggered output buffer in the layout. For the most part, the simulation showed that the heat dissipation is highly localized in specific transistors such as output drivers. There appears to be no substantial cross-heating due to the layouts.

8.3. *Package design and characterization*

Figure 20 shows a photo of the package used to house a 10 Gb/s 12 × 12 switch. The package housing is fabricated from aluminum, houses coaxial connectors, and functions as the module shield. The transition between the die and the housing is accomplished with an Alumina substrate. The higher unit cost of the package

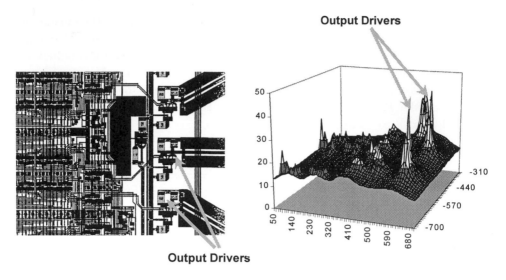

Fig. 19. Spatial thermal distribution of the switch IC near the output buffer which shows that the output buffer heating is localized near the output buffer and does not significantly affect the high-speed core.

Fig. 20. Prototype switch package for the 12 × 12 crosspoint switch showing the package housing, Alumina substrate, and switch die.

is well justified for prototype purposes when compared to the high NRE cost for conventional packages.

The inputs to the switch are made through GPO connectors. These connectors were chosen for their small size and high performance at the 10 Gb/s data rate. Due to the large number of I/O, the use of conventional SMA connectors would have resulted in a switch substrate diameter larger than 4 inches which is the maximum diameter handled by a large variety of substrate vendors. Power connections are made through DC feedthrough which contain an LC filter to ground to prevent high-frequency noise from entering and leaving the package.

Figure 21 shows a close up photo of the die/substrate for a 8 × 8 switch. The substrate is polished 99.5% Alumina that is 25 mils thick. This substrate allows for the 18 mil signal to signal spacing with a 4 mil gap between at the die I/O. For the connectors, the center to center signal spacing is 265 mils with a 6.7 mil gap between centers. The package was designed with a total of four transmission line width changes to minimize the loss. Taper transitions are used for low return loss. For low cross-talk and jitter due to ground bounce, the GSGGSGGSG pad I/O pattern is used. Conformal mapping combined with odd/even modal analysis was used to study the CPW coupling. The simulated results suggest that with the narrowest CPW lines nearest the die (where the adjacent channel spacing is the closest) can

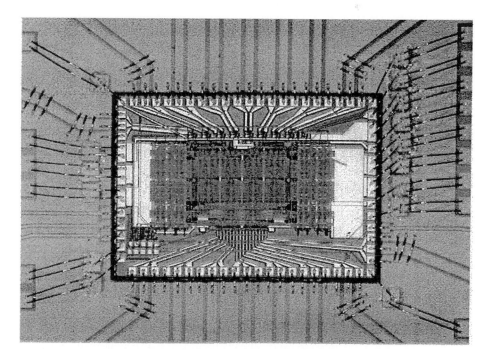

Fig. 21. A close up photo showing the transition between the die and the substrate. Chip capacitors are used to bypass the power. An 8 × 8 switch is shown as an example. The transitions for the 12 × 12 are similar.

have a worst case coupling of 25 dB. The simulated results for the worst case coupling for the widest CPW is 51 dB. In both cases, the worst case occurs when the transmission line length is near 1/4 wavelength. The transmission line calculations were used to form a lump element model for coupling simulations, with the four CPW width transformations, for the entire package. Figure 22 shows the measured and simulated coupling. The results shows that the worst case peak coupling in this package is −40 dB and the return loss at 5 GHz is better than 20 dB which is quite good for the crosspoint switch. All high-speed measurements presented for the 12 × 12 example were measured through this package which demonstrates the time-domain large signal performance. The results show that when the key packaging concerns for crosspoint switches are addressed, it is possible to design low cross-talk packages operating at 10 Gb/s with over 120 Gb/s data throughput.

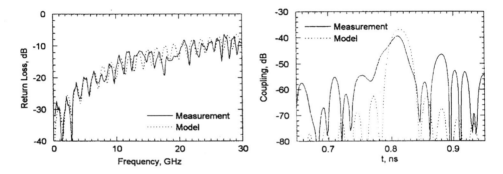

Fig. 22. Measured versus simulated coupling and return loss for the four CPW transitions used in the package. The models were based on a combination of conformal mapping and SPICE simulations.

8.4. Switch testing

Due to the large number of high-speed channels, the testing of the switch is quite challenging. In this work, on-wafer tests were used to verify the DC performance and the speed of a few selected channels were measured. Once potentially fully functional switches were determined, the switches were packaged for full high-speed characterization.

DC on-wafer testing

The primary evaluation of the functionally of the 12 × 12 switch was done with on-wafer tests at low frequencies. A low speed 128 pin probe card was custom manufactured by Rucker and Kolls. A Hewlett Packard pattern generator and logic analyzer (16 500B mainframe) was used to supply test vectors and confirm proper outputs. When the control lines were driven from a pattern generator with a fast rising edge, such as the HP 16522A, series resistance were used in series with the output pod to prevent excessive signal overshoot through the long unterminated control lines.

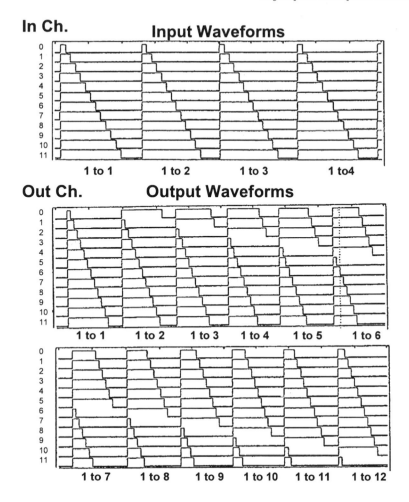

Fig. 23. Rolling stair case pattern used to verify full DC functionality on-wafer and in-package. The input pattern does not change and with each cycle, the switch program state is changed.

Functionality was examined with the "rolling stair-case" pattern shown in Fig. 23, in which a fixed input pattern (with pulses of different duration for each of the inputs) is directed to each output. By observing the output pulse, it is possible to confirm that the proper input is sent to the proper output. By observing 12 channels at the same time, all possible switch states can be verified with 12 settings. The first setting is channel 1 to 1, 2 to 2, ..., 12 to 12. The next setting is 1 to 2, 2 to 3, 3 to 4, ..., 12 to 1. The last setting is 1 to 12, 2 to 1, 3 to 2 and 12 to 11.

Although the transistor count in the circuit is relatively high, preliminary dc yield was excellent (approaching 100% on selected wafers). It was also shown that the circuit was functional over a fairly wide power supply range — control data could successfully be read in as VEE was varied from −6.4 V to about −8 V.

High-speed on-wafer tests

Initial high speed tests were conducted on a microwave probe station and made use of high frequency Ground-Signal-Ground (GSG) style probes. This ensured that a 50 Ω impedance match would be maintained. Power supply connections were supplied from multi-pin DC probes. Except for a necessary power line connection on the output side, the pad layout had been arranged purposely so that GSG probes would be able to contact the multiple high speed input and outputs. The remaining power line was contacted with a dc cantilever probe. High-speed pseudo random inputs were supplied from a 12 Gb/s and/or 3.3 Gb/s pattern generator or bit error rate tester (BERT) and the switch output was viewed on a 50 GHz digital sampling oscilloscope.

Unfortunately, probe geometry considerations limited the probable on-wafer inputs and outputs to a few select channels. For this test, the switch was configured by pulling the control input "TEST_CONFIG" to +3 V. This automatically configured the switch to send input 1 to output 1, input 2 to output 2, and so forth so the switch would not have to be programmed with additional equipment to simplify high-speed testing. The high-speed test setup is illustrated in Fig. 24.

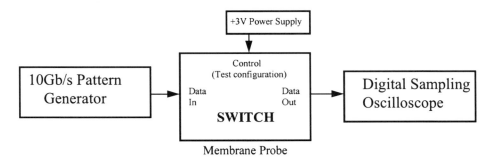

Fig. 24. Test Configuration for high-speed test with a pre-programmed switch stage.

The use of isolated grounds on the high speed inputs and outputs caused some difficulty and probe grounds had to be interconnected with external connectors. Because of this, on-wafer measurements proved not to be as successful as later in-package tests where the grounds are tied to a low inductance path resulting in a RMS jitter of 4 ps. With on-wafer probe testing, it was shown that there was some frequency dependence of the jitter that can be attributed to resonance in the power line bypass structures of the probes.

High-speed Packaged Switch Test

In-package testing provided excellent results and facilitated interchannel interference testing by allowing for simple multiple channel operation tests. Low speed functionality and high speed tests were done with the switch inside a completed switch

module. The success of the package tests could be attributed to the low inductance package ground provided by the CPW ground plane and the low-inductance
resonance free bypassing on the power lines. The package housing also provided
isolation from external noise sources and interference.

A representative output eye diagram is shown in Fig. 25 and the jitter histogram
is shown in Fig. 26 for the packaged switch with 10 Gb/s data (single channel active).
RMS jitter was determined to be in the range of 3 to 3.5 ps for the switch with
a single channel throughput. Some of the measured jitter is due to inter-symbol
pattern interference. With the addition of a parallel RC network in series with the
output, the "dispersion" characteristic of the switch was improved which reduced
the jitter to 2.5 ps (without the 1.5 to 2 ps source jitter correction). The measured
results did not demonstrate any resonance associated with the power by-pass.

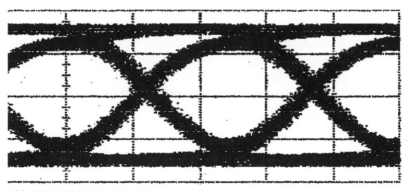

V: 120 mV/div H: 40 ps/div

Fig. 25. Eye diagram at 10 Gb/s with one channel active.

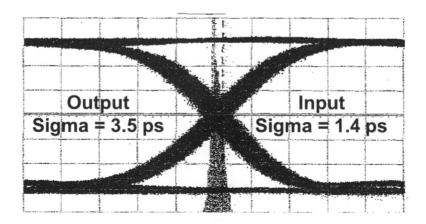

Fig. 26. Jitter histogram of the eye diagram shown in Fig. 25. The total measured output jitter
is 3.5 ps. The measured jitter of the input is 1.4 ps.

Multi-channel tests were performed by routing a 10 Gb/s signal through the switch in a loop-back fashion. This is accomplished by connecting outputs back to the input of a different channel with a random cable length delay to create interference. Another 10 Gb/s source is used to test the performance of the desired channel in the presence of interfering channels. Such re-routing with different length cables creates a quasi-asynchronous situation where a worst case cross-talk is expected with the test and interference switch edges are close in time. For the 12×12 switch, a 3.3 Gb/s asynchronous data (from a slower BERT) was added to the four 10 Gb/s loop-back channels for the tests to increase the number of interfering channels.

When additional channels were activated, the increase in jitter due to cross-talk was measured to be small, with a value of approximately 0.15–0.25 ps increase in RMS jitter per added channel. With three channels operating in addition to the test channel, the measured RMS jitter is only 4 ps. It is suspected that the main source of cross-talk was not through the multiplexing circuits, but rather, through power supply connections. These results show that at this time, this 12×12 crosspoint switch with the DEMUX/MUX high-isolation architecture may be the lowest jitter switch demonstrated for 10 Gb/s operation. The demonstrated fast rise times, high yield, and good uniformity should indicate that the 12×12 switches may make a viable component for a 10 Gb/s per channel WDM system crosspoint switch.

9. Future Possibilities

As capacity of data transmission systems expands, increasing demands are placed on crosspoint switch ICs for greater number of inputs and outputs, and higher data rates. Particularly for switching of small packets (such as ATM cells), shorter reconfiguration time is also desirable. The integrated circuits described here have been evolving to meet these requirements. However, they are reaching a variety of (time-varying) limits imposed by present technology, including power dissipation, interconnect speed and associated transmission line effects, I/O pad count, I/O power, and package complexity.

It may be possible in the future to bring about a significant change in the complexity and speed of crosspoint switches by employing optoelectronic interconnects. On the nearest term basis, it is likely that optical transmission of signals may allow easier connection of signals to critical regions of the chip. For example, read-in of the control signal bits (64 bits for a 16×16 switch) using a high degree of parallelism, achieved with a multiplicity of photodetectors distributed across the chip, would allow the overall switch reconfiguration time to be drastically shortened. Optical inputs and outputs for the high speed data can potentially alleviate problems of cross-talk among signals in the driver circuits or in the package, and, with the advent of low threshold current lasers, they may also reduce power dissipation. On a longer term basis, for optical fiber-based systems, high speed signal routing can be carried out entirely in the optical domain, with the electronics relegated to memory and control functions. This approach could allow extremely high data throughput if the optical signals are appropriately modulated. For example, signal routing can be done with "relational" devices which are insensitive to the bits propagating through them,[26] such as spatial light modulators, optical directional couplers and optical amplifiers. Optical signals with data rates above 100 Gb/s achieved with time division multiplexing or frequency division multiplexing can potentially be switched (albeit with long switch reconfiguration times). At a further level of sophistication, optical "logic" devices, which are sensitive to the propagating data, and are also very rapidly reconfigurable, can be employed for a general purpose switching fabric.[26]

10. Summary

This paper has discussed the issues associated with implementing very high speed digital crosspoint switch ICs, and has reviewed recent accomplishments in switch implementation. A detailed description was given of a recent 12×12 switch which provides 10 Gb/s operation with low jitter, implemented with AlGaAs/GaAs HBT technology. The aggregate data throughput of the crosspoint switches is among the highest of any IC reported to date.

Acknowledgements

The authors would like to thank Bert Hui and Bob Leheny at DARPA for funding on the 12×12 crosspoint switch; Al Price for helpful advice and experience in both the design and characterization of the switch; Jeff Liu for the package isolation characterization and modeling; both Derek Cheung and Jon Rode for support at the Rockwell Science Center; the processing staff at Rockwell Semiconductor Systems, Newbury Park for the fabrication of the switch; and April Collins of Wedgeway for assembly of the packaged switch.

References

1. C. Siller and M. Shafi, eds., *SONET-SDH: A Sourcebook of Synchronous Networking*, IEEE Press, New York, 1996.
2. C. E. Chang, K. D. Pedrotti, A. Price, A. D. Campana, D. Meeker, S. M. Beccue, D. Wu, K. C. Wang, A. Metzger, P. M. Asbeck, D. Huff, N. Kwong, M. Swass, S. Z. Zhang, and J. Bowers, "40 Gb/s WDM cross-connect with an electronic switching core: preliminary results from the WEST consortium", *1997 IEEE Lasers and Electron-Optics Society*, 1994, pp. 336–337.
3. T. Cloonan, "Applications of free-space photonic technology for ATM switching", *1994 IEEE Lasers and Electron-Optics Society*, 1994, pp. 228–229.
4. E. Munter, J. Parker, and P. Kirkby, "A high-capacity ATM switch based on advanced electronic and optical technologies", *IEEE Commun. Mag.*, Nov. 1995, pp. 64–71.
5. Triquent Semiconductor, Inc, *TQ8017 1.2 Gigabit/sec 16 × 16 Digital PECL Crosspoint Switch*, Data Sheet.
6. AMCC, *S2025 32 × 32 1.5 Gbit/s Differential Crosspoint Switch*, Data Sheet.
7. AMCC, *S2024 "Crossbow" 32 × 32 800 Mbit/s Differential Crosspoint Switch*, Data Sheet.
8. VITESSE, *VCS864A-2 200 Mb/s 64 × 64 Crosspoint Switch*, Data Sheet.

9. C. J. Anderson, J. H. Magerlein, G. J. Scott, S. Bermon, A. Callegari, J. D. Feder, J. H. Greiner, P. Hoh, H. Hovel, A. Pomerene, P. Roche, and M. Thomas, "A GaAs MESFET 16 × 16 crosspoint switch at 1700 Mbits/sec", *Tech. Dig. 1988 IEEE GaAs IC Symp.*, p. 91.

10. R. Savara and A. Turudic, "A 2.5 Gb/s 16 × 16 crosspoint switch with fast programming", *Tech. Dig. 1995 IEEE GaAs IC Symp.*, p. 47.

11. R. Savara, "A high-speed and high precision 64 × 33 crosspoint switch IC", *Tech. Dig. 1997 IEEE GaAs IC Symp.*, p. 105.

12. G. S. LaRue and T. A. Dao, "Gigabit complementary HFET communication circuits: 16:1 multiplexer, 1:16 demultiplexer and 16 × 16 crosspoint switch", *Tech. Dig. 1996 Int. Solid-State Circuits Conf.*, p. 124.

13. H. Shin and M. Immediato, "An experimental 5 Gb/s 16 × 16 Si bipolar crosspoint switch", *IEEE J. Solid-State Circuits* **27** (1992) 1812.

14. K. Lowe, "A GaAs HBT 16 × 16 10 Gb/s/channel crosspoint switch", *IEEE J. Solid-State Circuits* **32** (1997) 1297.

15. A. Metzger, C. E. Chang, P. M. Asbeck, K. C. Wang, K. Pedrotti, A. Price, A. Campana, D. Wu, J. Jiu, and S. Beccue, "A 10 Gb/s 12 × 12 crosspoint switch implemented with AlGaAs/GaAs heterojunction bipolar transistors", *1997 IEEE GaAs IC Symp.*, pp. 109–112.

16. Stratedge Corporation, San Diego CA, Package Catalog.

17. TriQuint High-Speed Packaging Catalog.

18. MSI Package Catalog.

19. G. Ghione and C. U. Naldi, "Coplanar waveguides for MMIC applications: effect of upper shielding, conductor backing, finite-extent ground planes, and line-to-line coupling", *IEEE Trans. Microwave Theory and Techniques*, **MTT-35**, No. 3, March 1987, pp. 260–267.

20. AMCC, *S2025 32 × 32 1.5 Gbit/s Differential Crosspoint Switch*, Data Sheet.

21. T. Saadawi, M. Ammar, and A. El Hakeem, *Fundamentals of Telecommunication Networks*, Wiley & Sons Inc, New York, 1994 pp. 399–442.

22. S. Ramo, J. Whinnery, and T. Van Duzer, *Fields and Waves in Communication Electronics*, John Wiley & Sons, Inc. New York, 1994.

23. P. Grey and R. Meyer, *Analysis and Design of Analog Integrated Circuits* (3rd ed.), John Wiley & Sons, Inc. New York, 1993.

24. P. Horowitz and W. Hill, *The Art of Electronics* (2nd ed.), Cambridge University Press, New York, 1989.

25. W. Wolf, *Modern VLSI Design: A Systems Approach*, Prentice Hall, Inc. Englewood Cliffs, 1994.

26. H. S. Hinton, "Photonic switching fabrics", *IEEE Commun. Mag.*, Apr. 1990, p. 71.

International Journal of High Speed Electronics and Systems, Vol. 9, No. 2 (1998) 549–566
© World Scientific Publishing Company

HBT ICs FOR OC-192 EQUIPMENT

JOHN SITCH and ROBERT SURRIDGE

Nortel, P.O. Box 3511, Station C, Ottawa, Ont. K1Y 4H7, Canada

Heterojunction bipolar transistor (HBT) integrated circuits are just beginning to appear on the commercial market, mostly as small-scale microwave and broadband parts. The high-speed portion of Nortel's OC-192 fiber communications product makes extensive use of this new technology, and this article describes the rationale for using HBT ICs in such a system, and the philosophy behind HBT IC introduction.

1. Introduction

In this paper we describe the development and transition to production of the Nortel GaAs/GaInP HBT IC technology, and its first application in ICs for OC-192 optical fiber communication equipment. In this introduction, we will set the stage by briefly introducing OC-192 systems and their marketplace, optical fiber and the high-speed portion of a system block diagram, to put the rest of the article in context. Subsequent sections cover the topics of technology choice, HBT development strategy, process, design, fabrication, test, reliability and packaging.

Many of today's trends point to dramatic and continual increases in demand for communication bandwidth, so the potential sales volume for OC-192 equipment should be high. "High" in this context still means orders of magnitude less than the volumes of consumer goods such as personal computers or cellular phones, since no-one will sell millions of OC-192 systems unless there is an explosive upturn in the market. Development costs must be scaled to suit the market size; on the other side of the coin it means that the asymptotic chip cost at infinite volume is far from the whole story, leading to interesting technological trade-offs. One might assume that with system data rates quadrupling every five years, equipment designed for a twenty-year life will be obsolete long before it wears out. But, considering the huge amounts of revenue earning data on each fiber, customers need the system to be very dependable, little different from a submarine system. In fact, this leads to high reliability requirements right across the system and a need to establish, for example, conservative junction temperatures in a compact assembly.

OC-192 is a SONET[1] optical fiber transmission system with a data rate of 9.953 Gb/s. The equivalent SDH system is STM-64. SONET signals are made up of interleaved bytes of tributaries at the base rate (about 51.8 MHz). The OC-designation strictly refers only to the optical format, while the electrical, often byte-parallel, form of a SONET signal is denoted as STS-N. Data is not restricted to rates lower than STS-1, higher rate data is dealt with as a concatenated stream; thus OC-12C (with the C suffix denoting concatenation) is 622 Mb/s and cannot be broken down, although it can, of course, be combined with other data in, for example, OC-48. A SONET signal consists of the header, sent every 125 μs, followed by the payload. The payload and most of the header are scrambled to improve the signal statistics, but the header definition bytes are left plain. Byte multiplexing results in N repeats of each byte, which make for easy frame finding but also leads to noticeable spectral content at 8 kHz.

Single-mode optical fiber is notable for its low loss, less than 0.25 dB/km at wavelengths around 1550 nm. A second, slightly higher, loss minimum occurs for wavelengths around 1300 nm. Dispersion, the variation of propagation velocity with wavelength, arises from two causes, first the guided mode is non-TEM, which gives rise to dispersion even with ideal materials, and the material from which the fiber is made is itself dispersive. Standard fiber, also known as non-dispersion shifted fiber (NDSF), which makes up the bulk of installed fiber in North America, has zero dispersion in the 1310 nm window. But, despite this advantage 1550 nm is more popular for long haul applications owing to the lower loss and the availability of erbium doped fiber amplifiers (EDFA) in this window. Dispersion-shifted fiber (DSF) has its dispersion zero near 1550 nm, and is being increasingly used for new installations. A third type of fiber, dispersion compensating fiber (DCF) has dispersion opposite to that of standard fiber in the 1550 nm window, so a suitable length can be used to null the dispersion of a standard fiber span at one wavelength. DCF is not presently laid in the ground, it is installed at amplifier sites or terminals, increasing the length of fiber in the span without increasing the span. The main drawback to DCF is its fairly high loss, which increases the number of amplifiers needed in a compensated system. Fiber itself is not very expensive, but the cost of putting it in the ground is high enough to justify going to some lengths to maximize the data carried by any fiber. Wavelength division multiplex (WDM) is one way to increase the total data rate on a fiber, but as each wavelength needs its own set of terminal equipment (pending the development of integrated WDM systems[2]) it still makes sense to have the highest data rate per wavelength.

The practical result of the foregoing is that fiber systems equipment must operate effectively with dispersive fiber. Directly modulated lasers have chirp: the lasing wavelength changes during optical transitions, and even during pulses as the carrier and photon populations change and the laser temperature varies; dispersion applied to such waveforms leads to inter-symbol interference (ISI), closing the data eye as the fiber span is increased. At OC-48 (2.488 Gb/s), for example, using a good laser in the 1550 nm window, we can reach about 80 or 90 km on standard fiber with good margins; the smaller bit period and faster transitions reduce this to an unacceptably short distance for OC-192. So in an OC-192 system, we use a CW laser followed by an electro-optic modulator to reduce the width of the optical spectrum to close to the minimum required to carry the information. Commercial fiber systems use the simplest possible modulation scheme — AM, double sideband with carrier. Those readers familiar with radio systems will recall that this, coupled with envelope detection, offers the least protection from dispersive effects, as the two sidebands get folded into one by the square-law detector and interfere with one another. We await with interest the technological advances needed to achieve in fiber the noise and distortion immunity, not to mention spectral efficiency, of present-day wireless and radio systems.

Figure 1 shows the block diagram of the high-speed portion of a fiber system, it shows functions rather than chips — some functions can be split across two or more chips, or more than one function may be combined on one chip. Thanks to SONET, all the digital processing can be done in silicon at a convenient low data rate (chosen by trading off speed against complexity), leading to straightforward multiplexer and demultiplexer (mux and demux) ICs. Even though the mux and demux are digital functions, great stress is placed on their analog parametrics as the closer they come to ideal, the better the optical link performs. As we have discussed, we need a modulator to meet the chirp requirements, so after the mux comes the modulator driver, whose job is to deliver data with the correct voltage swing and the right pulse shape to the modulator.

After the fiber link, we come to the receiver. Optical amplifiers give us a choice of where to put the gain, and how sensitive to make the receiver. But looking at the cost of an optical amplifier compared to ICs, we will put a low-noise preamplifier followed by a main amplifier after the photodetector (the required noise and overload characteristics of the preamplifier are slightly changed by the choice of a PIN compared to an APD detector). Next comes clock and data recovery where the noisy and distorted received signal is turned back into digital form. The

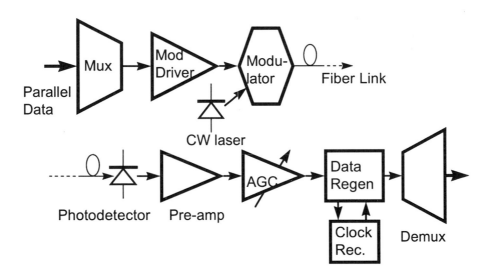

Fig. 1. Fiber system block diagram, showing the high-speed functions.

recovered clock, when used to retime the data, makes each bit period equal to the average bit period of the incoming signal, an operation similar to (but more difficult than) using the voltage limiting nature of logic gates to give all the bits constant amplitude. Recovered data and clock are passed on to the demux and converted into parallel data (and a lower rate clock). Partitioning this block diagram is crucial to all aspects of the system — how well it works, how much it costs to make, how long it takes and how much it costs to develop, how easy is rework or repair, and so on. The right answers are not fixed, they vary with the product life-cycle, volume and many other factors, and have great bearing on IC strategy and design.

2. Partitioning and Technology Choice

No development is without risk. The risks faced all end up as financial, but aside from having a product that costs so much to make that it cannot be sold at a profit, there are a couple of common types of risk that are readily defined. Performance risk is where we risk the thing not working, or having such small margins that yield is low and extensive testing is required at every stage, while a long development cycle involves a schedule risk — that the finished product will be too late for the market. Risk management is a large component of development project management, with system partitioning and technology selection important tools in this endeavor.

10 Gb/s fiber communication requirements are so stringent that before starting development of its OC-192 system, Nortel conducted a technological audit of high-speed technologies. The contenders were all the available or potentially available technologies that could perform some or all of the high-speed functions. As a rule of thumb for the digital parts, we need a technology capable of making a flip-flop that runs at about twice the bit rate, which translates into F_t at least four times the bit rate — 40 GHz in our case.

40 GHz silicon processes are hard to find, so the main silicon contender is heterostructure bipolar transistor (HBT) technology using SiGe. Not yet fully developed at the beginning of the program, SiGe is certainly promising for the digital parts. At the other end of the spectrum we have III-V HEMT technologies. These have been around for several years, with gate lengths down to 0.15 μm and F_{max} values that can exceed 200 GHz available from a number of foundries. The analog functions, particularly where low noise is required, seem well within reach of HEMTs. III-V HBT technology occupies a middle position in almost all respects, with $F_t > 40$ GHz devices available from a few vendors. Fast logic and high linearity microwave amplifiers are the perceived strengths of III-V HBTs.

Bipolar transistors have some important advantages over FETs (including HEMTs) when it comes to high-speed logic. The first one is high transconductance (gm). To make a proper comparison we divide gm by the operating (drain or collector) current; this normalized gm is around 20 v^{-1} for HBTs and about one-tenth that for FETs, when biased for high F_t. We can view the inverse normalized gm as the voltage change needed to significantly change the current, so bipolars switch more abruptly than FETs, leading to a smaller logic voltage swing and lower currents in capacitances and loads. High gm helps in other ways, too; output drive is improved by using buffers such as emitter or source followers, and the high normalized gm of bipolars leads to emitter followers with voltage gain of almost unity, compared to significant losses in source followers. For devices with the same F_t, reduced gm corresponds to reduced capacitance — this advantage of FETs is nullified in digital circuits by the ever-present wiring parasitics. Another bipolar advantage is the uniformity of turn-on voltage, even for small transistors Vbe varies by only a few mV across a wafer, whereas in a FET process one has to work very hard to get 20 mV, this is particularly important for the decision circuit. Bipolar transistors have their drawbacks, of course. The finite current gain (β) means there is always input current, which becomes appreciable for low β; and Vbe, highest in GaAs-based HBTs at about 1.4 V, leads to higher supply voltages. The number of transistors that can be yielded on a single chip is higher for FET than bipolar technologies, although the yield difference in III-V compounds is probably a lot smaller than in silicon. The chips considered here, ranging from 27 to 2500 transistors are well within the complexity limitations of HBT technology.

Summing up, bipolar technologies have more advantages for high-speed, as opposed to high complexity logic, and GaAs-based HBTs are favored over SiGe or InP-based HBTs as the most mature, despite the extra dissipation implied by their higher Vbe.

A priori decisions concerning analog circuits are rather harder to make, as each circuit design will need to be optimized considering all the specified parameters, and the different technologies might use completely different configurations. F_t and $F_{\max}$ are useful, but in a restricted sense. F_t refers to current gain, the usefulness of which depends on impedance levels. $F_{\max}$ predicts the gain per stage of a tuned amplifier (worse, a tuned unilateralised amplifier if Mason's unilateral gain is used instead of maximum available or stable gain when calculating $F_{\max}$), a condition hard to achieve over the six decades of bandwidth that we require. One thing that is clear right from the start is that a modulator driver will be far easier to design if the breakdown voltage is more than the output swing. As lithium niobate modulators need about 5 V (III-V modulators need less), SiGe and many HEMT technologies are virtual non-starters for this role. The receive side amplifiers need low noise, which tends to favor HEMTs, plus good matching and high linearity, which is a characteristic of bipolar amplifiers.

Our need to manage risk by changing the system partitioning leads us to favor the use of one technology over the use of several — note that this is not the case when our main goal is to optimize the system cost. Having a single technology means that we can repartition the system by moving functional building blocks from chip to chip without extensive redesign, apart, of course, from input and output structures. Another reason to minimize the number of new technologies is to reduce the amount of technology qualification that needs to be done. GaAs-based HBT technology is a good choice in these respects; apart from the noise figure of the preamplifier and possibly the power dissipation of the digital circuits, it is as good as any other technology for each of the required functions. In addition, repeatability and ease of design characteristic of bipolar circuits make HBTs the prime contender. There are local reasons why Nortel has pursued HBTs, too. Our policy in III-V fabrication is to operate a "minifab", with the entire operation scaled to be cost-effective when running at relatively low volume. Nortel's history in optoelectronics has made crystal growth a strength, while the minifab concept applied to capital has led us away from the very fine photolithography needed for PHEMTs. An in-house, as opposed to an external technology, was chosen because of its quicker turn-round and greater responsiveness to customer needs.

3. HBT Strategy

Ideally one should never introduce a new technology as part of a new system, as the development risks are combined. But new systems have new needs, development of a new technology is greatly focused by having a targeted first application. With the product development starting from scratch, there are no pre-existing constraints: a more practical approach would be to only use new technology in a new system if it cannot be avoided.

HBT process development at Nortel started a couple of years before the OC-192 system development, but 10 Gb/s optical fiber communication was selected as the first application. With a new technology, we are confronted by two options: (1) Do we design the electrical characteristics we need for the circuits we want to make and then develop the technology to fit the designs? (2) Do we make and evaluate some devices and see what circuits we can design with them? Clearly no one would pursue one of these strategies exclusively, but the circumstances surrounding the development can cause us to lean one way or the other. In our case, the desire to utilize existing fabrication equipment and to evaluate several epitaxial growth techniques led to a realization that IC designs would be easier to change than the process, which biased the strategy towards making repeatable devices and then designing circuits with them. F_t and $F_{\max}$ targets were set at 60 GHz, with breakdown (BVceo) of at least 6 V. Current gain was given a deliberately loose target of > 20. The first mask set contained only generic circuits based on models built using the output from physical and analytical device simulations,[3] but even that allowed the first 10 Gb/s operation of circuit building blocks.

Since that time, a set of basic devices, the process control monitor or pcm, has been measured on every wafer that has completed fabrication, while the process has been refined, with the only major change being the move from AlGaAs to GaInP as the wide bandgap emitter compound. Customized SPICE models have been developed along with the process,[4,5] with the parameters evaluated for transistors from representative wafers. DC and microwave measurements on two transistor sizes provide data for the scaleable models, which is then extended to arbitrarily sized 2 by n and 3 by m μm devices, plus multiple parallel transistors, within an allowed range. Parameters are extracted in a step-by-step fashion, avoiding optimizer use, with basic junction and transport properties taken from Gummel and resistance measurements, followed by junction capacitances measured at 1 MHz versus voltage and finally small-signal S-parameters from 45 MHz to 40 GHz at a number of bias points. Temperature effects are very important in HBT designs, so the models include self-heating, as well as a way of approximating the effects of adjacent devices. A thermal simulation is used, once the design is laid out, to check temperature rises and allow the designer to move things around to optimize the temperature distribution.

4. Epitaxy

Epitaxial growth for our standard device consists of a 300 nm thick $N+$ sub-collector, a 400 nm thick collector, a 50 nm thick $p+$ base, a 50 nm thick GaInP n-type emitter, and a 200 nm thick $n+$ emitter cap layer. The doping levels are 5×10^{18} cm^{-3} for the sub-collector and emitter cap, 4×10^{16} cm^{-3} for the collector, 4×10^{19} cm^{-3} for the base, and 4×10^{17} cm^{-3} for the emitter. The n type dopant is silicon, and the base dopant is carbon. All material used for production devices have been grown on an in-house MOCVD reactor, using a rotating suscep-tor for improved layer uniformity. Devices have also been successfully fabricated on in-house MOMBE, and MOCVD material from an external vendor. Devices from these material systems are currently undergoing extensive reliability tests, prior to their utilization as a second material source.

In addition to our standard process, we have also produced devices from material with collector layers between 550 nm and 1.0 μm for higher power (higher breakdown voltage) applications.

All wafers are inspected for flatness, resistivity, particle density, and material composition; using a Tropel wafer sorter, a non-contact resistivity mapper, a Surfs-can and photo-reflectance respectively. The photo-reflectance is an especially valu-able measurement as it provides quantitative data concerning the doping level in the emitter and collector layers, as well as composition (band-gap) of the GaInP emitter layer. Hall, resistivity, X-ray and SIMS measurements are also routinely performed on calibration wafers, ensuring that out-of-spec wafers do not undergo processing.

5. Processing Sequence

The Nortel HBT process is based on the use of a dummy dielectric emitter which is self aligned to the base ohmic contact metal that completely surrounds it. A thick sidewall formed on the dummy emitter provides ample alignment tolerance for the emitter ohmic metal. The base-emitter metal separation is precisely controlled by the magnitude of the lateral etching during the formation of the dummy emitter. To minimize base-collector capacitance, the area beneath the majority of the base metal is rendered isolating by a deep He+ implant. Square and rectangular emitters, as

243

well as multiple emitter structures, can be fabricated. The minimum allowed emitter size is 2×2 μm.

The process was originally developed for HBTs with AlGaAs emitters,[6] but was modified to allow fabrication of devices based on GaInP, as it had been shown that this alloy system gave HBTs with improved reliability.[7] An important feature of the process is a selective dry etch to the GaInP emitter, followed by the deposition of the base metal which is then alloyed through to the base. This facilitates contacting the thin base layer. The GaInP layer between the edge of the alloyed base contact and the edge of the emitter mesa is left in place in order to reduce the surface recombination, and increase the current gain of the devices, especially at low collector currents.[8]

In order to obtain a good ohmic contact the p-metal must be alloyed through the GaInP into the p+ base. For reliability reasons it is very important that the metal penetration be self limiting and uniform, with no tendency to spike through to the collector layer. The contact metallization employed in this process consists of Pd-Pt-Au-Pd and gives good morphology and uniform penetration of the GaInP. The penetration depth during alloying can be controlled by the thickness of Pd below the Pt barrier. Originally we incorporated zinc in this ohmic contact to ensure that good contact resistance to the base layer was achieved. However this proved to be unnecessary provided the base layer doping exceeded 2×10^{19}; and as zinc was a potential source of contamination and a potential source of poor reliability, it was removed.

The emitter and collector layers are contacted using a sintered PdGe layer, which we have shown to give improved reliability over the NiGeAu contacts previously used; as well as minimal in-diffusion and no evidence of spiking, which would seriously impact device performance and reliability. A thick sidewall, deposited on the dummy emitter prior to its selective removal, provides alignment tolerance for the location of the emitter contact.

In addition to HBTs the process sequence allows for MIM capacitors using SiON as the dielectric, and 50 ohm/sq. nichrome resistors. In the standard process interconnection is provided by a 600 nm thick first level, and a 1.2 μm thick second level metal. In both cases the interconnect metal is e-beam evaporated TiPtAu, patterned using a bi-layer lift-off technique. Interconnection is facilitated by 2×2 μm gold posts 2.5 μm high, or large area vias. A lift-off technique is also used to deposit the posts.

The interlevel dielectric used in our process is BCB (Dow Chemicals product name Cyclotene), which was chosen for its excellent dielectric and planarization properties, as well as low shrinkage and cure temperature. Interconnection is achieved by planarizing a thick BCB layer by plasma etching, and revealing the tops of the posts prior to second level metal deposition. A polished cross section of a completed device is shown in Fig. 2, where most of the salient features of the process can be identified. The complete process, as described, requires 13 reticles. All photolithography is performed on a Canon MkII stepper, which has an

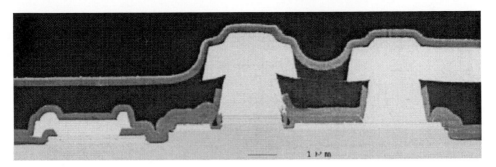

Fig. 2. Polished cross-section of a finished transistor. Collector, contacted by first metal, is on the left; emitter, center, and base, right, are contacted by second metal using posts. The emitter is completely surrounded by base ohmic metal.

alignment accuracy of better than 0.2 μm, and a minimum linewidth of 0.8 μm, which has proven to be adequate for this process.

6. Design

As described above, our strategy has been to design circuits based on measured transistor parameters. Formal best- and worst-case models do not exist; each design is analyzed for sensitivity to parameter variations, using variances related to measured values, it is then tested by being simulated with the range of extracted models. The magnitudes of the parasitics extracted from the layout depend on selecting the extraction rules file. SPICE is the simulator of choice for this work; the digital circuits are not complex enough to warrant sophisticated digital simulators, but the performance is critical, so a transistor level simulation of each chip is vital.

Emitter-coupled logic (ECL) is used for the digital circuits. In its fully differential form it offers high noise immunity with a minimum of power supply noise generation, good output drive when buffered by emitter followers, and fast transitions. One drawback associated with ECL is the relatively poor performance of combinational logic, especially with a high number of inputs. Looking at our applications, we see that robust mux and demux designs consist almost entirely of flip-flops, elements at which ECL excels, due to the series gating inherent in its current-steering tree structure. As the logic elements were designed while the process was still being developed, full advantage was taken of ECL's robustness. A typical gate schematic is shown in Fig. 3. The exclusive OR (XOR) function between the first level input, A and the second level input, B can be verified by inspection. Outputs are buffered by emitter followers, with diode-connected transistors providing level shifts to lower level outputs. The logic swing is 350 mV per side. In the 10 Gb/s cells, this is obtained with 1.4 mA switch current and 250 Ω loads. Lower power, lower speed cells have higher value resistors to scale the current, but the transistors remain the same size, operating at lower current density. High fanout clock buffers use double emitter followers to provide second level outputs with low impedance drive.

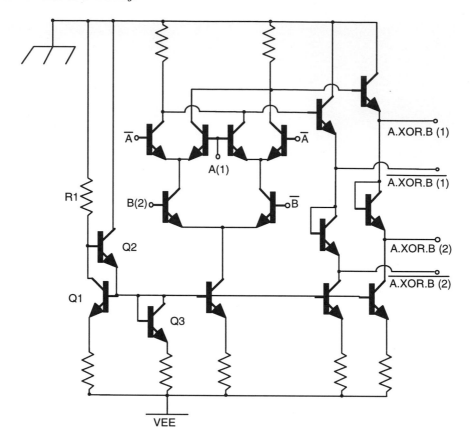

Fig. 3. Circuit schematic of a 2-level XOR gate.

To minimize the effects of on-chip voltage drop and prevent noise transmission, each cell has its own current mirror bias generator. Referring to Fig. 3, the reference current flows through R1 and Q1, while emitter follower Q2 increases the fanout of the mirror for low β, with Q3 drawing enough current to keep Q2 properly biased, even at high β. Using current mirror bias maximizes the common-mode rejection ratio of the differential switches, and when used on the emitter followers it helps to cut power supply noise. Nominal power supply voltages have been chosen to give plenty of headroom for all the devices; -6.5 V is used with two level logic and -8 V with three levels. Output drivers can use lower voltages to save power. For low speed and low fanout, the so-called zeroth logic level, with no emitter followers, is also available. System constraints often lead us to choose one supply voltage, rather than using the range that will minimize the chip power consumption. Then we have to compare the designs that result from using only two logic levels with those where three are allowed. We need to compare the current saved where one three level gate replaces two two-level ones with the voltage saved by using two levels. We then choose the realization that gives the best result overall. In the OC-192 case, the use of flip-flops with multiplexed data inputs has led to the mux being designed with three-level logic, whereas the demux, with fewer opportunities for three-level logic, uses only two levels.

Low output jitter and ample operating margins characterize the multiplexer design which has input flip-flops followed by successive 2:1 multiplexing gates, with flip-flop retiming after each stage. Retiming the data after the final 2:1 is essential to minimize the output jitter. To time-center the data for multiplexing, the second data stream going into each 2:1 stage is delayed by half a cycle relative to the first by putting it through a three latch (master-slave-master) flip-flop instead of the regular master-slave. The clock is ripple divided, driving the selectors and flip-flops, with clock divided by eight output to time the intermediate mux. The phase of the intermediate clock output is selectable and the phase of the incoming data is measured with respect to the internal clock for automatic alignment of the intermediate data. A tracking reference voltage for the ecl inputs is also output.

The modulator driver uses a differential limiting distributed output stage following a conventional input buffer,[9] giving high voltage swing, excellent signal waveform and good output impedance match at the same time. The design trade-offs involved here, for example extra power dissipation to avoid having to use external chokes, and a limiting structure with adjustable output amplitude to avoid sensitivity to interconnect losses, show the importance of close coupling between circuit design and system design.

Use of just one technology and its effects and benefits are best exemplified by the receive circuitry. During development, different configurations were explored as increasing experience led to new ideas being taken up, and this process continues as the product evolves through successive hardware releases. Having a working system mitigates the development risk of more aggressive, lower cost parts (and also packaging technologies). At the feasibility study stage, for example, the adaptive decision circuit and demultiplexer were fabricated as two chips as well as a single IC. Comparison of performance, power dissipation and yield show the clear advantage of the single chip as well as the value of careful design to keep transients generated by the output cells away from the decision circuit. Figure 4 shows the "superdemux" die. The use of a semi-custom cell-based approach, as in the mux, results in the characteristic digital LSI chip appearance. This technology, running at this speed, does not need transmission lines or inductors, for example, the signal rise and fall times are still equivalent to more than 1 mm of track.

A transimpedance preamplifier with a distributed output stage followed by a two-chip AGC amplifier make up the analog portion of the receiver. More circuit details are published elsewhere.[10] The uniformity and repeatability of the In-GaP/GaAs HBT process has made optimization of these components to minimize signal impairments a relatively straightforward matter.

The very small active volume of HBTs makes them vulnerable to electrostatic discharge (ESD). Measurements on minimum size (2×2 micron) transistors show an ESD damage threshold as low as 50 V for reverse-direction discharges through the

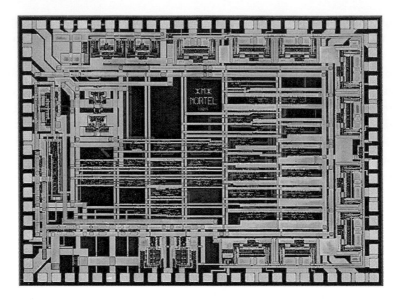

Fig. 4. Die photograph of a "superdemux" — integrated decision circuit and demultiplexer.

base-emitter junction. As normal laboratory precautions cannot guarantee values less than four times this, we must use built-in protection circuits in the ICs. ESD voltages are too high to be held off by the components available in HBT ICs, so our strategy is to shunt the ESD to ground or a convenient power supply. The shunt path is made up of diode-connected transistors (i.e., the base and collector are shorted together to form the anode and the emitter is the cathode) while resistors provide damping and help isolate the active circuits. The schematic of a typical ESD protection circuit is shown in Fig. 5. Thin film resistors are also susceptible to

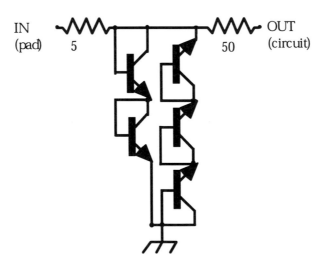

Fig. 5. ESD protection circuit for low-speed input.

ESD; in this case protection is provided in the form of a design rule which stipulates the minimum width of a resistor connected to a pad. High speed I/O is problematic, as the parasitics introduced by the protection cells are too large for the required frequency response. Some protection is afforded by using suitably stout termination resistors, along with multiple transistors.

The HBT test lab in Nortel has a static-dissipative floor enabling those working there to be grounded through shoe straps. When a footstrap tester measures an acceptable impedance, it opens the lab door. The room has controlled humidity (50%). People engaged in critical tasks use wrist straps as well as shoe straps. Electrical overstress can also destroy devices under test, and although HBT ICs can survive power supply overvoltages of about 40%, precautions such as avoiding "live" probing and contact with charged capacitors are needed to avoid yield dropout.

7. Test

All completed wafers undergo extensive electrical testing of the process control monitor (PCM) prior to circuit testing. The PCM has been designed around a common four pad configuration to facilitate automatic probe testing. Each of the probe pads is oversized to permit access from two probes to each pad, allowing full force and sense measurements. A significant amount of information is derived from Gummel plots measured on HBTs with emitter dimensions of 2×2 and 6.5×3 μm. These devices include stabilizing circuitry to ensure oscillation does not occur during measurement. Among the parameters monitored are beta at both 4×10^4 and 10^3 A/cm^2, gm at 4×10^4 A/cm^2 base-emitter turn on voltage (at 10^3 A/cm^2), current at unity gain (the point at which Ic = Ib), and the ideality factors of both Ib and Ie. A typical Gummel plot is shown in Fig. 6.

Breakdown voltages BVebo, BVcbo and BVceo, junction capacitances and isolation, are measured on separate devices. The s-parameters of the two sizes of HBT are also measured up to 40 GHz by wafer probing, and from these we estimate the F_t and F_{max} of the devices by extrapolating at 20 dB/decade from the derived H$_{21}$ and maximum available gain (MAG) data. Strictly speaking F_{max} should be derived from the unilateral gain, which would actually give a higher value. In practice the MAG figure is more relevant, especially for broadband circuits, and so we usually report this more conservative value. Typical measured gains of a 6.5×3 HBT biased at 4×10^4 A/cm^2 and Vce = 2 volts, are plotted in Fig. 7. The derived F_t and F_{max} values for this device are $F_t = 66$ GHz and $F_{max} = 84$ GHz. The F_{max} value based on the unilateral gain would be 111 GHz.

In addition to DC, capacitance and s-parameter tests on HBTs, the contact and sheet resistances of the emitter, base and collector and nichrome resistor layers are determined. Other test structures allow MIM capacitors and contact chains of all possible interconnect combinations to be measured.

The probe station, test equipment and extraction procedures are all controlled by a custom designed program running under HP-VEE. The extracted PCM

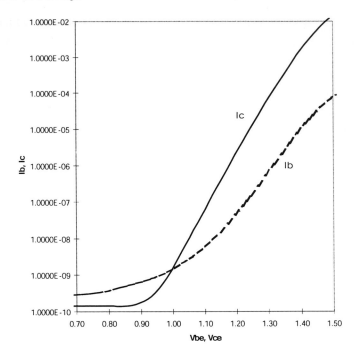

Fig. 6. Gummel plot — log scale Ib & Ic versus Vbe with Vcb = 0 — for 6.5 × 3 μm transistor.

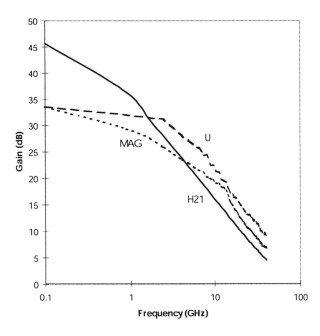

Fig. 7. Microwave characteristics for a 6.5 × 3 μm transistor, derived from S-parameters measured from 45 MHz to 40 GHz. The line labeled MAG is maximum available gain for frequencies for which the device is unconditionally stable, and maximum stable gain elsewhere; the line labeled U is Mason's unilateral gain.

parameters are logged into the database to allow correlation with material and process parameters. Median, upper and lower quartile values, and maximum and minimum values, as well as yield against upper and lower specification limits, are routinely charted against run number and epitaxial growth number to reveal any trends. Wafers with any parametric yield less than 50% are rejected, or waivered in the case of non-critical parameters. The criteria for rejection are continuously evaluated to ensure that all good wafers, and only good wafers, proceed to circuit testing.

All circuits on PCM good wafers are probe tested at speed using Cascade diaphragm probes. Wherever possible all specified parameters are tested, including bit error rates. A custom built test setup has been built specifically to test OC-192 circuits, and is currently being duplicated to accommodate increased volume. The control system for this setup is also based on HP-VEE. All circuit measurements are tracked against the unique wafer location number printed on each die, allowing wafer level traceability of measurements. These die numbers are ultimately used to select known good die for packaging.

8. Reliability

Because of the extreme reliability requirements for telecommunications applications, discussed earlier, a large amount of effort has been devoted to reliability testing. To avoid data being skewed by premature failure due to ESD or oscillation a special reliability test coupon has been designed. Each coupon contains eight individual devices, all with input and output ESD protection, and series base resistors and feedback capacitors designed to stabilize the transistors.

The most fundamental test has been multi-temperature accelerated aging of devices on these coupons. Three and four temperature studies have been performed at junction temperatures up to 300°C. The majority of these tests have been performed at a bias current density of 4×10^4 A · cm^{-2}, with the base and collector held at the same potential. High temperature biasing is usually performed in air, although tests have also been performed on devices packaged with nitrogen or hydrogen environments, the results of which showed no discernible differences. All the parameters derivable from Gummel plots have been examined to determine failure modes. Standard Arrhenius manipulation of the data derived from these tests, as well as similar experiments on unbiased devices, has revealed the presence of at least three different failure mechanisms.[11] These mechanisms are described below.

The most well-behaved and benign failure mechanism is associated with transconductance (gm) degradation, with an activation energy of approximately 1.75 eV, and a median time to failure (MTTF) of greater than 10^8 hours at a bias current density of 4×10^4 A·cm^{-2}, and 125°C junction temperature. This degradation mode is accelerated by heating alone, is not affected by applied bias, and is attributed to degradation of the emitter ohmic contact. This type of degradation is not typically observed in devices with refractory metal contacts to InGaAs contact layers, commonly employed in GaAs HBT fabrication. However, because of the well-behaved nature of this failure mode, and long expected life, these results obtained from our standard devices with sintered ohmic contacts are not considered to be a cause for concern.

The second failure mode is characterized by current gain (beta) degradation, with an activation energy of approximately 1.5 eV, and an MTTF of 10^8–10^9 hours at a collector current density of 4×10^4 A·cm^{-2}, 125°C junction temperature. This degradation mode has been observed in all GaInP/GaAs HBTs we have tested to date. The degradation is accelerated by both current density and temperature. We believe this is the characteristic failure mode of our GaInP/GaAs heterojunctions, and its severity may be dependent upon the exact epitaxial growth conditions used. It is speculated that this failure mechanism may be the result of current enhanced diffusion of carbon from the base into the emitter, resulting in a reduction of the acceptor concentration in the base layer immediately adjacent to the emitter. A similar effect was postulated for AlGaAs/GaAs HBTs,[12] which various groups have reported to have an activation energy closer to 0.6 eV. Unlike AlGaAs/GaAs HBTs where carbon can cause type conversion of the emitter layer, carbon is known not to dope GaInP p-type under most circumstances; and this factor, along with different diffusion rates, may be instrumental in causing the higher activation energy observed for GaInP/GaAs devices.

A second beta degradation mode, with a lower activation energy of approximately 0.3 eV, and a low MTTF (possibly as low as 10^4 hours at 4×10^4 A·cm^{-2} and 125°C junction temperature) has also been observed, and could lead to early failure of circuits. Like normal beta degradation, this mode is accelerated by bias, but as it has only been unambiguously identified in a handful of wafers there is a lack of statistical information and we have not yet been able to positively determine its cause. Eliminating its occurrence, and confirming its elimination, is difficult due to its rarity. To avoid any possibility of using devices with limited life due to the presence of this failure mode, we have instituted an elevated temperature screen (200 and 250°C) of all product wafers. The elimination of this failure mode is now a priority as there is an increasing demand for HBT circuits, and full screening is becoming increasingly undesirable.

To ensure that an unidentified failure mode does not limit operational life, and as part of the circuit qualification exercise, we are also studying the long-term reliability of individual devices and complete circuits. This is carried out under bias and at slightly elevated ($150 - 170$°C) junction temperatures (circuits operate normally up to at least 175°C). So far we have accumulated many thousands of operational hours under these conditions, as well as thousands of circuit-years, corresponding to billions of transistor-hours, in the field, without any significant failures.

9. Further Applications of HBT ICs

HBT ICs, with their attributes of high speed, excellent device matching, good linearity and fair noise figure are suitable for a number of uses, apart from fiber communications systems, ranging from high-volume "commodity" chips to highly specialized custom ASICs. High gain and linearity in a very small die size are selling points for wideband gain blocks, which are now commercially available.[13] Mixed analog-digital functions such as analog-to-digital converters,[14] or automatic data synchronizers[15] take advantage of HBT's unique combination of speed and precision, while large space switches, such as the 16×16 operating at 10 Gb/s described by Lowe[16] show their ability to give high yield in 5000 transistor circuits. HBT ICs are also competitive in the microwave regime, particularly in the field of power amplifiers,[17] where simultaneously high efficiency and linearity are selling points; while the low 1/f noise corner frequency gives low phase noise in microwave oscillators.[18]

10. Conclusions

Introducing a new and untried technology is a challenge, and the risks are magnified when the new technology is incorporated in a new type of product. OC-192 has shown that with a technology that is well-suited to the application, and proper risk management, new technology introduction can lead to significant systems advantages. Broadband fiber systems need very fast logic, precise and repeatable analog circuits, good output drive, high reliability and reasonable noise. GaAs/GaInP HBT ICs, with their combination of bipolar characteristics and III-V speed, enable us to fabricate all the chips with one technology, simplifying system partitioning and reducing the amount of technology qualification that needs to be performed. The volumes are such that a small team can fabricate and test the parts, leaving scope to explore other HBT IC applications while running products in the same fabrication facility.

Acknowledgments

We are proud to acknowledge the contributions of the GaAs IC development group and crystal growth and analysis departments in the Ottawa labs of Nortel Advanced Components, as well as the OC-192 team in Nortel Broadband and Nortel Microwave Modules.

References

1. Bellcore GR-253-CORE *Synchronous Optical Network (SONET) Transport Systems: Common Generic Criteria* and GR-1377-CORE *SONET OC-192 Transport System Generic Criteria.*
2. J. Sitch, C. Drèze, D. Pollex, K. Warbrick, K. Lowe, E. Best, T. Wilson, P. Corr, and G. Weston, "The use of III-V ICs in WDM optical network equipment", *IEEE GaAs IC Symp. Tech. Dig.*, 1995, pp. 177–180.
3. J. J. X. Feng, D. L. Pulfrey, J. Sitch, and R. Surridge, "A physics-based HBT SPICE model for large-signal applications", *IEEE Trans. Electron. Dev.* **ED-42** (1995) 8–14.
4. Q.-M. Zhang, J. Hu, J. Sitch, R. Surridge, and J.-M. Xu, "A new large-signal HBT model", *IEEE MTT-S Int. Microwave Symp. Dig.*, 1994, pp. 1253–1256.

5. Q.-M. Zhang, J. Hu, J. Sitch, R. Surridge, and J.-M. Xu, "A new large-signal HBT model", *IEEE Trans. Microwave Theory & Techniques* **MTT-44** (1996) 2001–2009.

6. T. Lester, R. Surridge, S. Eicher, J. Hu, G. Este, H. Nentwich, B. MacLaurin, D. Kelly, and I. Jones, "A manufacturable process for HBT circuits", *Inst. Phys. Conf. Ser. No.* **136** (GaAs & related compounds) (1993) 449–451.

7. T. Takahashi, S. Sasa, A. Kawano, T. Iwai, and T. Fujii, "High-reliability InGaP/GaAs HBTs fabricated by self-aligned process", *IEEE IEDM Tech. Dig.*, 1994, pp. 191–194.

8. R. J. Malik, L. M. Lunardi, R. W. Ryan, S. C. Shunk, and M. D. Feuer, "Submicron scaling of AlGaAs/GaAs self-aligned thin emitter heterojunction bipolar transistor (SATE-HBT) with current gain independent of emitter area", *Electron. Lett.* **25** (1989) 1175–1177.

9. T. Y. K. Wong, A. P. Freundorfer, B. C. Beggs, and J. Sitch, "A 10 Gb/s AlGaAs/GaAs HBT high power fully-differential limiting distributed amplifier for a III-V Mach-Zehnder modulator", *IEEE GaAs IC Symp. Tech. Dig.* 1995, pp. 201–204.

10. J. Sitch, "HBTs in telecommunications", *Solid-State Electron.* **41** (1997) 1397–1405.

11. C. Beaulieu, B. Beggs, J. Bennett, J. P. D. Cook, L. Hobbs, T. Lester, B. Oliver, and R. Surridge, "Degradation modes of GaAs heterojunction bipolar transistors and circuits fabricated in a GaInP emitter technology", *Proc. GaAs Rel. Workshop*, 1996, pp. 11–14.

12. W. A. Hagley, R. Rutyna, R. K. Surridge, and J.-M. Xu, "Investigation of emitter degradation in heterostructure bipolar transistors", *Inst. Phys. Conf. Ser. No.* **145** (Int. Symp. Compound Semiconductors) 1995, pp. 667–672.

13. See, for example, TRW or RF micro devices catalogs.

14. K. R. Nary, R. Nubling, S. Beccue, W. T. Colleran, J. Penny, and K.-C. Wang, "An 8-bit, 2 gigasample per second analog to digital converter", *IEEE GaAs IC Symp. Tech. Dig.*, 1995, pp. 303–306.

15. T. Y. K. Wong, J. Sitch, and S. McGarry, "A 10 Gb/s ATM data synchronizer", *IEEE GaAs IC Symp. Tech. Dig.*, 1995, pp. 49–51.

16. K. Lowe, "A GaAs HBT 16×16 10 Gb/s/channel cross-point switch", *IEEE J. Solid-State Circuits* **32** (1997) 1293–1298.

17. T. Yoshimasu, M. Akagi, N. Tanba, and S. Hara, "A low distortion and high efficiency HBT MMIC power amplifier with a novel linearisation technique for $\pi/4$ DQPSK modulation", *IEEE GaAs IC Symp. Tech. Dig.*, 1997, pp. 45–48.

18. H. Wang, K. W. Chang, L. Tran, J. Cowles, T. Block, D. C. W. Lo, G. S. Dow, A. Oki, D. Streit, and B. R. Allen, "Low phase noise millimeter-wave frequency sources using InP based HBT technology", *IEEE GaAs IC Symp. Tech. Dig.*, 1995, pp. 263–266.

International Journal of High Speed Electronics and Systems, Vol. 9, No. 2 (1998) 567–593

PRESENT STATUS AND FUTURE PROSPECTS OF HIGH-SPEED LIGHTWAVE ICS BASED ON INP

EIICHI SANO and KAZUO HAGIMOTO

NTT Optical Network Systems Laboratories,
1-1 Hikari-no-oka, Yokosuka-shi, Kanagawa 239-0847, Japan

YASUNOBU ISHII

NTT System Electronics Laboratories, 3-1 Morinosato Wakamiya,
Atsugi, Kanagawa 243-0122, Japan

High-speed integrated circuits (ICs) are essential for expanding the capacity of light-wave communications. InP-based heterostructure field effect transistors (HFETs) and heterojunction bipolar transistors (HBTs) are very promising for producing high-speed digital and analog ICs. This paper reviews the current status of InP-based lightwave communication ICs in terms of device, circuit, and packaging technologies. A successful 40-Gbit/s, 300-km optical fiber transmission using InP HFET ICs demonstrates the feasibility of the ICs. Furthermore, we estimate future IC performance based on the relationship between electron device figures-of-merit and IC speed. To keep up with the performance trend, technological problems, like inter- and intra-chip interconnections, have to be solved.

1. Introduction

Much effort is being devoted to expanding the transmission bit rates of backbone networks to realize multimedia communications systems. We have developed a 10-Gbit/s time division multiplexing (TDM) system and brought it into commercial use.[1] TDM systems based on fully digital technology are the most practical and the most mature in terms of accurate phase matching, reliability, and cost. On the other hand, optical time division multiplexing (OTDM) and wavelength division multiplexing (WDM) techniques potentially offer greatly improved performance: OTDM techniques can increase the transmission bit rate to 400 Gbit/s while WDM techniques yield throughputs beyond 1 Tbit/s.[2-5] However, these remarkable increases in transmission capability are based on high-speed electronic IC technology, and some breakthroughs are needed before practical systems can be built: OTDM systems have difficulty in phase matching, and WDM systems require significant advances in photonic devices. Therefore, at present it is very useful to increase the single-channel bit rate through progress in TDM techniques. One of the major issues in implementing an ultra-high speed TDM system is the development of ultra-high speed electronic ICs. Recently, several 40-Gbit/s class ICs using Si bipolar, GaAs MESFETs, GaAs HBTs, GaAs HFETs, InP HFETs, and InP HBTs have been reported.[6–11] Among them, InP-based devices are very promising because of the excellent electron transport properties of the material systems and the capability of monolithic integration with photonic devices.

This paper reviews the current status of InP-based lightwave communication ICs in terms of device, circuit, and packaging technologies. Successful 40-Gbit/s, 300-km optical fiber transmission using InP HFET ICs demonstrates the feasibility of these ICs. Furthermore, we estimate future IC performance based on the relationship between the electron device figure-of-merit and IC speed. Technological problems that have to be overcome to reach the 100 Gbit/s region, like inter- and intra-chip interconnections, are also addressed.

2. Requirements for Lightwave Communication ICs

Figure 1 depicts the basic transmitter and receiver configuration for lightwave communications systems. The transmitter block consists of a laser diode (LD), an optical modulator (MOD), a modulator driver (DRV), and multiplexers (MUXs). The receiver block consists of a photodiode (PD), a preamplifier (Pre), a baseband amplifier (Base), a decision circuit (DEC), demultiplexers (DEMUXs), and a clock recovery circuit that includes a differentiator (DIF), a rectifier (REC), a resonator (RES), and a limiting amplifier (Limit). Er-doped fiber amplifiers (EDFAs) are used as a booster amplifier in the transmitter and a preamplifier in the receiver. The transmitter and receiver ICs, except for the clock recovery block, require broadband operation from near DC to the maximum bit rate with good eye openings. The advent of EDFA has drastically changed the design methodology of the equalizing amplifiers including the Pre and Base. Minimum detectable optical power is not determined by Pre noise but by EDFA noise. This means that low-noise characteristics are not critically important. Of course, the gain requirement of the amplifiers has been relaxed. Nevertheless, a regenerating function, performed by DEC and

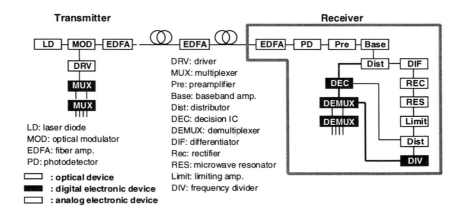

Fig. 1. Basic transmitter and receiver configuration for lightwave communication systems.

the clock recovery circuit, is still needed to reduce the timing jitter produced by the cascaded EDFAs.

Emitter-coupled logic (ECL) and source-coupled FET logic (SCFL) are widely used in high-speed digital ICs because of their advantageous series-gate configuration. The propagation delay of an ECL inverter is approximated by

$$t_{pd} \approx \frac{1}{2\pi}\left[\left(\frac{1}{f_T} + \sqrt{\frac{2V_{SW}}{0.15}}\frac{1}{f_{max}}\right) + \left(2 + \frac{V_{SW}}{0.15}\right)\frac{f_T}{4f_{max}^2}\right], \qquad (1)$$

where f_T is the unity current gain cutoff frequency, f_{max} is the maximum oscillation frequency, V_{SW} is the logic swing.[12] For simplicity, we assume the figure-of-merit for bipolar transistors can be given by

$$f_{fom} = 2\left(\frac{1}{f_T} + \frac{1}{f_{max}}\right)^{-1}. \qquad (2)$$

On the other hand, the propagation delay of an SCFL inverter is approximated by

$$t_{pd} \approx \frac{C_{gs}}{g_m} + (2 + 3A)\frac{C_{gd}}{g_m} \propto \frac{C_{g0}}{g_m} \propto \frac{1}{f_T}, \qquad (3)$$

where C_{gs} is the gate-to-source capacitance, C_{gd} is the gate-to-drain capacitance, g_m is the transconductance, A is the inverter gain, and C_{g0} is the zero bias gate capacitance.[12] Then we apply f_T to the figure-of-merit for FETs.

D type flip-flops (D-F/Fs) and baseband amplifiers are critical digital and analog circuit components, respectively, for the lightwave communication ICs shown in Fig. 1. Figure 2 shows the maximum operating speeds of D-F/Fs reported in the literature as a function of device figures-of-merit.[13] A clear dependence on the device figure-of-merit is obtained. The difference between Si bipolar and HBT circuits might be due to the differences in the logic swings and the circuit optimization levels.[14] The evolution of device technologies enhances the IC performance following

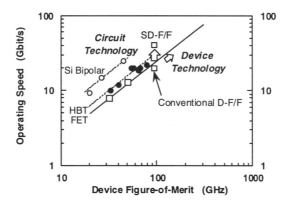

Fig. 2. Relationship between device figures-of-merit and maximum operating speeds of D-F/Fs.

the trend line, while circuit design technologies can boost it in the vertical direction. When the same figures-of-merit are used for the baseband amplifier bandwidths, we get the relationship shown in Fig. 3. An f_T and f_{max} of over 150 GHz for HBTs and an f_T of over 200 GHz for FETs are required for 40-Gbit/s operation in D-F/Fs and baseband amplifiers. Such a device performance can be obtained only by using state-of-the-art device and process technologies. Our 0.1-μm gate InP HFETs can attain 40 Gbit/s operation in on-wafer measurement. Considering speed loss due to IC packaging, and temperature and supply-voltage deviations, however, faster operation is needed. Therefore, circuit design technologies, like a super-dynamic flip-flop (SD-F/F) and a distributed baseband amplifier (DBA), are very important for achieving 40-Gbit/s class ICs.[15,16]

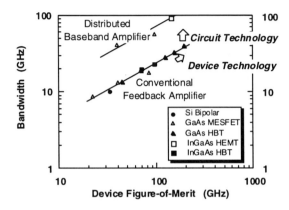

Fig. 3. Relationship between device figures-of-merit and bandwidths of baseband amplifiers.

3. Device Technology

3.1. *HFET*

Shortening the gate to reduce the electron transit time is an effective way to boost FET speed. Figure 4 shows the relationship between gate length L_g and unity current gain cutoff frequency f_T for FETs. Effective electron confinement and the epitaxially grown thin channel layer enable us to shorten the gate to around 50 nm while keeping a sufficient aspect ratio for InP-based HFETs. Because of superior electron transport properties and the ultra-short gate, a high f_T of more than 340 GHz has been attained for an InP HFET.[17] In the longer gate region, f_T linearly increases with decreasing L_g, and as L_g becomes shorter, f_T tends to saturate when parasitic effects are not negligible. The saturating points are 0.1 μm for InP HFETs and 0.2 μm for GaAs HFETs and MESFETs.

    Figure 5 shows a schematic cross-sectional view of the InP HFET developed by NTT.[18] The epitaxial layers were grown by metalorganic chemical vapor deposition (MOCVD). The most important feature of the heterostructure is the InP recess etch stopper introduced in the InAlAs barrier layers. In a conventional etching process,

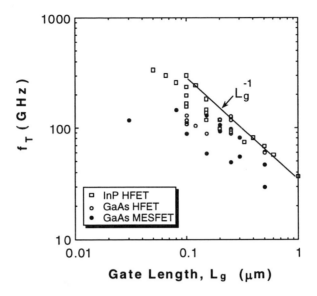

Fig. 4. Relationship between gate length and unity current gain cutoff frequency for FETs.

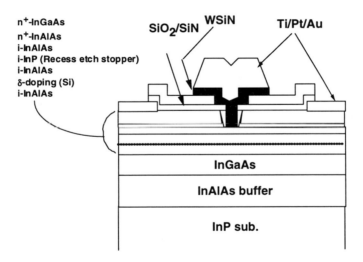

Fig. 5. Cross-sectional view of the InP HFET used in high-speed ICs.[18]

the gate recess depth is controlled by means of etching time, which results in a large threshold voltage scattering. InGaAs and InAlAs are etched more than 400 times more easily than InP, so recess etching is almost automatically terminated at the surface of the InP etch stopper. The standard deviations of the threshold voltages in 2-inch wafers were improved from 100 mV to 25 mV.[18] The T-shaped gate was made using an electron beam direct writer for the gate footprint and a g-line stepper for the top part of the gate electrode. Typical f_T and $f_{\max}$ for 0.1-μm gate HFETs are 200 GHz and 250 GHz, respectively.

3.2. *HBT*

InP-based HBTs have been widely investigated because the material systems provide potential advantages, like short electron transit times, low base resistances,
and low voltage operation owing to the narrow InGaAs energy gap, in addition to
the high current drivability inherent to bipolar transistors. Furthermore, the base
and collector regions in InP-based HBTs are similar to those of pin photodiodes
(*pin*-PDs), so a monolithically integrated photoreceiver can be obtained without
modifying the HBT fabrication process.[19]

Figure 6 shows a schematic cross-sectional view of the InP HBT developed by
NTT.[20] The epitaxial layers were grown by MOCVD. The InGaAs/InP composite
collector enables us to shorten the electron transit time while keeping the breakdown
voltage high (over 8 V). The n^+-InP subcollector is indispensable for reducing the
thermal resistance of the HBTs.[21] The thermal conductivity of InGaAs is one-
order lower than that of InP. A thermal simulation indicated that the junction
temperature of an HBT with an InGaAs subcollector is 50 degrees higher than that
of an HBT with an InP subcollector.[21] This might result in a poor device lifetime
of HBTs with an InGaAs subcollector.

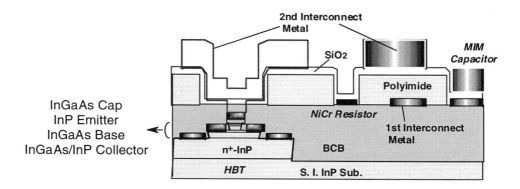

Fig. 6.   Cross-sectional view of the InP HBT.[20]

A self-aligned fabrication process was adopted to reduce the external base resistance and base-to-collector capacitance.[22] We started the fabrication process by
lifting off the emitter metal evaporated on the InGaAs cap layer. The emitter metal
was used as a mask for the emitter mesa etching, which is performed with a combination of electron cyclotron resonance (ECR) dry etching and wet chemical etching.
The wet etching formed an undercut approximately 0.2-μm deep around the emitter
mesa. Base electrode metal was then evaporated over the region including the entire
emitter mesa. Owing to the undercut, the base metal, emitter mesa, and emitter
metal all self-aligned without short-circuiting the emitter and the base electrodes.
Next, base-collector mesa etching was performed, where the mesa was defined by
the base metal. Then, collector metal was evaporated onto the subcollector, and

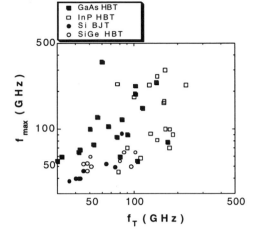

Fig. 7. f_T versus $f_{\max}$ characteristics for bipolar transistors.

a third mesa etching was carried out to isolate the devices. A hexagonal emitter geometry enabled us to produce an HBT with a submicrometer emitter width and to obtain both f_T and $f_{\max}$ of over 220 GHz.[23] Benzocyclobutene (BCB) films were used for device passivation and planarization of the mesa structure. Photosensitive polyimide films were used as the insulator to reduce the parasitic capacitances between the two-level interconnections. Figure 7 shows f_T versus $f_{\max}$ characteristics for InP HBTs along with other bipolar transistors reported in the literature. The advantage of our InP HBTs is clearly illustrated.

4. Circuit Technology

4.1. *Digital ICs*

A flip-flop (F/F) and 2:1 selector (SEL) are basic circuit components for constructing high-speed digital ICs. F/Fs are employed in DEC, DIV, MUX, and DEMUX shown in Fig. 1, while SELs are used in MUX. The maximum operating speed of D-F/Fs is lower than that of SELs when the same device technology is used. In fact, the SEL fabricated by 0.2-μm gate GaAs MESFETs can be operated at 25 Gbit/s, while the D-F/F can operate only at 13 Gbit/s.[14] Although an excellent operation at 50 Gbit/s has been obtained for a Si bipolar SEL, the maximum operating speed of Si bipolar D-F/Fs remains at 25 Gbit/s.[24] This is due to the fact that a clock frequency for SELs is a half that for D-F/Fs when the same bit rate is achieved. Improving the F/F speed is the key to obtaining faster digital circuits. An F/F with double emitter followers and a high-speed latching operation (HLO) F/F were used to do this.[24,25] To further improve speed, we devised the super-dynamic F/F (SD-F/F) shown in Fig. 8.[15] The circuit features are: (1) a series-gate connection to separate the current paths of the reading and latching circuits; (2) a latching current that is smaller than the reading current; and (3) a

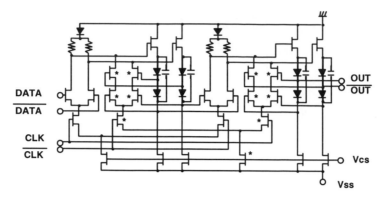

* FET with smaller gate width

Fig. 8. Circuit configuration of super-dynamic D-F/F.[15]

source-coupled negative feedback pair (SCNFP) inserted in the first-level latching differential pair in a cascade manner. The SCNFPs drastically reduce the effective logic swing without any degradation in the signal transition slew rate. This is the key to increasing F/F speed. Details of the circuit operation are described in another paper in this issue.[26] An SD D-F/F can operate 100% faster than a conventional master-slave (MS) D-F/F.[15,26] To compensate for source follower loss in the high frequency region, capacitive peaking was used in the source followers.

The basic circuit configuration of a selector is shown in Fig. 9. Data 1 and data 2 are alternatively selected by the clock signal, which results in a 2:1 multiplexing. A cascade connection and a combination of inductor peaking and parallel feedback techniques have been proposed to improve the maximum operating speed of SELs.[26,27]

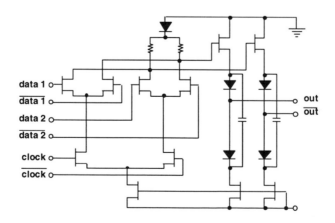

Fig. 9. Circuit configuration of SEL.

Once the speed of the core circuits has been improved, input and output interface circuits can still limit the maximum operating speed of the entire IC. Simple emitter followers and source followers are often used for input buffers under the differential input condition.[24,28] In the frequency region of over 40 Gbit/s, however, it is very difficult to realize an IC-to-IC interconnection in the differential mode without a phase shifter. Therefore, we adopted a single-ended interconnection scheme and used several analog circuit techniques for widening the bandwidths of the input amplifiers. The wideband data buffer we used in InP HFET digital ICs consists of a two-stage differential buffer. Figure 10(a) shows the first stage of the data buffers.[29] The data buffers employ capacitive feedback at the differential buffer and capacitive peaking at the source follower.[30] Capacitive feedback cancels the influence of the gate-to-drain capacitance, and capacitive peaking compensates the source follower loss in the high-frequency region. As the feedback capacitance, gate-to-drain and -source capacitance in an HFET was used. The wideband data amplifier used also employs the same configuration except for the source follower at the input stage. A performance comparison of this wideband buffer and a conventional buffer is shown in Fig. 10(b). The bandwidth of the conventional buffer is insufficient for 40-Gbit/s

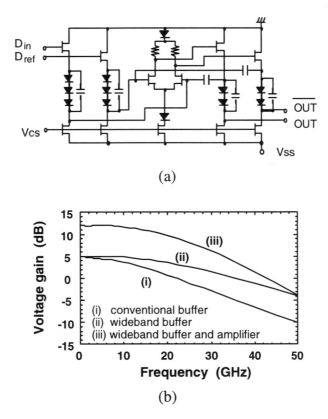

Fig. 10.   Wideband data buffer.   (a) Circuit configuration.   (b) Simulated gain-bandwidth characteristics.[29]

operation and its voltage gain is not so high. On the other hand, the wideband buffer offers adequate bandwidth; the gain is still not so high. The data buffer with wideband amplifiers increases the gain to 12 dB, and its bandwidth is 40% wider than the conventional ones.

The circuit configuration of the clock buffer is shown in Fig. 11(a).[29] The clock buffer consists of a two-stage inductor peaking differential buffer. A capacitively coupled resistive divider was introduced as a low-loss, passive RF level shifter instead of source followers. The simulated frequency characteristics of the clock buffers are shown in Fig. 11(b). The clock buffer using a meander inductor has a wider bandwidth and is suitable for higher bit-rate operation. Another buffer using a spiral inductor has a higher voltage gain and is expected to achieve shorter rise and fall times at the buffer output around 40 GHz. The meander inductors were used in the DEC and DIV for high speed operation, and the spiral inductors were used in the MUX and DEMUX for better retiming performance. Due to the capacitive

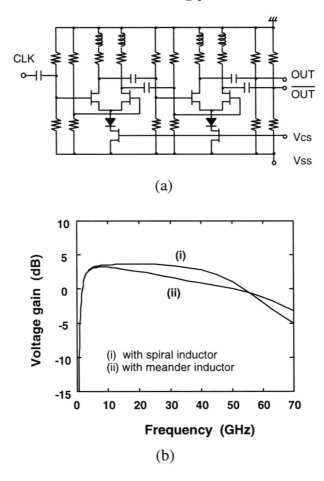

(a)

(b)

Fig. 11. Wideband clock buffer. (a) Circuit configuration. (b) Simulated gain-bandwidth characteristics.[29]

coupling, the minimum clock frequency offering adequate amplitude is about 1 GHz. It is found from the comparison between Figs. 10(b) and 11(b) that this clock buffer gives 10 dB higher gain than a conventional buffer at 50 GHz.

Other circuit configurations, like a parallel feedback amplifier and a source follower with a bypass capacitor, are employed to enlarge the bandwidth of input buffers.[26,31] The data buffer shown in Fig. 10 was also used as an output buffer in our HFET ICs.

At frequencies over 50 GHz, interconnection is a severe problem, even in an IC chip. The maximum interconnection length in our HFET MUX IC is 400 μm, which corresponds to around two-tenths of the wavelength at 50 GHz.[32] This means that the distributed element nature should be taken into account even in IC design. To reduce the interconnection propagation delay and to obtain an impedance-matched microstrip line, we introduced a high-speed interconnection process using two metal layers consisting of Au and a 2-μm-thick BCB film as an insulator.[32]

4.2. *Analog ICs*

As described in Sec. 2, it is very difficult to achieve a stable 40-Gbit/s operation using a conventional feedback amplifier. To overcome the problem, we introduced a distributed baseband amplifying (DBA) methodology.[16] Distributed amplifiers are commonly used in millimeter wave applications. Their great merit is that they can significantly widen the bandwidth. The parasitic capacitance of FETs is effectively utilized as a circuit element to enable velocity-matched signal transmission between the gate input and drain output lines. However, we run into a serious problem when we try to apply this design to baseband amplifiers; namely, the drain conductance of the FETs degrades the DC gain. To cope with this, a novel frequency-dependent bias termination was devised to compensate for the DC gain (Fig. 12).[16] Another wideband design we devised is the drain peaking line and the loss compensation circuit. The effects of these circuit design techniques are schematically depicted in Fig. 12.

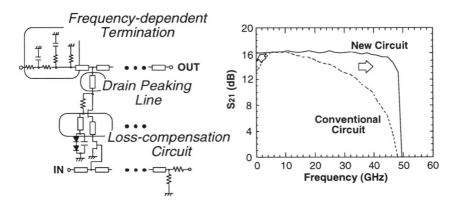

Fig. 12. Circuit configuration of distributed baseband amplifier and effects of the circuits.[16]

4.3. *Packaging*

Packaging is another key technology for attaining ultra-broadband signal transmission without undesirable losses due to cavity and/or parasitic resonance and coupling. The basic structure we adopted is the so-called "chip-size cavity" package that minimizes the inner cavity.[33] We developed an improved version of this package.[34] Figure 13 shows the package structure for analog ICs. The RF ports are made with V-band coaxial connectors. The digital-type package can accommodate up to 6 RF ports. A die is mounted on an inner thin-film multilayer interconnection substrate by means of flip-chip bonding or ribbon bonding. The upper and lower metal lids reduce the size of the inner cavity so as to shift undesirable cavity resonance out of the transmission band. The module with metal lids had a high isolation of more than 60 dB up to 50 GHz. The result agrees well with electromagnetic field analysis.[34] A microphoto of the ribbon-bonded InP HFET MUX chip is shown in Fig. 14.

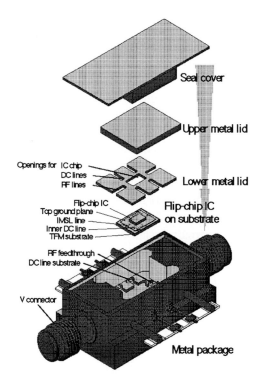

Fig. 13. A view of broadband chip-size cavity package.[34]

5. IC Performances

High-speed lightwave communication ICs were designed and fabricated using the technologies described above. In this section, we present our IC performances compared with other InP-based ICs reported in the literature.

Fig. 14. Microphoto of ribbon-bonded HFET MUX chip.

Fig. 15. Chip microphoto of InP HFET DBA.[35]

5.1. *Ultra-high speed ICs*

Figure 15 shows a chip microphoto of fabricated HFET distributed baseband amplifier (DBA). The chip size is 1×2 mm^2. Scattering (S) parameters for the DBA were measured on a wafer using a network analyzer up to W-band. Figure 16 shows the results. The DBA has a gain of 10 dB with an 89.2 GHz bandwidth. Figure 17 shows the relationship between device figures-of-merit and bandwidths of baseband amplifiers. The advantage of DBAs over conventional feedback amplifiers are clear. A 1-50-GHz bandwidth was attained for an InP HBT distributed amplifier.[11] DC characteristics of the amplifier can be improved by using our frequency-dependent termination circuit.

Figure 18(a) shows the circuit configuration of a fabricated HFET DEC.[29] The DEC consists of the circuit components described in the previous section and was mounted in the digital-type package. Figure 18(b) shows observed eye diagrams for

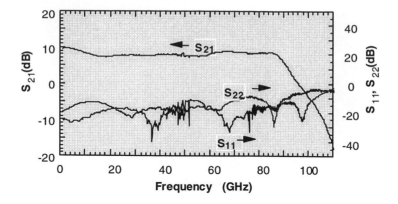

Fig. 16. Measured S parameters for DBA.[35]

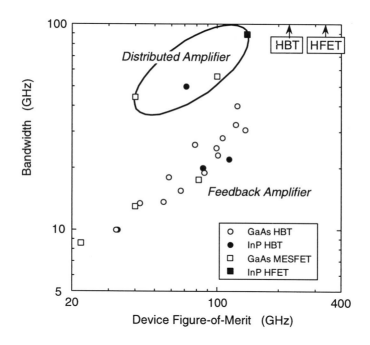

Fig. 17. Relationship between device figures-of-merit and bandwidths of baseband amplifiers.

a non-return-to-zero (NRZ) 46-Gbit/s, $2^{15} - 1$ pseudo-random bit stream (PRBS). The input PRBS was generated by an HFET MUX module. An error-free operation was confirmed from 15 Gbit/s to 46 Gbit/s.[29] The maximum speed of the DEC is limited by the measurement setup. Figure 19 shows the relationship between device figures-of-merit and maximum operating speed of D-F/Fs. The advantage of SD D-F/Fs over conventional master-slave (MS) D-F/Fs is clearly demonstrated. The maximum operating speed reported for InP HBT MS D-F/Fs still remains at 20 Gbit/s.[11] This might be limited by the measurement setup bandwidth.

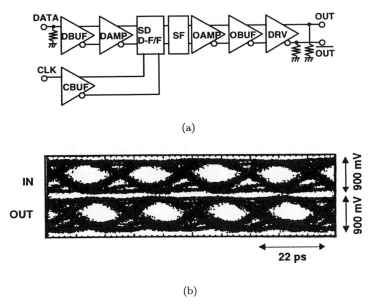

(a)

(b)

DBUF: data input buffer, DAMP: data input amplifier, CBUF: clock input buffer, SD D-F/F: super-dynamic D-F/F, SF: source follower, OAMP: output amplifier, OBUF: output buffer, DRV: driver.

Fig. 18. (a) Circuit configuration of fabricated HFET DEC. (b) Eye diagrams of DEC at 46 Gbit/s.[29]

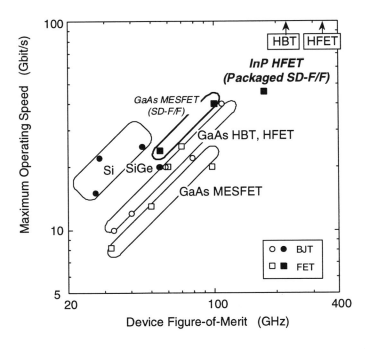

Fig. 19. Relationship between device figures-of-merit and maximum operating speeds of D-F/Fs.

Exclusive ORs (EXORs) are used for REC in the clock recovery circuit shown in Fig. 1. We introduced an inductor peaking technique in the core circuit to get higher EXORs.[36] Figure 20 shows the experimental result for a 40-GHz timing extraction from an NRZ 40-Gbit/s PRBS by using fabricated HFET EXOR.[36] This is the first successful demonstration of 40-GHz timing extraction using InP devices.

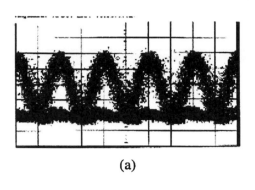

(a)

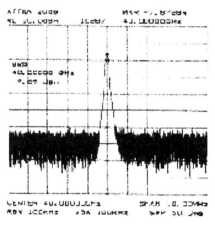

(b)

Fig. 20. Experimental results for a 40-GHz timing extraction.[36] (a) Output waveform from EXOR. (b) Frequency spectrum.

Figure 21(a) shows the circuit diagram of fabricated HFET 1:2 DEMUX.[29] A combination of an MS D-F/F and a tristage (Tri) D-F/F enables us to enlarge the phase margin.[14] The DEMUX was also designed using the concept as described in the previous section and was mounted in the digital-type package. Figure 21(b) shows observed DEMUX operation for an NRZ 40-Gbit/s, $2^{23} - 1$ PRBS.[29] Error-free operation was confirmed up to 40 Gbit/s. Although a true experiment has not been performed, the possibility of the 40-Gbit/s DEMUX operation was indicated for an InP HBT IC.[37]

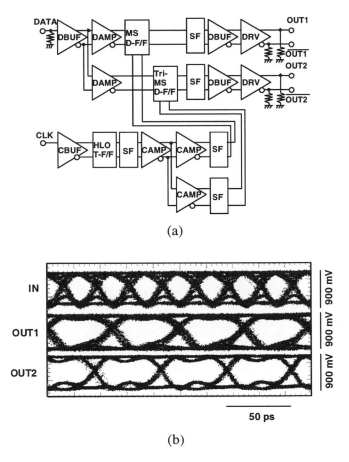

(a)

(b)

HLO T-F/F: high-speed latching operation T-F/F, Tri-MS D-F/F: tristage MS D-F/F.

Fig. 21. (a) Circuit configuration of fabricated HFET DEMUX. (b) Eye diagrams of DEMUX at 40 Gbit/s.[29]

Figure 22 shows the circuit diagram of a fabricated HFET 2:1 MUX.[29] The MUX was mounted in the package described above and a clear eye-opening of 52-Gbit/s output was obtained.[29] This MUX employs a conventional interconnection methodology that does not consider the distributed nature of the IC chip. To improve the maximum operating speed, we introduced impedance-matched microstrip lines from the data input buffer to the SEL core and from the clock input buffer to the SEL core.[32] Figure 23 shows observed eye diagrams for the HFET 2:1 MUX on wafer.[32] Although the bandwidth of the measurement setup is limited to 40 GHz, clear eye openings are obtained for 80-Gbit/s output data. This is the highest bit rate ever reported among all digital electronics. Figure 24 shows the relationship between device figures-of-merit and maximum operating speed of MUX ICs. A clear eye opening at 40-Gbit/s was obtained for an InP HBT MUX IC.[11]

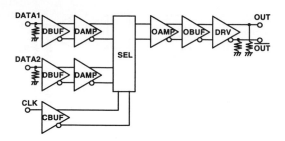

Fig. 22. Circuit configuration of fabricated HFET 2:1 MUX.[29]

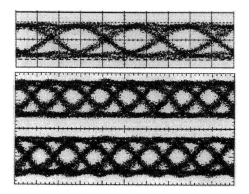

Fig. 23. Eye diagrams of HFET MUX at 80 Gbit/s.[32] Upper: input data, middle: OUT, lower: /OUT.

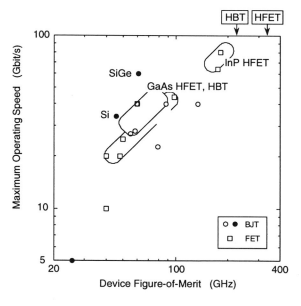

Fig. 24.   Relationship between device figures-of-merit and maximum operating speeds of MUX ICs.

Table 1. Maximum operating speed and power dissipation of InP-based ICs.

	HFET on wafer		HFET module		HBT on wafer	
baseband amplifier feedback amp.	10 dB, 90 GHz	—	—	—	12.7 dB, 37 GHz	54 W[35]
DBA	—	0.86 W[35]	9 dB, 60 GHz	0.66 W[33]	> 20 Gbit/s	—
decision	—	—	46 Gbit/s	1.7 W[29]	—	—
exclusive OR	40 GHz	1.7 W[36]	—	—	—	—
limiting amplifier	40 GHz	70 mW[39]	—	—	—	—
1:2 DEMUX	40 Gbit/s	3.8 W[29]	40 Gbit/s	3.8 W[29]	40 Gbit/s ?	2.5 W[37]
divider	46 GHz	1.1 W[29]	45 GHz	1.1 W[29]	39.5 GHz	—[40]
2:1 MUX	80 Gbit/s	2.7 W[32]	52 Gbit/s	2.2 W[29]	> 40 Gbit/s	1.6 W[11]
modular driver	—	—	—	—	30 Gbit/s, 2.2 V	1.8 W[37]

273

Table 1 summarizes the speed performances of InP-based ICs reported to date.

5.2. *Applications of InP HBTs to low-power ICs and OEICs*

As described in Sec. 3, InP HBTs are promising candidates for low-power ICs and monolithic photoreceivers. Figure 25 shows the circuit configuration and chip microphoto of a regenerative receiver IC fabricated by our HBTs as shown in Fig. 6.[41] The IC was constructed with a pre-amplifier, post-amplifier, automatic offset controller (AOC), PLL-based timing recovery circuit, and D-F/F. The timing recovery circuit includes a 90-degree delay, phase detector, low pass filter (LPF), and voltage-controlled oscillator (VCO). To reduce the power dissipation, HBTs with a submicrometer effective emitter width were employed. The chip is 1.6×1.6 mm^2. A 20-Gbit/s error-free operation for an input dynamic range of 13 dB was achieved with a power dissipation of only 0.6 W. This indicates that InP HBTs are very attractive for relatively large scale integration.

A *pin*-PD can be obtained without modifying our HBT fabrication process. A monolithically integrated photoreceiver, constructed with a *pin*-PD and a transimpedance preamplifier, has demonstrated the feasibility of 40-Gbit/s optical-to-electrical conversion.[37] Figure 26 shows observed eye diagrams of 40-Gbit/s RZ

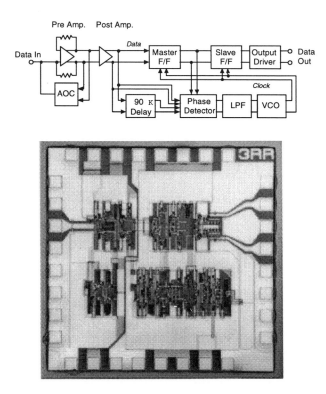

Fig. 25. Circuit configuration and chip microphoto of regenerative receiver IC.[41]

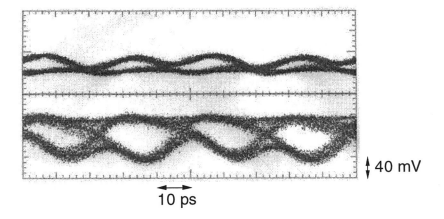

40 mV

10 ps

Fig. 26. Eye diagrams from photoreceiver and commercial PD. Upper: commercial PD, lower: photoreceiver.

PRBS from the photoreceiver and a commercial 40-GHz PD. This is the first observation of an eye opening at 40 Gbit/s for OEICs. The bandwidth of the *pin*-PD was 36 GHz. The preamplifier had a gain of 12.7 dB with a bandwidth of 37 GHz and consumed only 54 mW. Furthermore, a decision circuit with a *pin*-PD was successfully fabricated with InP HBTs.[42] A 17-Gbit/s error-free operation was confirmed with a power dissipation of 620 mW. This digital photoreceiver is suitable for use as a simple optical receiver using a high-power EDFA.[43]

6. Application of InP HFET ICs to 40-Gbit/s Optical Fiber Transmission

InP HFET IC modules were successfully used for the first 40-Gbit/s electrically multiplexed/demultiplexed optical fiber transmission.[44] Figure 27 shows the experimental setup.[29] Four-channel 10-Gbit/s, $2^7 - 1$ PRBS signals from a PPG were multiplexed to 40 Gbit/s using two GaAs MESFET MUXs and the HFET MUX module. The 40-GHz optical pulse train was directly generated by a monolithic mode-locked laser diode (ML-LD) module and was encoded by a Mach Zehnder modulator.[45] The modulated 40-Gbit/s RZ optical signal was amplified by an EDFA and injected into the transmission line. The transmission line comprised four 75-km long dispersion shifted fibers (DSFs) connected by three EDFAs. Each EDFA accommodated a dispersion compensator to cancel the dispersion in each section. The transmitted signal was amplified by another EDFA, followed by a dispersion equalizer module.[46] The signal was received by the newly developed optical receiver module which comprised a waveguide type *pin*-PD and a GaAs MESFET DBA.[47] The received signal was fed into the HFET DEC module, then demultiplexed to 10 Gbit/s by another HFET DEC module and a Si bipolar DEC. The bit error rate (BER) performance of the four 10-Gbit/s channels in the 40-Gbit/s data stream was measured by adjusting the clock timing of the DECs and is shown in Fig. 28. The average receiver sensitivities at the BER of 10^{-9} before and after transmission

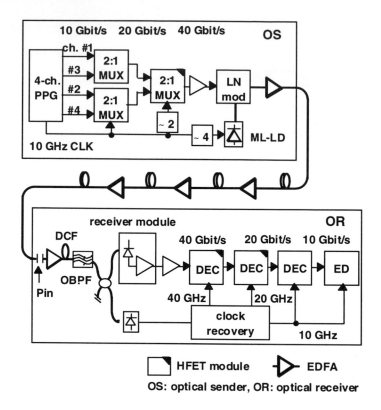

Fig. 27. Experimental setup for 40 Gbit/s, 300 km transmission.[29]

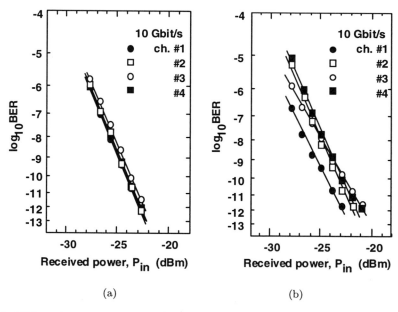

(a)   (b)

Fig. 28. BER performance at 40 Gbit/s. (a) Back-to-back. (b) After 300-km transmission.[29]

were −24.8 dBm (standard deviation $\delta = 0.24$ dB) and −24.3 dBm ($\delta = 0.6$ dB), respectively. This experiment demonstrated that InP HFET ICs are very effective for enlarging the optical transmission capacity.

7. Future Prospects

The high potential of InP-based devices has been confirmed, as described in the previous section. Compared with Si and GaAs-based devices, however, InP-based heterostructure devices are still immature, and there remain some issues to be settled for system application. Epitaxial growth, which is the key to making the heterostructures, needs to progress further. Devices reliability should also be investigated and confirmed for practical use.

As can be found from Table 1, progress in optical modulator drivers having a large output voltage swing lags behind that of other lightwave communications ICs. This is because of the tradeoff between breakdown voltage and RF performance of transistors. Also, Eq. (1) clearly indicates that the propagation delay increases with increasing voltage swing. Figure 29 plots modulator bandwidth f_{mod} and driver speed f_{drv} versus driving voltage V_{drv}. Modulator and driver figures-of-merit are expressed as f_{mod}/V_{drv} and $f_{drv}V_{drv}$, respectively. The two lines in Fig. 29 indicate the highest figures-of-merit for modulators and drivers reported to date. Their intersection shows the highest obtainable bit rate for optical modulation. The most reasonable way to increase the bit rate would seem to be to develop devices using a

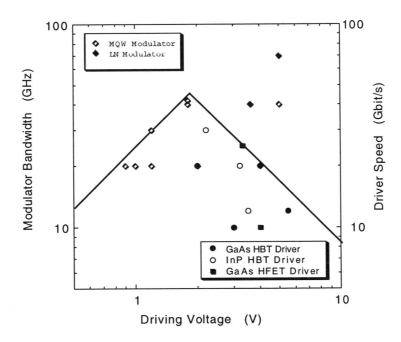

Fig. 29. Modulator bandwidths and driver speeds versus driving voltage characteristics.

driving voltage of around 2 V. InP-based devices are candidates in terms of driving voltage and extensive work should be conducted.

In Figs. 17, 19, and 24, the maximum device figures-of-merit obtained by InP-based HFETs and HBTs are indicated by the arrows. From these figures, one might predict that a 100-Gbit/s operation could be achieved by using devices with the highest figures-of-merit. To keep up with the estimation, however, several technological problems have to be solved.

(a) Almost all the data in Figs. 17, 19, and 24 were obtained for on-wafer measurements using RF probes. IC packaging degrades the IC performance obtained on wafer. In fact, the maximum speed of our divider module is slightly lower than that for the on-wafer measurement as listed in Table 1.

(b) Our digital-type can be applied to a frequency region up to 50 GHz, which corresponds to a cutoff frequency of the higher-order mode of the coplanar waveguide (CPW) on the outer substrate shown in Fig. 14. A finer pattern formation technology for CPWs is required to shift the cutoff frequency to a higher one. The cavity resonance becomes serious with increasing operating speed. If the concept of "chip-size cavity" package continues to be adopted, the chip size must be reduced to below 1 mm at 100 GHz.

(c) In general, the circuit delay time determining the maximum operating speed includes the intrinsic delay time inherent to circuit elements and the propagation delay time of the electromagnetic field in the interconnection line between the elements. The latter is not taken into account in Eqs. (1) and (3) because it can be neglected for ordinary small-scale integrated circuits in the frequency region below 40 GHz. If the interconnection technology is unchanged, the trend lines in Figs. 17, 19, and 24 tend to saturate in a higher frequency region where the propagation delay cannot be neglected. Transistor sizes and interconnection line widths/spacings should be reduced without increases in the parasitic resistance of the transistors and the line resistance.

The question then is whether these problems can be solved by adopting a new approach. The answer is unclear, but an attempt is being made to produce high-speed, smart ICs having optical-to-electrical and serial-to-parallel (O/E&S/P) conversion function by using InP-based uni-traveling-carrier photodiodes and resonant tunneling diodes.[48] Ultra-fast optoelectronic transmitters are the key to the success of 100-Gbit/s-class TDM lightwave communications systems.

8. Conclusions

We have reviewed the current status of InP-based lightwave communications ICs in terms of device, circuit, and packaging technologies. All the ICs, except the modulator driver, have achieved 40-Gbit/s operation by using 0.1-μm gate InP HFET, high-speed design, and broadband packaging technologies. 40-Gbit/s, 300-km optical fiber transmission was successfully carried out using InP HFET ICs. The feasibility of high-speed, low-power ICs and OEICs has been confirmed with InP HBTs. Furthermore, we estimated future IC performance based on the relationship between electron device figures-of-merit and IC speeds. Technological problems that have to be overcome to reach 100 Gbit/s region, like inter- and even intra-chip interconnections, have been addressed.

Acknowledgments

The authors would like acknowledge I. Kobayashi, S. Horiguchi, and K. Hirata for continuous support and encouragement throughout this work.

References

1. K. Hagimoto, "Experimental 10 Gbit/s transmission systems and its IC technology", *GaAs IC Symp. Tech. Dig.*, 1993, pp. 7–10.
2. S. Kawanishi, H. Takara, T. Morioka, O. Kamatani, K. Takiguchi, T. Kitoh, and M. Saruwatari, "400 Gbit/s TDM transmission of 0.98 ps pulses over 40 km employing dispersion slope compensation", *OFC'96 Tech. Dig.*, 1996, PD 24.
3. M. Nakazawa, K. Suzuki, E. Yoshida, E. Yamada, T. Kitoh, and M. Kawachi, "160 Gbit/s soliton data transmission over 200 km", *Electron. Lett.* **31** (1995) 565–566.
4. Y. Yano, T. Ono, K. Fukuchi, T. Itoh, H. Yamazaki, M. Yamaguchi, and K. Emura, "2.6 terabit/s WDM transmission experiment using optical duobinary coding", *Proc. ECOC'96*, 1996, pp. 5.3–5.6.
5. H. Onaka, H. Miyata, G. Ishikawa, K. Otsuka, H. Ooi, Y. Kai, S. Kinoshita, M. Seino, H. Nishimoto, and T. Chikama, "1.1 Tb/s WDM transmission over 1 150 km, 1.3 μm zero-dispersion single-mode fiber", *OFC'96 Tech. Dig.*, 1996, PD19-1.
6. A. Felder, M. Moller, J. Popp, J. Bock, and H.-M. Rein, "46 Gb/s DEMUX, 50 Gb/s MUX, and 30 GHz static frequency divider in silicon bipolar technology", *IEEE J. Solid-State Circuits* **31** (1996) 481–486.
7. K. Murata, T. Otsuji, M. Yoneyama, and M. Tokumitsu, "A 40 Gbit/s super-dynamic Decision IC using 0.15-μm GaAs MESFET", *MTT-S IMS Tech. Dig.*, 1996, pp. 465–468.
8. K. Runge, P. J. Zampardi, R. L. Pierson, P. B. Thomas, S. M. Beccue, R. Yu, and K. C. Wang, "High speed AlGaAs/GaAs HBT circuits for up to 40 Gb/s optical communication", *GaAs IC Symp. Tech. Dig.*, 1997, pp. 211–214.
9. M. Berroth, M. Lang, Z.-G. Wang, Z. Lao, A. Thiede, M. Rieger Motzer, W. Bronner, G. Kaufel, K. Kohler, A. Hulsmann, and J. Schneider, "20-40 Gbit/s 0.2 μm GaAs HEMT chip set for optical data receiver", *GaAs IC Symp. Tech. Dig.*, 1996, pp. 133–136.
10. T. Otsuji, E. Sano, Y. Imai, and T. Enoki, "40-Gbit/s ICs for future lightwave communications systems", *GaAs IC Symp. Tech. Dig.*, 1996, pp. 14–17.
11. H. Suzuki, K. Watanabe, K. Ishikawa, H. Masuda, K. Ouchi, T. Tanoue, and R. Takeyari, "InP/InGaAs HBT ICs for 40 Gbit/s optical transmission systems", *GaAs IC Symp. Tech. Dig.*, 1997, pp. 215–218.
12. E. Sano, Y. Matsuoka, and T. Ishibashi, "Device figure-of merits for high-speed digital ICs and baseband amplifier", *IEICE Trans. Electron.* **E78-C** (1995) 1182–1188.
13. E. Sano, Y. Imai, and H. Ichino, "Lightwave-communication ICs for 10 Gbit/s and beyond", *OFC'95 Tech. Dig.*, 1995, pp. 36–37.
14. H. Ichino, M. Togashi, M. Ohhata, Y. Imai, N. Ishihara, and E. Sano, "Over-10-Gb/s IC's for future lightwave communications", *J. Lightwave Technol.* **12** (1994) 308–319.
15. T. Otsuji, M. Yoneyama, K. Murata, and E. Sano, "A super-dynamic flip-flop circuit

for broadband applications up to 24-Gbit/s utilizing production-level 0.2-μm GaAs MESFETs", *GaAs IC Symp. Tech. Dig.*, 1996, pp. 145–148.

16. S. Kimura, Y. Imai, Y. Umeda, and T. Enoki, "A DC-to-50-GHz InAlAs/InGaAs HEMT distributed baseband amplifier using a new loss compensation technique", *GaAs IC Symp. Tech. Dig.*, 1994, pp. 96–99.
17. L. D. Nguyen, A. S. Brown, M. A. Thompson, and L. M. Jelloian, "50-nm-self-aligned-gate pseudomorphic AlInAs/GaInAs high electron mobility transistors", *IEEE Trans. Electron Devices* **39** (1992) 2007–2014.
18. T. Enoki, T. Kobayashi, and Y. Ishii, "Device technologies for InP-based HEMTs and their applications to ICs", *GaAs IC Symp. Tech. Dig.*, 1994, pp. 337–340.
19. Y. Matsuoka and E. Sano, "InP/InGaAs double-heterostructure bipolar transistors for high-speed ICs and OEICs", *Solid-State Electron.* **31** (1995) 1703–1709.
20. S. Yamahata, K. Kurishima, H. Nakajima, and E. Sano, "InP/InGaAs DHBTs technology for single-chip 20-Gbit/s regenerative receiver circuits with extremely low power dissipation", *OSA TOPS on Ultrafast Electron. Optoelectron.* **13** (1997) pp. 135–140.
21. H.-F. Chau, W. Liu, and E. A. Beam, III, "InP-based HBTs and their perspective for microwave applications (invited)", *Proc. 7th Int. Conf. on InP and Related Materials*, 1995, pp. 640–643.
22. Y. Matsuoka, S. Yamahata, S. Yamaguchi, K. Murata, E. Sano, and T. Ishibashi, "IC-oriented self-aligned high-performance AlGaAs/GaAs ballistic collection transistors and their applications to high-speed ICs", *IEICE Trans. Electron.* **E76-C** (1993) 1392–1401.
23. S. Yamahata, K. Kurishima, H. Ito, and Y. Matsuoka, "Over-220-GHz-f_T-and-f_{max} InP/InGaAs double-heterojunction bipolar transistors with a new hexagonal-shaped emitter", *GaAs IC Symp. Tech. Dig.*, 1995, pp. 163–166.
24. H.-M. Rein, "Design considerations for very-high-speed Si-bipolar IC's operating up to 50 Gb/s", *IEEE J. Solid-State Circuits* **31** (1996) 1076–1090.
25. K. Murata, T. Otsuji, M. Ohhata, M. Togashi, E. Sano, and M. Suzuki, "A novel high-speed latching operation Flip-Flop (HLO-FF) circuit and its application to a 19 Gb/s decision circuit using 0.2 μm GaAs MESFET", *GaAs IC Symp. Tech. Dig.*, 1994, pp. 193–194.
26. T. Otsuji, K. Murata, K. Narahara, K. Sano, E. Sano, and K. Yamasaki, "20-40-Gbit/s-class GaAs MESFET digital ICs for future optical fiber communications systems", this issue.
27. H. Hamano, T. Ihara, I. Amemiya, T. Futatsugi, K. Ishii, and H. Endoh, "20 Gbit/s AlGaAs/GaAs-HBT 2:1 selector and decision ICs", *Electron. Lett.* **27** (1991) 662–664.
28. Z. Lao, U. Npwotny, A. Thiede, V. Hurm, K. Kaufel, M. Rieger-Motzer, W. Bronner, J. Selbel, and A. Hulsmann, "45 Gbit/s AlGaAs/GaAs HEMT multiplexer IC", *Electron. Lett.* **33** (1997) 589–590.
29. M. Yoneyama, A. Sano, K. Hagimoto, T. Otsuji, K. Murata, Y. Imai, S. Yamaguchi, T. Enoki, and E. Sano, "Optical repeater circuit design based on InAlAs/InGaAs HEMT digital IC technology", *IEEE Trans. Microwave Theory Tech.* **45** (1997) 2274–2282.
30. M. Vadipour, "Capacitive feedback technique for wide-band amplifiers", *IEEE J. Solid-State Circuits* **28** (1993) 90–92.
31. Z.-G. Wang, M. Berroth, A. Thiede, M. Rieger-Motzer, P. Hofmann, A. Hulsmann, K. Kaufel, K. Kohler, B. Raynor, and J. Schneider, "Low power data decision IC for 20-40 Gbit/s data links using 0.2 μm AlGaAs/GaAs HEMTs", *Electron. Lett.* **32** (1996) 1855–1856.
32. T. Otsuji, K. Murata, T. Enoki, and Y. Umeda, "An 80-Gbit/s multiplexer IC using InAlAs/InGaAs/InP HEMTs", *GaAs IC Symp. Tech. Dig.*, 1997, pp. 183–186.

33. T. Shibata, S. Kimura, H. Kimura, Y. Imai, Y. Umeda, and Y. Akazawa, "A design technique for a 60 GHz bandwidth distributed baseband amplifier IC module", *IEEE J. Solid-State Circuits* **29** (1994) 1537–1543.

34. S. Yamaguchi, Y. Imai, S. Kimura, and H. Tsunetsugu, "New module structure using flip-chip technology for high-speed optical communication ICs", *Tech. Dig. MTT-S IMS*, 1996, pp. 243–246.

35. S. Kimura, Y. Imai, Y. Umeda, and T. Enoki, "0-90 GHz InAlAs/InGaAs/InP HEMT distributed baseband amplifier IC", *Electron. Lett.* **31** (1995) 1430–1431.

36. K. Murata, T. Otsuji, T. Enoki, and Y. Umeda, "An exclusive OR/NOR IC for over 40 Gbit/s optical transmission systems", *Electron. Lett.* (to be published).

37. J. Godin, P. Andre, J. L. Benchimol, P. Desrousseaux, A. M. Duchenois, A. Konczykowska, P. Launay, M. Meghelli, and M. Riet, "A InP DHBT technology for high bit-rate optical communications circuits", *GaAs IC Symp. Tech. Dig.*, 1997, pp. 219–222.

38. E. Sano, K. Sano, T. Otsuji, K. Kurishima, and S. Yamahata, "Ultra-high speed, low power monolithic photoreceiver using InP/InGaAs double-heterojunction bipolar transistors", *Electron. Lett.* **33** (1997) 1047–1048.

39. M. Nakamura, Y. Imai, S. Yamahata, and Y. Umeda, "Over-30-GHz limiting amplifier IC's with small phase deviation for optical communication systems", *IEEE J. Solid-State Circuits* **31** (1996) 1091–1099.

40. J. F. Jensen, M. Hafizi, W. E. Stanchina, R. A. Metzger, and D. B. Rensch, "39.5-GHz static frequency divider implemented AlInAs/GaInAs HBT technology", *GaAs IC Symp. Tech. Dig.*, 1992, pp. 101–104.

41. E. Sano, K. Kurishima, and S. Yamahata, "20 Gbit/s regenerative receiver IC using InP/InGaAs double-heterostructure bipolar transistors", *Electron. Lett.* **33** (1997) 159–160.

42. M. Yoneyama, E. Sano, S. Yamahata, and Y. Matsuoka, "17 Gbit/s pin-PD/decision circuit using InP/InGaAs double-heterojunction bipolar transistors", *Electron. Lett.* **32** (1996) 393–394.

43. K. Hagimoto, Y. Miyamoto, T. Kataoka, H. Ichino, and O. Nakajima, "Twenty-Gbit/s signal transmission using a simple high-sensitivity optical receiver", *OFC'92 Tech. Dig.*, 1992, p. 48.

44. K. Hagimoto, M. Yoneyama, A. Sano, A. Hirano, T. Kataoka, T. Otsuji, K. Sato, and H. Noguchi, "Limitation and challenges of single carrier full 40-Gbit/s repeater systems based on optical equalization and new circuit design", *OFC'97 Tech. Dig.*, 1997, pp. 242–243.

45. K. Sato, I. Kotaka, Y. Kondo, and M. Yamamoto, "Active mode-locked strained InGaAsP lasers integrated with electroabsorption modulators for 20 Gbit/s pulse generation", *OFC'95 Tech. Dig.*, 1995, pp. 37–38.

46. A. Sano, T. Kataoka, H. Tsuda, A. Hirano, K. Murata, H. Kawakami, Y. Tada, K. Hagimoto, K. Sato, K. Wakita, K. Kato, and Y. Miyamato, "Field experiments on 40 Gbit/s repeaterless transmission over 198 km dispersion-managed submarine cable using a monolithic mode-locked laser diode", *ECOC. Tech. Dig.*, 1996, pp. 2.207–2.210.

47. K. Tsuda, Y. Miyamoto, A. Sano, K. Kato, Y. Imai, and K. Hagimoto, "40 Gbit/s baseband-type optical receiver module using a waveguide photodetector and a GaAs MESFET distributed amplifier IC", *OECC'96. Tech. Dig.*, 1996, pp. 506–507.

48. K. Sano, K. Murata, T. Akeyoshi, N. Shimizu, T. Otsuji, M. Yamamoto, T. Ishibashi, and E. Sano, "Ultra-fast optoelectronic circuit using resonant tunnelling diodes and uni-travelling carrier photodiode", *Electron. Lett.* **34** (1998) 215–217.

International Journal of High Speed Electronics and Systems, Vol. 9, No. 2 (1998) 595–630
© World Scientific Publishing Company

InP-HBT ICs FOR 40 Gb/s OPTICAL LINKS

MEHRAN MOKHTARI, URBAN WESTERGREN and BO WILLÉN

Photonics and Microwave Engineering, Department of Electronics,
Royal Institute of Technology,
Isafjordsgatan 22-26 S-164 40 Stockholm, Sweden

THOMAS SWAHN

Ericsson Microwave Systems, Microwave and High Speed Electronics,
Research Center, Bergfotsgatan 2, S-431 84 Mölndal, Sweden

ROBERT WALDEN

HUGHES Research Laboratory,
3011 Malibu Canyon Road, Malibu, CA 90265-4799, USA

InP-based HBT technology has proven to be a strong candidate for ultra high speed electronic as well as optoelectronic integrated circuits. The cut-off frequencies of the available devices exceed 100 GHz. To challenge the technology, a variety of circuits, suitable for a demonstrator for the 40 Gb/s fiber optical transmission system have been designed, fabricated, and tested. All the circuits show potential for 40 Gb/s applications. The electrical parts have been implemented in MSI/LSI AlInAs/InGaAs-HBT technology, while the monolithic optoelectronic receivers were realized in SSI- InP/InGaAs HBT technology. The verification of performance of the circuits have been mainly limited by available measurement equipment. All the electronic parts were operational with 3 volt supply voltage. All the circuits were fully functional after the first processing round. No redesign was necessary.

1. Introduction

High-speed digital electronics (> 10 Gb/s) is a key enabling technology in telecommunication and information technology. Its role comprises and penetrates most areas such as transmission, switching and processing. Progress in the area has been extremely rapid, fuelled by technology achievements as well as the request for ever higher communication and processing speed, and increased circuit complexity. InP-based Heterojunction Bipolar Transistor (HBT) technology is one of the candidates for applications where very high speed and electrooptical compatibility with long distance fiberoptical links may be combined. Bipolar devices in general, and HBT in particular offer high performance devices with no necessity for advanced photolithography steps. Field Effect Transistors utilize deep sub-micron feature sizes (gate length), to achieve better performance. In bipolar devices the performance is determined mainly in the vertical direction and thereby defined by epitaxy technology rather than horizontal layout geometries.

However, when targeting integrated circuits for ultra high speed such as 40 Gb/s operation, the trade off between circuit complexity from the yield and uniformity point of view and circuit flexibility with respect to possibilities of external adjustments becomes a major issue. On one hand, a high degree of integration minimizes the high frequency I/Os which, in turn decreases problems with signal degradation due to bonding wires and packaging. It also lowers the power consumption. But on the other hand, a high degree of integration limits the possibility to externally adjust various circuit parts for best bias conditions during operation and requires thorough interconnect modeling. Experimental work on both HEMT- and HBT-based technologies is proceeding in research and development units around the world.[1–4]

The overall system solution for a fiber optical transmission system, presented in Fig. 1, describes the basic functional blocks which are commonly utilized. Parallel digital inputs are time multiplexed to a serial bit stream. The output of the multiplexer controls a driver circuit which in term modulates the light source, generating an optical bit stream. On the receiving end the optical bit stream is converted to electronic pulses by a photo diode and one or more front-end amplifier(s). The output of the amplifier(s) is further magnified to achieve satisfactory logical levels by using limiting- or Automatic Gain Control (AGC) amplifiers. The signal, assuming Non Return to Zero (NRZ) data, is then partly fed into a clock recovery unit to synchronize the local oscillator to the frequency and phase of the incoming data, while a decision circuit retimes the input to the clock phase of the demultiplexer.

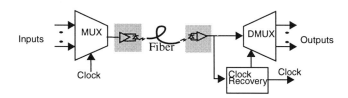

Fig. 1. Principle diagram of a Fiber optical link.

2. Heterojunction Bipolar Transistors (HBT)

The idea of a heterojunction bipolar transistor (HBT) is as old as the transistor itself,[5] but, it took about 20 years until the development of liquid phase epitaxy made it possible to grow the first device in 1972[6] and ten years later, the theory behind the HBT was extensively explained by Kroemer,[7] initiating an international effort in HBT-development. Most interest is today concentrated on fabrication of HBTs on Si, GaAs, and InP.

Silicon was initially the material of choice for electronics because of the excellent passivation layer offered by silicon dioxide. Silicon bipolars have always improved enough to satisfy the requirements for higher speed, and for a long time it was said that "GaAs is the material for the future, and it will always be". Looking at extreme high-speed devices, bipolar transistors suffer from the Johnson limit, i.e., that the transit frequency times the breakdown voltage is constant.[8] This is an effect of the device speed being limited by the transit time for the charge carriers, which requires a thin collector layer, while a high breakdown voltage is obtained through a thick layer. SiGe-based HBTs therefore suffer from the same basic limitation, as in Silicon bipolars, whereas GaAs and InP-based HBTs increase the maximum value considerably through the high electron velocity. Silicon seems finally to have reached its limit with a typical value of about 100 GHz V, and GaAs is no longer a technique for the future: There are a few million GaAs-HBT circuits delivered every month, limited by the fabrication capacity only.

In theory, InP-HBTs offer every advantage, compared with GaAs, but suffer from some practical drawbacks, i.e., a less mature fabrication technology and lack of 4″-substrates. Three main areas have been the driving force for developing both InP and GaAs-based HBTs: High power at high speed, high-speed integrated circuits, and optoelectronic integrated circuits (OEIC). Although the first area today is the main reason for the success of GaAs-HBTs, a higher power density as well as a higher power efficiency has been proven for InP-HBTs.[9] The properties of InP-HBTs are superior with regards to speed, current gain, low power dissipation, and thermal conductivity of the substrate, which all are important when making high-speed integrated circuits. The advantage in the OEIC-area is that InP-HBTs are made from the same material system as optoelectronic devices for fibre-optical communication at 1.3 to 1.55 μm wavelengths.

2.1. *The InP-HBT*

Concentrating on devices for integrated circuits, a dense packing is desirable, which requires a low power consumption to reduce heat dissipation. The dimensions of the interconnections may also be narrower when operating at a lower current. A low bias voltage is necessary to reduce the power consumption, and the turn-on voltage for an HBT is proportional to the sum of the bandgap in the base layer and the conduction band discontinuity ($E_{g,\text{base}} + \Delta E_C$). Both InP and AlInAs are in common use as materials for the wide bandgap emitter, but the alignment of InP to the InGaAs-base is advantageous, although the energy bandgap is smaller. An InP emitter offers a significantly lower turn-on voltage, and GaAs-based HBTs are clearly an inferior choice regarding power consumption. High-frequency operation with f_T over 100 GHz has been shown possible for InP/InGaAs-HBTs at a collector-emitter voltage of 0.60 V.[10]

Today, both MOCVD and different types of MBE-techniques are used to grow the HBT-structures. Conventionally, MBE-growth has been limited in the choice of material for the wide bandgap emitter to AlInAs, but the use of valved cracker cells loaded with elemental red phosphorus (SSMBE), or high-temperature cracking of phosphine (GSMBE), has made it possible to use InP for the emitter layer. For a long time, the low temperature used in MBE-growth was regarded as favorable, reducing diffusion broadening of the heavily doped base layer and by that facilitating both uniform characteristics and excellent high-speed performances. The establishment of carbon as p-type dopant in MOCVD has bypassed the problem, since it is almost immobile. All techniques have proven successful and in particular SSMBE has a potential for large-scale production.

The structures of high-speed bipolar transistors are very similar, independent of the material, but some differences are necessary when fabricating integrated circuits. Proton implantation is employed in GaAs-HBT fabrication to create a high-resistivity layer as a base for the metal interconnects, where the silicon technology uses silicon dioxide. Proton isolation is not effective for the narrow bandgap material InGaAs, so all epitaxial material must be removed outside the active areas. A necessary step in fabrication of high-speed InP-HBT circuits is therefore to achieve a reliable planarisation layer, as shown in Fig. 2. The conventional technique is to spin polyimide over the entire surface, use dry etching to thin the layer until the emitter contacts are revealed, and etch via holes to the remaining contacts.[11] Good results have also been obtained using spin-on glass[12] and silicon nitride.[13]

The sheet resistivity of the base layer in a high-speed HBT is on the order of 1 kΩ per square. A self-aligned base contact metal is therefore important, since it reduces the base resistance and also the base-collector capacitance. The technique also improves the uniformity of these parameters. A critical step in self-aligned HBT fabrication is the necessary undercutting of the emitter metal, requiring an excessive etching of one crystal plane to be sufficient for the other since the etch-rate is strongly crystal orientation dependent for III/V-materials. This is easily solved in the GaAs-HBT process, where the emitter overlaps the active area, reducing the need for self-alignment to two parallel sides only, but when making small area InP-HBTs, this effect conventionally limits the minimum emitter width. Using the SiN-planarization technique it was possible to let the emitter metal overlap the active area, very similar to the GaAs-process,[13] and in another approach, the conventional rectangular emitter mesa was replaced by a hexagonal to replace one crystal plane with a triangular shape.[14]

High frequencies are conventionally obtained at high collector current densities for bipolar transistors, since it reduces the influence of the dynamic emitter

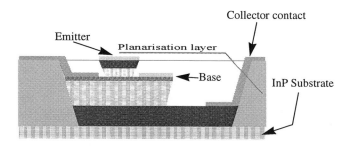

Fig. 2. Schematic cross-section of planarized HBT using silicon nitride.[13]

resistance. A reduction of the active device area is therefore necessary to reduce the current at a preserved high current density. Doing this, the external parasitics become more important, so the surface leakage current reduces the current gain and the base-collector capacitance reduces the cutoff frequencies, especially f_{max}. A high surface recombination velocity obstructs downscaling of GaAs-HBTs, while high-frequency, low current operation has been shown possible for InP/InGaAs-HBTs: Both cutoff frequencies were more than 100 GHz and the current gain over 20 at less than 0.5 mA for a nominal emitter area of 2.5 μm^2.[13]

The high-speed properties of InP-HBTs are expected to improve when using InP for the collector instead of InGaAs, since it increases the electron speed. The resulting structure is called a double heterojunction transistor (DHBT). High frequencies are often obtained by optimising one cutoff frequency, i.e., f_T or f_{max}, but the potential for DHBTs has been proven by more than 200 GHz for both figure-of-merits simultaneously.[14] A second advantage for the DHBT is an improved breakdown voltage. Conventional HBTs with a single heterostructure are limited to about two volts due to avalanche breakdown in the InGaAs collector. InP can sustain a higher voltage, and a breakdown voltage of 5.5 V has been shown at high current densities.[15] The product of the breakdown voltage and the transit frequency (f_T) is thus in the range of 500 to 1000 GHzV for InP-HBTs even in the on-state, meaning five times better than silicon, or more.

3. Monolithic optoelectronic receivers

There are (at least) two different approaches to high-sensitivity receivers for 40 Gb/s being investigated in several laboratories around the world; one uses an optical amplifier before the detector and the other uses an electrical preamplifier after the detector. The scope here will be limited to the latter, monolithic solution.[16-20]

The purpose of a monolithic optoelectronic receiver consisting of a detector and a preamplifier is to act as an active low-noise interface between modulated light and high-speed electronics. The receiver should detect and amplify the signal with a minimum of distortion and added noise. The most basic figure-of-merit used to measure the performance of the receiver is the sensitivity, which is the minimum average input optical power required for a certain bit-error rate (usually 10^{-9} for telecommunications, 10^{-12} or lower for computer communication) at a given bitrate. The sensitivity may be measured directly using a bit-error rate test set (BERTS) consisting of a pulse-pattern generator and an error detector using a pseudo-random binary sequence (PRBS) data pattern, usually with a non-return-to-zero (NRZ) code. The PRBS pattern can also be used with an oscilloscope to produce eye diagrams for inspection of the pulse shape. However, in order to analyze what causes the errors in the data, the behavior of a receiver is often expressed in terms of small-signal characteristics such as the responsivity or gain, the magnitude (-3 dB) bandwidth, the phase linearity and the total noise referred to the input.

The gain of the preamplifier in the receiver should be high enough so that the noise from the following electronics may be neglected. The exact minimum figure required for the gain will depends on the noise level of the circuit that is connected at the output of the receiver, but a reasonable level of gain in the preamplifier would be at least 10 dB, which means a current gain of more than 3 times or a transimpedance of more than 150 Ω with a 50 Ω load. If the gain is lower it may not be meaningful to try to establish a total noise for the receiver since this will not be determined by the preamplifier alone.

The magnitude (−3 dB) bandwidth (in Hz) should be at least 50% of the bitrate (in b/s), but in practice about 70% appears to be the most common figure used for high-speed applications. For 40 Gb/s this means that a receiver with at least 20 GHz magnitude bandwidth could be used in some cases, but that about 28 GHz is preferable.

It is quite common in early published work on receivers for a new and higher bitrate that only the magnitude bandwidth is investigated. However, the signal passing through the receiver is not a narrowband signal but a sequence of pulses with a frequency spectrum from very low (a few kHz or MHz) to very high (several GHz) frequencies, and it is therefore important that the delay for any part of the spectrum is independent of the frequency. This means that in addition to the magnitude bandwidth the phase must be linear, which is equivalent to a constant group delay. The allowed range of variation for the delay is system dependent, but a reasonable level could be a group delay spread of less than +/−10% of the period length at any frequency up to the −3 dB limit.

The noise added to the signal by the receiver may be caused by a number of separate noise sources. The dominating sources for 40 Gb/s receivers may be divided into two categories: thermal or shot noise caused by the transistors in the preamplifier and thermal noise from any resistor connected to the input of the preamplifier for detector biasing, preamplifier feedback or matching (termination). The detector has a dark current that also causes shot noise at the input, but this is usually relatively small in receivers for 40 Gb/s. All of these sources may be referred to the input for calculation of the signal-to-noise ratio in order to estimate the sensitivity for a given bit-error rate.

At least five research groups[16–20] have reported monolithic receivers for 1.55 μm wavelength with magnitude bandwidths of more than 20 GHz and a few more are at or close to 20 GHz. The results have been reached with two different technologies and preamplifier designs: traveling-wave amplifiers (TWA) with heterojunction field-effect transistors (HFET or HEMT)[16,17] or transimpedance amplifiers (TIA) with heterojunction bipolar transistors (HBT).[18–21] The capacitive character of the input impedance of a FET makes it suitable for use in a TWA since the capacitance then becomes part of the input transmission line. In a TIA with feedback through a resistor, the input capacitance adds to the capacitance of the detector which limits the bandwidth and requires low feedback resistor values. The resistive character of the input impedance of a bipolar transistor such as the HBT will load the input transmission line of a TWA which makes the design more difficult than with FETs, but in a TIA it becomes relatively easy to reach large bandwidths. The work on monolithic optoelectronic receivers, presented in, among others, the above mentioned publications, have therefore been concentrated on HEMT-TWA and HBT-TIA solutions.

To our knowledge, the highest magnitude bandwidth published for a monolithic receiver at 1.55 μm light wavelength is 46 GHz for a receiver with a side-illuminated waveguide detector and a InP-based HEMT-TWA preamplifier with 8 transistors.[16] The detector responsivity was 0.5 A/W and the preamplifier gain was about 5 dB, or 1.8 times current gain with a 50 Ω load. The TWA configuration uses a resistor to terminate the input transmission line, which may add a relatively large amount of thermal noise to the input. Eye-diagrams with 40 Gb/s return-to-zero (RZ) code were presented. The HEMTs had $f_T = 47$ GHz and $f_{max} = 100$ GHz.

Another HEMT-TWA receiver with four transistors has previously been reported to yield a 27 GHz magnitude bandwidth with a 0.36 A/W detector responsivity and 6 dB gain (two times current gain), with an estimated sensitivity at 40 Gb/s of -10 dBm.[17] The noise level is concluded to be determined largely by the input line termination resistor which is as low as 25 Ω, causing an input referred noise current spectral density of 20–30 pA/Hz$^{\frac{1}{2}}$ at moderate frequencies. Open eye diagrams with 20 Gb/s NRZ-code are presented and later also measurements with 40 Gb/s RZ-code have been performed.[21] This receiver has also been packaged, whereas the other results presented here are primarily probed measurements. The HEMTs had $f_T = 37$ GHz and $f_{max} = 90$ GHz.

For an InP-based HBT technology a magnitude bandwidth of 33 GHz[18] has been measured using an electrical rather than optical input signal, thereby improving an earlier figure from the same laboratory of 26 GHz.[22] The receiver had a top-illuminated pin-diode detector with a responsivity of 0.14 A/W and a preamplifier gain of 12 dB (four times current gain). Open eye diagrams for 40 Gb/s RZ-code were demonstrated. The noise level was calculated from simulations in the earlier work[22] to be 20–30 pA/Hz$^{\frac{1}{2}}$ resulting in an estimated sensitivity of -10.5 dBm for 40 Gb/s. The preamplifier circuit was a DC-coupled TIA with five transistors and resistive feedback to the input. Values of f_T and f_{max} above 150 GHz have been reported earlier[23] for the HBT technology at this laboratory.

A 24 GHz magnitude bandwidth for an InP HBT technology has also been reported, with 0.3 A/W detector responsivity and 16 dB preamplifier gain (six times current gain).[19] Figure 3 (left) illustrates the schematic of the receiver and Fig. 3 (right) presents the artist's impression based on the SEM-micrograph of the chip. Figure 4 shows the equivalent input noise current spectral densities for a receiver with a TIA having two different feedback resistors, providing bandwidths approximately suitable for 10 Gb/s (Rfb = 1800 Ω) and 40 Gb/s (Rfb = 400 Ω).

The equivalent input noise current spectral density was measured to be 8 pA/Hz$^{\frac{1}{2}}$ at low frequencies and the sensitivity was estimated from the noise measurements to be -12 dBm at 40 Gb/s. The preamplifier was a DC-coupled TIA with three

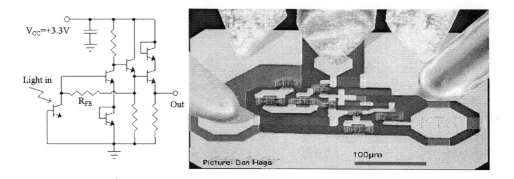

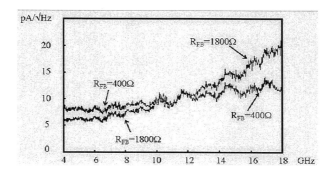

Fig. 3. The schematic and the artist's impression of the chip.

Fig. 4. Equivalent input noise current spectral densities for a receiver with a TIA having two different feedback resistors, providing bandwidths approximately suitable for 10 Gb/s (Rfb = 1800 Ω) and 40 Gb/s (Rfb = 400 Ω).

transistors and a 400 Ω feedback resistor. The noise level is primarily determined by the shot noise of the base current to the first transistor, and of the thermal noise of the feedback resistor. It is predicted that the sensitivity may be further improved by relatively simple changes in the lateral dimensions of the detector and the first transistor. The phase bandwidth was measured to be about 15 GHz and open eye diagrams to at least 25 Gb/s were observed. The HBTs were measured to have both f_T and $f_{\max}$ above 100 GHz.

Recently an InP HBT technology aimed for production has been reported to yield 24 GHz magnitude bandwidth with 10 dB gain.[20] This indicates that the technology even for 40 Gb/s monolithic receivers is maturing rapidly.

4. The Electronic Demonstrator

In order to set up a demonstrator to investigate the performance of electronic parts in system applications, the blocks presented in Fig. 5 are necessary.

To verify the potential of InP-HBT technology for high speed applications, all the necessary parts for the above structure have been realized in the same technology.

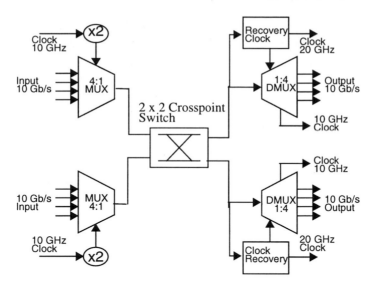

Fig. 5. Structure of the electronic demonstrator.

Common for all the parts are the following:

- All ICs were operational to the specifications, or better, after the first processing round (no redesign was necessary). However, a complete verification of operation, at speed, has not been possible due to lack of proper measuring equipment (BER etc.).
- All parts were operational with 3 volts power supply.
- All parts were designed only using "SPICE"-based simulators.
- The smallest feature size in all the circuits was 2 μm.

All the circuits were measurements on-wafer. Uniformity and yield of the fabricated chips were far above expectations. The received wafers were of two different categories, one with thin collector layer (0.3 μm) and the other with a thick collector layer (0.7 μm). Random measurements performed on all InP-wafers show better than 90% functional circuits. Within each category, good agreement on transfer-functions and bias conditions was observed during high frequency measurements. The limitations in characterizing the circuits were more due to available measurement equipment than uniformity and yield. Based on simulations, the packaging procedures will set the final specifications of circuit performance.

5. The InP-HBT IC Technology

AlInAs/GaInAs single heterojunction bipolar transistor technology[24] was used to realize the circuits. The technology utilized here, offers maximum values of more than 95 GHz for f_T and 100 GHz for $f_{\max}$. The cut-off frequencies are measured at about 5 mA collector current on the 2×5 μm^2 (as drawn) devices. The current gain and base-emitter forward diode-voltage drops are 25 and 0.7 V respectively. The smallest device used in the designs is 2×5 μm^2 in emitter area, however, emitter sizes as low as 2×2 μm^2 were available. A variety of emitter sizes, two layers of interconnect isolated by polyimide, MIM-capacitors and TaN-resistors with 50 Ω/sq were the available devices. The MIM-capacitors are created by Metal- Si_3N_4-Metal, corresponding to 0.3 $fF/\mu m^2$. The largest circuit designed in this technology is the 1:4 demultiplexer with a transistor count of about 500 and a similar number of resistors. The consumed area for this circuit is 2.5×2.5 mm^2.

6. Interconnect Modeling Considerations

As far as the HF-I/O problems are concerned, the benefits of a high degree of integration are overwhelming when high speed is targeted. This fact leads to complex geometrical structures interacting in various manners, degrading the signal integrity.[25] In order to deal with this problem, simplified and accurate models must be developed, taking higher order effects into account. There exist two aspects:

- The actual models for the interconnects with respect to "parasitics to ground" and "cross-talk".
- The capabilities of the CAD-software in recognizing various types of geometries and generating necessary models from the actual layout database.

The three models,[25] (capacitive loading, RC-equivalent, and third order LRC-filters), have been implemented for the InP-HBT technology using a commercial software's full extraction facilities. Extraction provides the parasitic circuit elements. SPICE simulations can be performed with or without interconnect parasitics. The model for the parasitics is chosen through the software prior to the extraction.

Numerous simulations were performed to investigate the behavior of various critical parts of different circuit-layouts when the influence of the interconnects was introduced. The agreement between simulation and measurements on deciding the allowed phase difference between clock and input-data was excellent.

7. InP-HBT-Based Differential Stage

The differential amplifier was thoroughly investigated as the unit is repeatedly used in various parts of the circuits. Such a stage is used as signal regeneration amplifier, limiter and output-stage for 10 and 20 Gb/s bitstreams. The amplifier is primarily used at the inputs of the 4:1 and 2:1 40 Gb/s multiplexers and outputs of 1:4 and 1:2 demultiplexers, for signal-regeneration. In a 20 Gb/s version of the system, the amplifier is used as gain stage prior to the decision circuit.

The characteristics of the amplifier are investigated at low power supply voltages mainly to achieve low power consumption and maintain power supply compatibility with the low speed parts of the complete system.

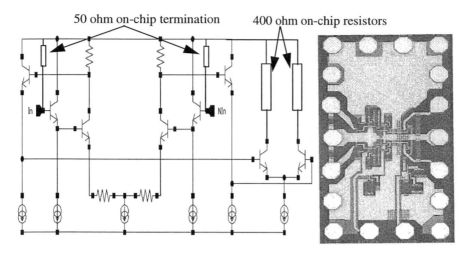

50 ohm on-chip termination   400 ohm on-chip resistors

Fig. 6. Schematics and photograph of the amplifier.

7.1. *Amplifier structure*

The amplifier was designed to produce a Bessel-transfer function which eliminates introduction of ringing at the data transitions.[26] The basic schematic and SEM-micrograph of the amplifier are presented in Fig. 6.

The first differential stage acts as linear amplifier whereas the second stage provides driving capability of the external 50 Ω load. Emitter-degeneration resistors are used in order to raise the input impedance of the input differential stage as well as the bandwidth. The two emitter follower pairs function partly as level shifts to introduce a suitable collector-emitter bias voltage for the following differential stages, and partly as current-gain stages (especially for the output stage) to drive the $2 \times 20 \ \mu m^2$ npn transistors. The layout was carefully carried out to preserve the constant-delay character of the amplifier. Both microstrip and coplanar structures were used to carry the signal through the amplifier. Massive on-chip decoupling of the DC-sources was utilized to minimize the HF-disturbance. 50 Ω on-chip terminations were provided to the inputs of the amplifier. Intensive work was put into introducing relevant interconnect models in order to observe the influence on signal integrity.[25]

Due to the fact that no precise models for the active or passive devices were available during the design period, the implementation was performed to be insensitive to device parameters. This has been achieved by allowing the current-source of the circuit to be accessible from pads for readjustment under various operating voltages.

7.2. *Measurements and results*

The amplifier was operational at 1.7 V with a current consumption of 40 mA including the load-current. The gain and bandwidth could be optimized for high gain or wide-band operation at supply voltage values ranging from 1.7 V to 3 V. Gain and bandwidth characteristics were optimized by readjusting the on-chip current mirrors through external pads.

7.2.1. *Frequency domain measurements*

Specially made probes were used for on-wafer measurements.[27] The available calibration procedures only allowed correction of unbalanced inputs and outputs. Therefore some inaccuracy was introduced due to the fact that one of the balanced-inputs was terminated by 50 Ω. One input was fed with HF-signal while one of the outputs was measured. Special Power-Signal-Ground-Signal-Power wafer-probes were used. Figure 7 shows the frequency response and the group delay up to 30 GHz at 1.7 V supply voltage. As presented, an almost flat response is observed up to 12 GHz. Thereafter, the gain falls with approximately 1 dB/GHz. The transfer function thereby proves that the circuit is suitable for 10- and 20 Gb/s applications, e.g., signal regeneration at the inputs of the multiplexers and outputs of the demultiplexers. The linear decrease of the phase of the amplifier indicates a Bessel-transfer-function character. The group-delay in the same figure shows a mean-variation of about +/− 2 ps.

Figure 8 presents the amplifier transfer function (S21) at 2.5 V with 40 mA and 3.3 V with 60 mA operation conditions. With a 2.5 V supply voltage the amplifier biasing conditions were optimized for the same bandwidth as the 1.7 V case but at a higher gain (15 dB). 10 GHz bandwidth and 24 dB gain were achieved at 3.3 volts. Naturally in both cases, the bandwidth could be raised by lowering the gain. At 3.3 V and 10 dB the bandwidth was increased to about 20 GHz.

It is important to point out that operation at 1.7 V was achieved by almost forward biasing the base-collector diode. Also the collector-emitter voltage was put as close as possible to the knee between the linear and saturation characteristics of the transistor, but still having high enough resistance for the current source. At 3 V, the bias conditions are more similar to nominal operation.

Although the transistors in the gain-stage of the amplifier are operating at less than 4 mA with a collector-emitter voltage of about 1 volt, the cut-off frequencies

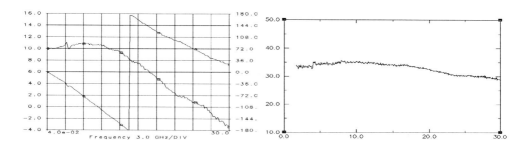

Fig. 7. The Frequency Response at 1.7 V (left) and the Group Delay (right) of the amplifier.

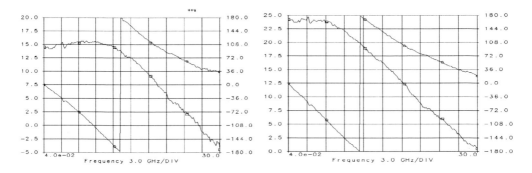

Fig. 8. The frequency response at 2.5 V (left) and 3.3 V (right).

of the InP-HBTs are high enough to achieve gain-bandwidth product of more than 150 GHz for the amplifier. By increasing the power supply voltage, both collector-emitter voltage and/or collector current of the devices may be increased resulting in a gain-bandwidth product of more than 240 GHz.

Figure 9 shows the input (S11) and output (S22) matching characteristics of the amplifier. The amplifier was designed with 50 Ω on-chip resistors at the HF-inputs. The output stage of the amplifier is an open-collector structure with on-chip 400 Ω resistors. S11 shows good 50 Ω matching at low frequencies. At higher frequencies, the influence of a capacitance component becomes more visible. The main portion of the capacitance originates from the equivalent input-capacitance of the input emitter follower stage.

S22 also shows a capacitive behavior. At low frequencies the on-chip 400 Ω resistor is the dominating component. As the frequency rises, an almost half-circle formed behavior is observed showing a parallel capacitance component. The parasitic capacitance is mainly based on the fringing and parallel-plate parasitic capacitance of the wires connecting the output transistors to the pads as well as output

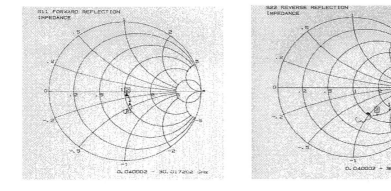

Fig. 9. Input (left) and Output (right) Reflection Characteristics.

transistor parasitics. Due to the fact that two different wafers with respect to collector thickness were at hand, a certain comparison of the behavior of the amplifiers with respect to influence of transit time and base-collector capacitance was possible.

Figure 10 presents S21 measurements for 0.3 μm and 0.7 μm intrinsic collector thicknesses at 1.7 V (upper) and 3.3 V (lower). A thin collector in the HBT results in a shorter transit time through the collector. However, it also results in larger base-collector capacitance, when the collector layer is fully depleted, which degrades the frequency response of the device.

As observed in Fig. 10 (upper), a higher gain for almost the same bandwidth for identical bias conditions is measured for the 0.3 μm case, indicating that transit time is the major limiting factor in performance of the amplifier under these conditions. At a 3 V power supply, in the same figure (lower) the gain difference is almost negligible. As shown in the figure (the 0.7 μm curves), by modifying the bias

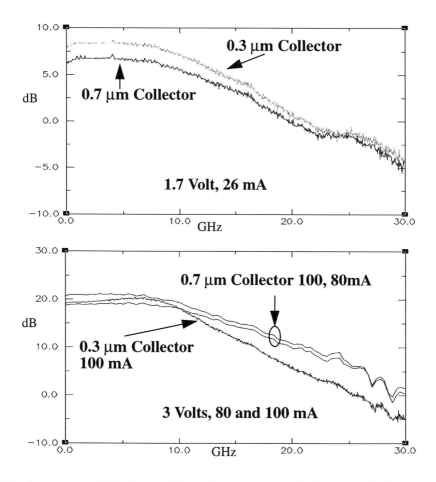

Fig. 10.   Comparison of S21 for amplifiers with transistors with 0.3 μm and 0.7 μm intrinsic collector thicknesses at various bias conditions.

conditions, the gain may be set to the desired value. However, the slope of the 0.3 μm case is sharper than the 0.7 μm one, indicating stronger parasitic capacitive influence in the amplifier behavior. The only difference between the amplifiers is the collector thickness, which then shows that the performance behavior of the amplifier has moved from largely transit time limited to base-collector capacitance limited. The weak peaking observed in the 0.3 μm case is due to the emitter degeneration feedback.

7.2.2. *Eye diagram measurements*

In order to investigate the wide-band characteristics of the amplifier, eye diagram measurements were performed, using a pseudo random bit sequence (PRBS) generator at 25 Gb/s with a sequence length of $2^{31} - 1$.

Probes, attenuators, cables and other components influence the eye opening. In order to be able to observe the actual waveform supplied to the amplifier ports, the output of the PRBS-generator was sent through the measurement set up by using an on-wafer-"thru" structure. Figure 11 presents the measurement results for "thru" (upper), at 1.7 V (middle) and 3 V (lower), respectively. The "thru"-measurement shows the actual rise and fall times at the input of the amplifier. The amplifier has a bandwidth of about 15 GHz (10 dB gain) at 1.7 V supply voltage. The middle figure shows that the rise- and fall-time of the input signal are kept after the gain stage.

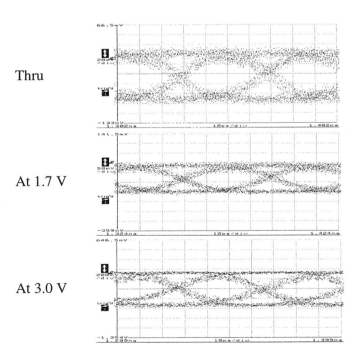

Fig. 11. Eye diagram measurements at 25 Gb/s: No amplifier, thru (upper), at 1.7 V (middle) and at 3 V (lower) power supply voltage.

At 3.3 V supply voltage the bandwidth is decreased to about 10 GHz (24 dB gain). The lower eye diagram presents the degraded rise- and fall-time of the amplifier compared to that of the 1.7 V case.

7.2.3. *Linearity measurements*

Linearity measurements were performed using a 20 GHz signal generator and a spectrum analyzer. Using a passive power-splitter and line-stretchers, 180° phase shifted input signals to the amplifier were generated and verified using a digital sampling oscilloscope. The input frequency for the measurements was chosen to be 10 GHz due to the following:

- Phase adjustment possibilities were limited to frequencies above 5 GHz due to the available line stretchers.
- 10 GHz is a good compromise for various bias conditions during which the amplifier was measured.

The attenuation due to power-splitter, line-stretchers, cables and probes was measured using the "thru"-structure on the calibration substrate. The measured value was then taken into account in the following measurements in order to compensate for the measurement set-up losses. Only one of the outputs of the amplifier was measured. The inverted output was terminated by 50 Ω. The output power was measured using the spectrum analyzer. The lowest input power, producing an output signal 10 dB above the noise floor measured by the spectrum analyzer, was observed to be around −130 dBm, while the contribution by the amplifier to the noise floor was less than 15 dB showing good noise characteristics due to InP-HBT based structure. The following table presents the 1 dB compression point of the amplifier under three different bias conditions:

Table 1. The amplifier's 1 dB compression point at different biasconditions.

Power supply [V]	Total current [mA]	Bandwidth [GHz]	output power @ 1 dB comp. point [dBm]
1.7	40	15	−10
2.5	40	15	−7
3.3	60	10	+1

The second harmonic (20 GHz) of the 2.5 V-case was measured to be more than 30 dB suppressed at the 1 dB compression point. In order to present the intermodulation characteristics of the amplifier, the third intercept point was measured. To perform the measurement, two independent signal sources were used to generate 8 GHz and 8.001 GHz. The two frequencies were summed by connecting the generators to the outputs of a passive power-splitter. In order to ensure that the spectrum analyzer used in this measurement does not have considerable contribution to the intermodulation, the sum of the frequencies were then observed at various power levels. No significant intermodulation due to the measurement equipment was noted. The output of the two signal sources were provided with isolators using 3-port circulators with the third port terminated with 50 Ω, to prohibit the two generators from modulating each other. The signal sources were then connected to the two inputs of the amplifier (balanced inputs). The summing operation was then performed by the differential amplifier operating at 3.3 V power supply voltage. The amplifier was adjusted to have about 20 dB gain and 12 GHz bandwidth during these measurements. Figure 12 presents the measurement results.

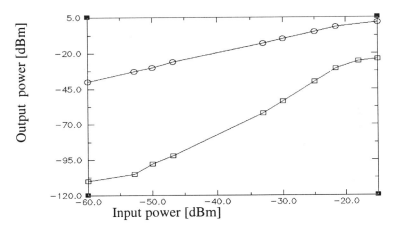

Fig. 12. IP3 measurement results for the differential amplifier. '□' represents the IP3- where '○' represent the basic tones (8 GHz and 8.001 GHz).

8. Three Stage Active Feedback Wideband Amplifier

In order to achieve higher bandwidths, active feedback was used in a three-stage amplifier structure.[28-31] The schematic principle of every stage is presented in Fig. 13. The first two stages have more than 3 dB gain each and the third stage acts as an active impedance transformer to adapt to the off-chip 50-Ω system.

8.1. *Measurements and results*

The complete amplifier has a gain of about 6 dB at 3 V power supply and 150 mA current consumption, with a bandwidth of more than 30 GHz. The schematic and the SEM-micrograph of the amplifier are presented in Fig. 13. As observed in Fig. 14, a peak in the transfer-function is observed at approximately 10 GHz. This was experimentally proven (by introducing a stand-alone-probe and pointing to the DC-probe which was connected to the current mirror of the first amplifier and observing a change in the position of the peak) to be due to parallel resonance

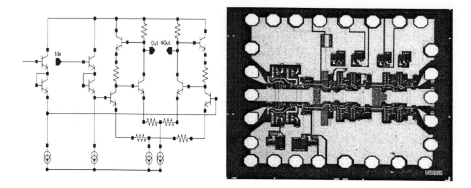

Fig. 13. Schematic and SEM-Micrograph of the active-feedback amplifier.

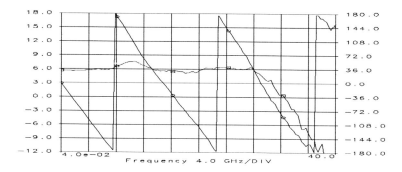

Fig. 14. S21 for the active feedback amplifier.

effects caused by the incomplete decoupling structure of the measurement setup consisting of first level on-chip decoupling, DC-probe series inductance and second level decoupling in approximately 2 mm distance from chip-pads. By introducing peaking at high frequencies (20 to 25 GHz) the bandwidth of the amplifier was increased to more than 35 GHz. This was performed by readjusting the current sources of the amplifier to increase the gain of the active feedback stage. Also, in this case, an almost constant group-delay was achieved (see Fig. 15).

A 20 Gb/s PRBS generator, run at 25 Gb/s ($2^{31} - 1$ sequence length), was used to investigate the pulse response of the amplifier. The eye diagrams were measured at 25 Gb/s by extending the range of the equipment mentioned above.

By raising the supply voltage to 4.0 V, the gain could be increased to about 10 dB with the same bandwidth.

Eye diagram measurement results at 25 Gb/s input are presented in Fig. 16. The upper diagram shows the signal form when transmitted directly through a "thru"-structure in order to give an understanding of the wave form degradation due to attenuators, cables, probes etc. The middle diagram presents the output of the amplifier. By readjusting the amplifier to peak more than 10 dB, sharper rise and fall times are achieved. This is illustrated in Fig. 16 (lower diagram).

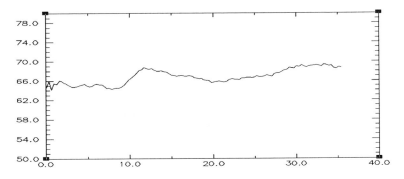

Fig. 15. Group-delay of the active feedback amplifier.

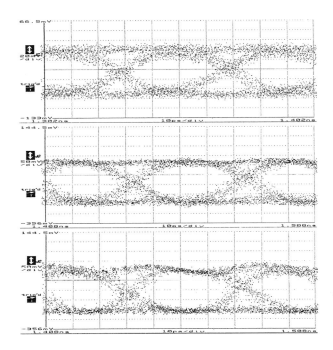

Fig. 16. Eye diagram for wide band amplifier, upper: waveform without amplifier, middle: amplifier output with out peaking, lower: amplifier output with peaking.

9. Narrowband Amplifier

In the narrowband amplifier implementation, the trans-impedance trans-admittance structure[31] was used according to Fig. 17. In this case, the narrowband characteristics of the structure have been exploited utilizing delta-type inductors, prior to the output-pads (see Fig. 17). In order to remove any DC off-set errors, the inputs were AC-coupled with 1 pF capacitance. The main reason for the choice of amplifier structure has been the achieved tuning possibilities. The center frequency is created by tuning the resonance between the base-collector capacitance of Q and L.

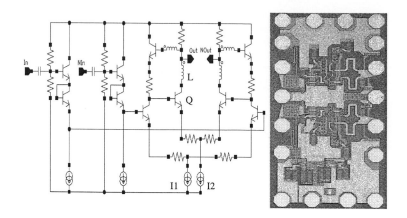

Fig. 17. Narrowband amplifier (schematic and die).

By varying the bias currents I1 and I2 of the amplifier the center-frequency may be changed as the base-collector capacitance of the transistors are bias dependent (see Fig. 17). Thereby, a robust structure is encountered where narrow-band behavior is achieved without the need for precise modeling of the inductors and transistors. The disadvantage of this approach is the lower Q-value compared to that of a fix structure. Nine different versions of the amplifier were implemented. The only difference among the amplifiers were the various inductor values used. By taking into account the dispersion and skin-effects, the inductors were designed between 100–200 pH assuming a 100 μm thinned substrate. The emitter area of Q was chosen in such a manner that the 40 GHz center frequency could be achieved even with unthinned substrate, i.e., with 2 to 2.5 times larger inductor values. The geometries were carefully laid out, considering the influence of the surroundings on the delta inductors. The T-structure at the output-pads gives rise to unwarranted effects which may be one of the mechanisms behind the phase behavior, observed during the measurements (see Fig. 18).

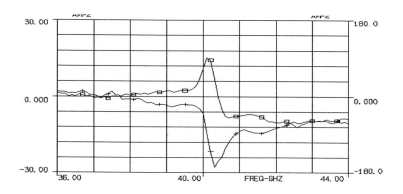

Fig. 18. S21 for the narrowband amplifier.

9.1. *Measurements and results*

Measurements were performed, using a commercial network analyzer. It is important to point out that at 40 GHz the power-signal-ground-signal-power wafer-probes contribute to phase error both during the calibration, which is performed assuming an unbalanced amplifier, and when measuring above 40 GHz. The center frequency of the amplifiers varied between 37.5 GHz and 43 GHz, at power supply levels between −3.0 V and −3.3 V, depending on the inductor values. Each amplifier showed tunability within 1 to 1.5 GHz range. No signs of oscillation were detected when changing Vee down to approximately −4 V without input signal. At higher power levels, signs of oscillation were observed on the frequency response of the amplifier, but the actual oscillation frequency has not yet been determined. The 180° phase-shift at the resonance frequency is proper, although the phase shift turns unexpectedly just after the center frequency. The circuit seems to have a combination of serial and parallel resonance (see Fig. 18, the immediate change of the phase after the 180° shift). The behavior may also be due to geometry effects, as mentioned earlier.

Figures 19 and 20 present amplifier behavior with the same inductor value, but under different bias conditions and with various inductor values.

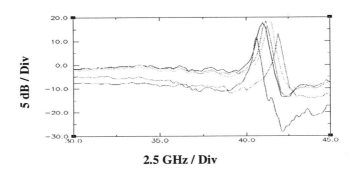

Fig. 19. S21 for different bias conditions.

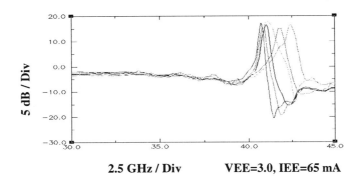

Fig. 20. S21 for different inductor values, under identical bias conditions.

As shown in Fig. 19, by changing the operating current of the amplifier, the resonance frequency may be tuned within a range of about 2 GHz. Comparing the measurements on amplifiers with different inductor values and under identical bias conditions, we noted that the resonance frequency is raised as the inductor value is decreased as shown (Fig. 20).

10. Edge Detector and Mixer

The edge detector and the mixer in the clock recovery unit[33,34] and the frequency multiplier in the transmitter block are realized using Gilbert-cell based structures (Fig. 21). Two cells have been designed, one fully DC-coupled and another with one differential input AC-coupled with a 1 pF input capacitance. The cells are otherwise identical. In the AC-coupled version, the correct input dc-level to the cell is created after the AC-capacitance, by resistors. The main reason for AC-coupling the cell is for usage as mixer in the sub-harmonic oscillator in the clock recovery block (see Sec. 11). The AC-coupled input will be connected to the branch coming from the narrowband amplifier in order to prohibit off-set errors created by the amplifier being propagated to the recurrent part (loop) of the block diagram, presented in Fig. 24.

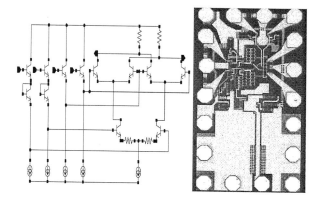

Fig. 21. The Gilbert-cell schematic and micrograph.

10.1. *Measurement results*

As a frequency multiplier, the circuit has been measured with 10 GHz, 300 mVp-p input-signal at one input and a similar signal delayed by 90° to the other input. Figure 22 presents the multiplier output wave form with the corresponding real-time fourier transform. The supply voltage is 3 V and current consumption is about 50 mA.

In order to evaluate the circuit as edge detector, a 40 Gb/s PRBS source is necessary, which is not at hand. Therefore, a single 20 GHz signal was connected directly to one input and after 90° phase delay, connected to the second input.

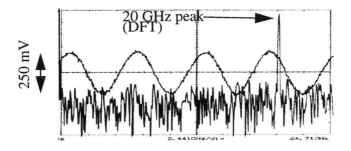

Fig. 22. 20 GHz output of the multiplier.

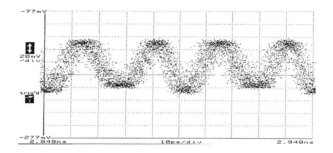

Fig. 23. 40 GHz output of the edge detector.

This experiment was done to observe the characteristics of the mixer's 40 GHz component at the output, if 40 Gb/s input-data would be provided. The resulting 40 GHz output is presented in Fig. 23. The output power of the generator was set to +6 dBm, which in turn (after power splitter, cable and probe attenuation) was about 1 dBm at every input. Please observe that the probes, cables and connectors were specified for 40 GHz bandwidth, which in turn puts the measurement conditions at the edge of specifications. These conditions together with triggering limitations, cause at least part of the jitter observed in the output as well as the low swing.

11. Clock Recovery Block

The principle used for clock extraction,[32] shown in Fig. 24, has the advantage of being independent of a Voltage Controlled Oscillator (VCO).

First, the input signal is split and amplified. One branch is connected to the decision circuit, while the other is sent to another power splitter stage. One of these two outputs is shifted by 90°, and thereafter both signals are fed to an edge detector. The 40 GHz component of the output is filtered with respect to phase and frequency through the bandpass filter and amplified by a narrow-band amplifier. The output locks the loop consisting of a mixer, a discrete 20 GHz band pass filter and an amplifier, to the phase of the input data. The discrete band pass filters are connected to the circuit off-chip by wire bonding.

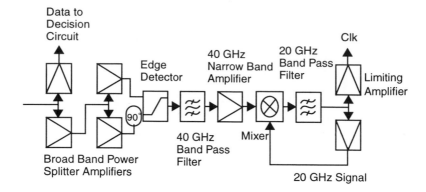

Fig. 24. The clock recovery block diagram.

The amplifiers and mixers described in the previous sections will be used to implement this block as soon as they are mounted and packaged.

12. 2 × 2 Crosspoint Switch

Figure 25 presents the basic circuit structure.[35] The inputs are provided by two signal regeneration amplifiers. Two selectors are then fed by the input amplifiers. The selection is made by a separate differential input with a pre-distortion amplifier stage. The output-50-Ω drivers are of open-collector type. The die-micrograph of the circuit is presented in Fig. 26. As observed, the circuit includes relatively long wires between the channels. In order to thoroughly investigate/simulate the influence of the interconnects on circuit performance, previously described interconnect extraction routines were used. The final circuit consists of about 70 active devices and consumes 1100×1100 μm^2 area.

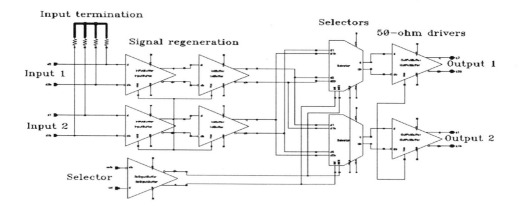

Fig. 25. Schematic of the 2 × 2 crosspoint switch.

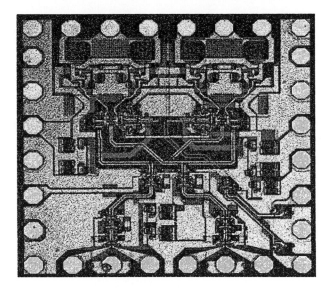

Fig. 26. Die microphotograph of the 2 × 2 crosspoint switch.

12.1. *Measurement and results*

12.1.1. *Frequency-domain measurements*

A network analyser was used to extract the small-signal behavior of the asynchronous switch. Due to the fact that the switch actually is a large-signal circuit, using "limiting"-behavior, two different power levels were chosen for S-parameter measurements. Once the system was calibrated for −30 dBm input power and once for −10 dBm. Specially made probes with S-G-S-S-G-S and P-S-G-S-P probe tip configuration were used for the on-wafer characterization. During the measurements, one channel was kept at constant DC-level while the other channel was evaluated. By changing the selector state, S21 for the "quiet"-channel was registered as a measure of cross-talk between the channels. The cross-talk between the probe-needles was not compensated due to lack of necessary calibration geometries. However, part of the cross-talk was measured to be due to "probe-coupling" by measuring on stand-alone pads.

Figure 27 presents the transmission on the active channel at −30 and −10 dBm input power, while Fig. 28 presents the same on the quiet channel.

As it can be seen, the cross-talk is more severe at higher frequencies, partially due to inevitable interconnect crossings on-wafer.

12.1.2. *Eye diagram measurements*

Using a 20 Gb/s PRBS-generator and BER-measurement equipment, eye diagrams and Bit Error Rate were measured. The PRBS-generator allows up to 25 Gb/s. Figure 29 shows the output of the active channel at 2 V and 20 Gb/s, (left) and

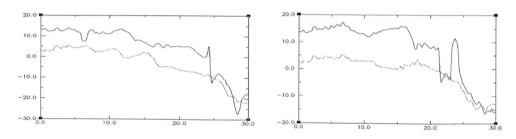

Fig. 27. Transmission at two input power levels in the active channel: Input1 to Output1 (left), and Input1 to Output2 (right).

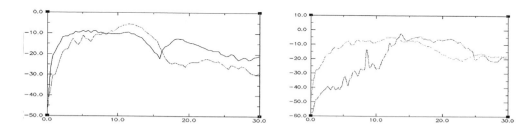

Fig. 28. Transmission at two input power levels in the quiet channel: Input1 to Output1 (left), and Input1 to Output2 (right).

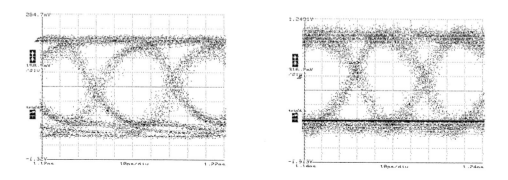

Fig. 29. The eye diagrams at 2 V and 20 Gb/s (left) and 3 V and 25 Gb/s.

3 V and 25 Gb/s (right). Given the clear eye-opening at 25 Gb/s and the above small-signal responses for the active channel, the circuit is a strong candidate for operation at even higher bitrates. However, the available measurement equipment does not allow characterization at higher bitrates.

BER-measurements were possible up to 20 Gb/s. Measuring both Input1 to Output1 and Input1 to Output2, while the unmeasured channel was provided with 12.5 Gb/s PRBS, neither at 2 V nor 3 V power supply, any errors that occurred were detected. The eye diagrams in Fig. 30 show the large signal coupling between the active and quiet channels in both selector states.

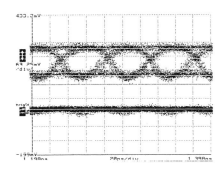 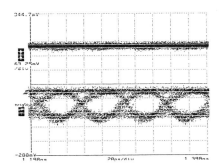

Fig. 30. The eye diagram of the active and quiet channels simultaneously Input1 to Output1 (left) and Input1 to Output2 (right) at 20 Gb/s.

The current consumption of the circuit at 2 V and 3 V were 125 mA and 300 mA, respectively. The externally adjustable current sources for various parts of the circuit were naturally tuned for best results at 2 V resp. 3 V.

13. Static Divider

The structure of the divider is based on two ECL-latches as presented in Fig. 31.[33,34]

"Out2" provides the 90° shifted signal to be used at system level on will. The divider is correctly operational at about 20 GHz. Figure 32 shows the result of the measurements of the divider at 20 GHz. The lowest necessary input power was measured to be around −15 dBm. The measurements were repeated with balanced and single ended inputs. No difference in behavior was observed.

14. Multiplexer

The 2:1 multiplexer is based on a 2:1 selector with ECL-latches at the inputs, as presented in Fig. 33.[33,34] In order to perform multiplexing at every clock edge, the input latches for D2 are triggered at different clock phases. In order to select data while the latch is in stable mode (hold mode), an additional latch is introduced to one of the inputs. The clock signal is predistorted through an input amplifier stage

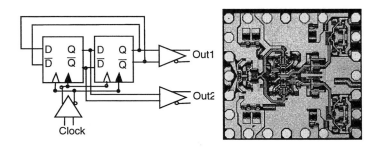

Fig. 31. The structure (left) and the micrograph (right) of the divider.

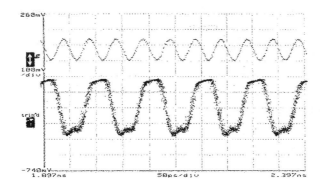

Fig. 32. The output of the static divider at 20 GHz.

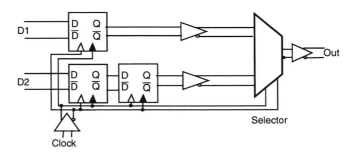

Fig. 33. The 2:1 multiplexer structure.

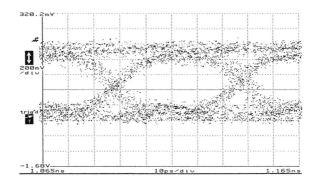

Fig. 34. Multiplexer eye diagram at 25 Gb/s output.

and the inputs of the selector are provided with an additional buffer to regenerate the signal after the latches.

Using a 20 Gb/s pseudo random bit sequence (PRBS) generator, run at 25 Gb/s, with a word length of $2^{31} - 1$, the eye diagram in Fig. 34 was measured. Only single-ended-input (data and clock) measurements were possible due to equipment limitations. The total current consumption of the multiplexer is about 280 mA at 3 V supply voltage. The circuit was actually operational at 2.7 V.

The allowed phase difference between data and clock was measured to be $+/-85°$ at 20 Gb/s input, which is in excellent agreement with simulation results on a back-annotated schematic including parasitics, extracted by earlier described routines. Externally re-adjustable current mirrors were placed in all the designs for each "function" on chip — i.e., one or several latches, the clock buffer(s) the output buffers etc. have their own current mirrors.

The idea is to introduce the ability to readjust the operation conditions of the structures, to compensate for operation in various supply voltage conditions. The approach has been verified to give similar behavior in simulation and measurement. However it is important to point out that due to limited possibility in probing, most of the bias-readjusting pads were left unconnected. Therefore the results achieved are not optimized for best conditions.

In order to investigate the 40 Gb/s response of the circuit, two independent 20 Gb/s sources are necessary. No such equipment was at hand during the measurements, therefore, by using the same data source for both inputs with a 50 ps time shift applied to one, a 40 Gb/s eye diagram was achieved which is shown in Fig. 35.

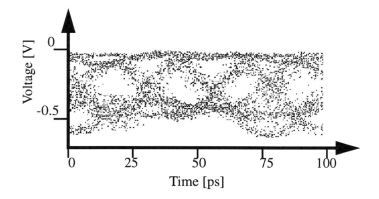

Fig. 35. 2:1 multiplexer eye diagram at 40 Gb/s output.

A 4:1 multiplexer has also been fabricated. The layouts for the 2:1 multiplexer- and the divider-units are identical to the respective circuits. Due to the large number of pads and lack of necessary probes the 4:1 multiplexer can only be fully characterized by packaged measurements. However, it has been possible to partially power up the circuit and as the outputs of the on-board divider are available on separate output-pads, the internal divider has been found to be operational to 21 GHz at −10 dBm input power. The micrograph of the 4:1 multiplexer is presented in Fig. 36.

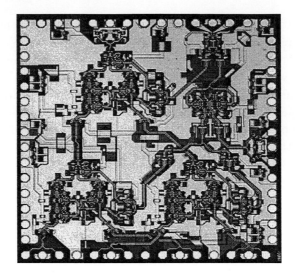

Fig. 36. The micrograph of the 4:1 multiplexer.

15. Demultiplexer

The 1:2 demultiplexer utilizes the principle shown in Fig. 37.[33,34] The input data is alternately read into the "latch-rows" at different clock edges. In order to minimize the data to clock phase alignment constraints and decrease signal leakage during the hold modes of the respective input latches, an extra latch has been added to every row.

Due to the fact that no true 40 Gb/s PRBS source has been available, the following measurements were done to verify 1:2 demultiplexer capabilities at 40 Gb/s.

First, a known sequence was used as input and the demultiplexer was clocked at twice the speed to "copy the input to both outputs". Figure 38 shows typical outputs where the solid line presents one output while the dashed line shows the other. Then a PRBS with a length of $2^{31} - 1$ was used as the input and the eye

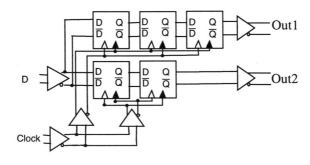

Fig. 37. The 1:2 demultiplexer structure.

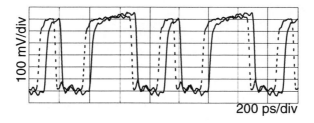

Fig. 38. Input to the demultiplexer copied to the outputs.

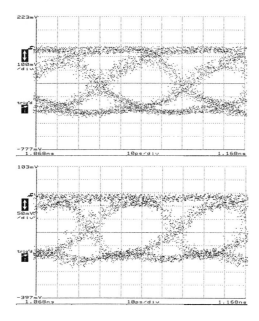

Fig. 39. Eye diagram of the demultiplexer clocked at 25 GHz with 25 Gb/s input.

diagrams at the outputs were registered. In this case probing the current source of the output buffer was possible, due to available wedge configuration.

From $-140°$ to about $-110°$ phase difference between clock and data at 10 Gb/s input, the demultiplexer was functioning erroneously. Otherwise no sensitivity to the phase relation was observed. Also in this case the measurement is in very good agreement with the back-annotated simulations.

Figure 39 shows the eye diagram of one of the outputs at two different bias conditions of the output buffer at 25 Gb/s input and 25 GHz clock. The upper diagram shows the output when the buffer is in the high-swing low-bandwidth mode, while the lower one presents the results at low swing high bandwidths.

The presented measurements were performed at 2.5 V supply voltage. The current consumption was measured to 280 mA.

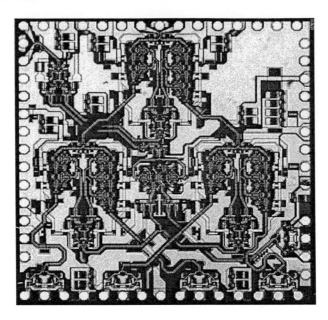

Fig. 40. Micrograph of the 1:4 demultiplexer.

The 1:4 demultiplexer is built using the layouts for the presented 1:2 demultiplexer and divider. Due to lack of proper wedge configuration, on-wafer measurements on this circuit have not been possible. Figure 40 presents the micrograph of the 1:4 demultiplexer.

16. Conclusions

InP-HBTs offer a combination of high cut-off frequencies and high current gain at a low bias voltage that makes them extremely interesting for high-speed integrated circuits. The Johnson limit is shifted towards higher values, making the technique about five times faster than conventional Si-bipolars. The required bias voltage for an InP-HBT is about the same as for a conventional Si-bipolar, making them exchangeable. Circuit designers therefore have the possibility to make an InP-HBT circuit as if it were to be realised in a conventional technique, but with a considerably higher bandwidth: Cut-off frequencies of more than 200 GHz have been proven for discrete InP-HBTs today. We believe that high frequency operation of InP-ICs is still limited by an immature fabrication process and insufficient models for the active devices as well as the interconnections.

With respect to monolithic optoelectronic receivers, in summary it may be noted that none of the results above are quite ideal for 40 Gb/s yet, even though some are close. HEMT-TWA receivers appear to be suitable for very large bandwidths, especially when comparing the bandwidth and the speed of the transistors, but the gain is low and the noise is high due to the input line terminations in the designs presented so far. The HBT-TIA receivers have less bandwidth compared to the transistor speed, especially concerning the phase, but the noise is lower since it is not limited by the thermal noise of a small resistor at the input.

In the future both the HEMT-TWA and the HBT-TIA approaches are predicted to be further developed, and there does no appear to be any obvious reason why all reasonable system requirements regarding magnitude and phase bandwidths at 40 Gb/s should not be possible to fulfil with realistic improvements of the existing designs in both types of technology. The result may be that the bipolar solution can provide slightly lower noise, but the sensitivity is determined not only by the noise but also by the responsivity of the detector. In the results above, the HEMT-TWA receivers have higher detector responsivity since they use side-illuminated detectors instead of the top-illuminated ones used for HBT-TIA. This is not a restriction for the latter technology, however, so the side-illuminated detectors may be used also in HBT-TIA solutions in the future.

With respect to VHSICs, a chip-set consisting of all the necessary high speed parts for 40 Gb/s fiber optical communication systems, except for the driver, have been designed, fabricated and characterized by on-wafer measurements. The chip-set is capable of 3 V single power supply operation. The AlInAs/GaInAs single heterojunction bipolar transistor technology in which the circuits were fabricated was found to be of excellent performance, uniformity and yield for current applications.

All the circuits were designed using SPICE-type simulators. Necessary routines for extraction of higher order interconnect parasitics were implemented in commercial software in order to enable prediction of circuit behavior with complex layouts by automatic extraction. All the circuits were operational after the first processing round. No redesign was necessary.

Acknowledgements

The authors wish to thank Professor Lars Thylen at Royal Institute of Technology, Department of Electronics (KTH/ELE), Photonics and Microwave engineering laboratories (FMI) for fruitful discussion and support. Also thanks to Dr. Boris Kerzar and Dr. Florian Sellberg for advice on high frequency circuit design and interconnect modeling aspects, and Miss Tarja Juhola and Mr. Gerd Schuppener for help with the simulations and layouts at FMI.

The authors would like to express their gratitude to Dr. Paul Greiling at Hughes Research Labs (HRL) for encouragement and support, Dr. William Stanchina, Dr. Joe Jensen and Dr. Ken Elliot for reviewing the designs, Mr. Michael Kardos for CAD-support, and Mr. Freddie Williams for helping with measurements at HRL.

The authors also wish to thank Dr. Thomas Lewin, Dr. Bjorn Rudberg and Dr. Ingmar Andersson at Ericsson Microwave Systems for sharing their experiences from the earlier 10 Gb/s project with us.

Finally the authors wish to thank the system manager at KTH/ELE, Mr. Hans Berggren for all the computer support.

References

1. M. Berroth, M. Lang, Z.-G. Wang, Z. Lao, A. Thiede, M. Rieger Motzer, W. Bronner, G. Koller, A. Husmann, and J. Schneider, "20-40 Gbit/s 0.2 mm GaAs HEMT chip set for optical data receiver", *IEEE GaAs IC Symp. 96*, pp. 133–136.
2. R. Yu, R. Pierson, P. Zampardi, K. Runge, A. Campana, D. Meeker, K. C. Wang, A. Petersen, and J. Bowers, "Packaged clock recovery integrated circuits for 40 Gb/s optical communication links", *IEEE GaAs IC Symp. 96*, pp. 129–132.
3. T. Otsuji, E. Sano, Y. Imai, and T. Enoki, "40-Gb/s ICs for future lightwave communications systems", *IEEE GaAs IC Symp. 96*, pp. 14–17.
4. A. Felder, M. Moller, J. Popp, J. Bock, and H.-M. Rein, "46 Gb/s DEMUX, 50 Gb/s MUX, and 30 GHz static frequency divider in silicon bipolar technology", *IEEE J. Solid-State Circuits* **31**, 4 (1996).
5. W. Shockley, "Circuit element utilizing semiconductive material", United States Patent 2,569,347 (1951).
6. W. P. Dumke, J. M. Woodall, and V. L. Rideout, "GaAs-GaAlAs heterojunction transistor for high frequency operation", *Solid-State Electron.* **15** (1972) 1339–1343.
7. H. Kroemer, "Heterostructure bipolar transistors and integrated circuits", *Proc. IEEE* **70** (1982) 13–25.
8. E. O. Johnson, "Physical limitations on frequency and power parameters of transistors", *RCA Review* **26** (1965) 163–177.
9. C. Nguyen *et al.*, "AlInAs/GaInAs/InP double heterojunction bipolar transistor with a novel base-collector design for power applications", *IEEE Electron Device Lett.* **17** (1996) 133–135.
10. Y. Matsuoka and E. Sano, "InP/InGaAs double-heterostructure bipolar transistors for high-speed ICs and OEICs", *Solid-State Electron.* **38** (1995) 1703–1709.
11. J. F. Jensen *et al.*, "AlInAs/GaInAs HBT IC technology", *IEEE J. Solid-State Circuits* **26** (1991) 415–421.
12. R. J. Malik *et al.*, "Self-aligned thin emitter C-doped base InP/InGaAs/InP DHBT's for high speed digital and microwave IC applications", *54th Ann. Device Research Conf. Dig.* (1996) 40–41.
13. B. Willén and D. Haga, "InP-HBTs with good high frequency performance at low collector currents using silicon nitride planarisation", *Electron. Lett.* **33** (1997) 719–720.
14. S. Yamahata, K. Kurishima, H. Ito, and Y. Matsuoka, "Over-200-GHz-f_T-and-f_{max} InP/InGaAs double-heterojunction bipolar transistors with a new hexagonal-shaped emitter", *Proc. IEEE GaAs IC Symp. 1995*, pp. 163–166.
15. H.-F. Chau and E. A. Beam III, "High-speed, high-breakdown voltage InP/InGaAs double heterojunction bipolar transistors grown by MOMBE", *Proc. 51th Ann. Device Research Conf.*, 1993.
16. Y. Muramoto, K. Takahata, H. Fukano, K. Kato, A. Kozen, O. Nakajima, and Y. Matsuoka, "46.5-GHz bandwidth monolithic receiver OEIC consisting of a waveguide pin photodiode and a distributed amplifier", *Proc. 23rd European Conf. Optical Communication (ECOC97)*, **5** pp. 37–40, IEE, 1997.
17. S. Ian Waasen, A. Umbach, U. Auer, H.-G. Bach, R. F. Bertenburg, G. Janssen, G. G. Mekonnen, W. Passenberg, R. Reuter, W. Schlaak, C. Schramm, G. Unterbörsch, P. Wolfram, and F.-J. Tegude, "27-GHz bandwidth high-speed monolithic

integrated optoelectronic photoreceiver consisting of a waveguide fed photodiode and an InAlAs/InGaAs-HFET traveling wave amplifier", *IEEE J. of Solid-State Circuits* **32**, 9 (1997).

18. E. Sano, K. Sano, T. Otsuji, K. Kurishima, and S. Yamahata, "Ultra-high speed, low power monolithic photoreceiver using InP/InGaAs double-heterojunction bipolar transistors", *Electron. Lett.* **33**, 12 (1997) 1047–1048.

19. U. Westergren, D. Haga, and B. Willén, "Monolithic optoelectronic receivers with up to 24 GHz bandwidth using InP pin-HBT technology", *Proc. 23rd European Conf. Optical Communication (ECOC97)*, **4**, pp. 105–108, IEE, 1997.

20. D. C. Streit, A. Gutierrez-Aitken, J. C. Cowles, L.-W. Yang, K. W. Kobayashi, L. T. Tran, T. R. Block, and A. K. Oki, "Production and commercial insertion of InP HBT integrated circuits", *19th Ann. IEEE GaAs IC Symp. Tech. Dig. 1997*, pp. 135–138, October 1997.

21. H.-G. Bach, R. M. Bertenburg, H. Bulow, G. Jacumeit, G. Mekonnen, A. Umbach, G. Unterbörsch, G. Veith, and S. van Waasen, "Ultrafast monolithic InP-Based photoreceiver module detecting a 40 Gbit/s optical TDM RZ modulated pulse sequence", *Proc. 23rd European Conf. Optical Communication (ECOC97)*, **4**, pp. 101–104, IEE, 1997.

22. E. Sano, M. Yoneyama, S. Yamahata, and Y. Matsuoka, "InP/InGaAs double-heterojunction bipolar transistors for high-speed optical receivers", *IEEE Trans. Electron Devices* **43**, 11 (1996) 1826–1832.

23. Y. Matsuoka and E. Sano, "InP/InGaAs double-heterostructure bipolar transistors for high-speed ICs and OEICs", *Solid-State Electron.* **38**, 9 (1995) 1703–1709.

24. W. E. Stanchina *et al.*, "An InP-based HBT fab for high-speed digital, analog, mixed-signal and optoelectronic ICs", *GaAs IC Symp. 1995*, pp. 31–34.

25. M. Mokhtari, T. Juhola, G. Schuppener, and F. Sellberg, "Automatic extraction of higher order interconnect parasitics for device level simulators for VHSIC applications", *IEEE-CAS Symp.* 44–49, Sep. 1996, Pavia, Italy.

26. M. Mokhtari, T. Juhola, B. Kerzar, G. Schuppener, U. Westergren, H. Tenhunen, T. Swahn, T. Lewin, W. E. Stanchina, and R. Walden, "Low voltage, broad and narrowband microwave differential amplifiers for 40 Gb/s demonstrator applications in InP-HBT", *ISCAS97*, pp. 149–152.

27. GGB Indistries' product manual for Picoprobe- "Multi-Contact Wedge".

28. Y. Kuriyama *et al.*, "DC to 40 GHz amplifiers using AlGaAs/GaAs HBTs", *Tech. Dig. GaAs IC Symp. 1994*, pp. 299–302.

29. E. M. Cherry and D. E. Hooper, "The design of wide-band transistor feedback amplifiers", *Proc. IEE* **110**, 2 (1963) 375–389.

30. W. Polman, "A silicon-bipolar amplifier for 10 Gbit/s with 45 dB gain", *IEEE J. Solid State Circuits* **29**, 5 (1994) 551–556.

31. M. Mokhtari, T. Juhola, B. Kerzar, H. Tenhunen, I. Andersson, T. Swahn, and W. E. Stanchina, "40 GHz narrow-band trans-impedance trans-admittance amplifier in InP-HBT technology", *ESSCIRC'96*, in Neuchatel, Switzerland, pp. 96–99.

32. A. Djupsjobacka, I. Andersson, and B. Rudberg, "A 10 Gb/s demonstrator", *Ericsson Rev.* **2** pp. 70, 1994, ISSN 0014-0171.

33. M. Mokhtari, T. Swahn, R. Walden, W. Stanchina, M. Kardos, T. Juhola, G. Schppener, H. Tenhunen, and T. Lewin, "InP-HBT chip-set for 40 Gb/s fiber optical communication systems operational at 3 volts", *IEEE J. Solid State Circuits* **32**, 9, (1997) 1371–1383.

34. T. Swahn, T. Lewin, M. Mokhtari, H. Tenhunen, R. Walden, and W. E. Stanchina, "40 Gb/s, 3 Volt InP HBT ICs for a fiber optic demonstrator system", Invited paper, *GaAs IC-Symp. 96*, pp. 125–128.

35. M. Mokhtari, B. Kerzar, T. Juhola, G. Schuppener, H. Tenhunen, T. Swahn, and R. Walden, "A 2 V, 120 mA 25 Gb/s 2 × 2 crosspoint switch in InP-HBT technology", *ISSCC'98*.

International Journal of High Speed Electronics and Systems, Vol. 9, No. 2 (1998) 631–642

A REVIEW OF RECENT PROGRESS IN InP-BASED OPTOELECTRONIC INTEGRATED CIRCUIT RECEIVER FRONT-ENDS

ROBERT H. WALDEN

HRL Research Laboratories, 3011 Malibu Canyon Road,
Malibu, CA 90265, USA

InP-based OEIC receivers look promising for high-speed ($\geq$ 10 Gb/s) optical communications systems and for WDM networks because of the inherent advantages of integration, and the intrinsic speed of the devices available. This paper reviews recent developments.

1. Introduction

The accelerating drive toward ever higher speed fiberoptic communications systems along with the increasing popularity of multiwavelength/multichannel network concepts, such as wavelength division multiplexing (WDM), has, over the past 3–6 years, raised the activity level in the already active field of optoelectronic integrated circuits (OEICs). A large part of this activity has been focused on OEICs fabricated using the InAlAs/GaInAs/InP material system because it is naturally suited for wavelengths in the 1.3–1.6 μm range, and, excellent optical (lasers, photodetectors) and electrical (JFETs, HEMTs, HBTs) devices are available in it. In fact, the first long-wavelength OEIC receiver ever reported was fabricated in InP-based technology.[1] This review concentrates on InP-based OEIC receiver front-ends.

OEIC receivers offer the advantages of lower parasitics, lower cost and higher reliability compared to hybrid approaches. Previously these advantages had been offset by the complexities of realizing high quality optical and electronic devices on the same wafer, but recent excellent results indicate that this is no longer the case. OEICs may now be the preferred choice particularly for: (1) receivers with data rates and bandwidths $\geq$ 10 GHz[2-12] and (2) OEIC arrays.[13-18] The latter, when combined with fiber-ribbons as inputs, permit the simultaneous alignment of all the fibers with all the OEIC photodetectors. This implies a more efficient packaging scheme relative to one-fiber-at-a-time alignment approaches and therefore lower assembly costs.

The following sections deal with key performance parameters of, and ingredients in, optical receiver front-ends. Section 2 discusses sensitivity, bit-rate and bandwidth. Sections 3 and 4 describe the devices used and the circuit architectures employed, respectively. Section 5 discusses OEIC arrays and Sec. 6 gives conclusions.

2. Receiver Sensitivity

Two of the most important performance characteristics of optical receiver front-ends are the receiver's sensitivity and its ability to handle the transmission (bit) rate, B, of the incoming data stream. The sensitivity, P_{min}, is the minimum average received optical signal power needed to achieve a specified bit-error-rate (BER). Often, a BER $\leq 10^{-9}$ bit-errors per second is required. The basic elements of a BER test setup are shown in Fig. 1 along with an example of the eye diagrams observed during the measurements. In this example, a 2.5 gigabit/s (Gb/s) pseudorandom bit sequence (PRBS) is generated and used to modulate a laser with wavelength, $\lambda = 1550$ nm.[16] The modulated light beam is sent through an optical attenuator to set the optical power level seen by the OEIC. Finally, the error detector receives the OEIC output datastream and compares it to the original PRBS. An error is recorded when a received bit differs from the corresponding original bit, e.g., a ONE instead of a ZERO.

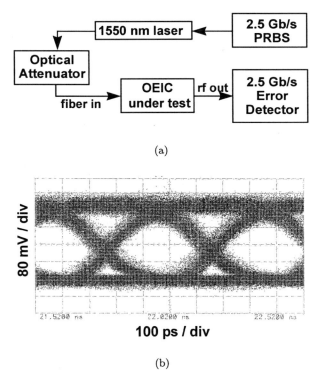

(a)

100 ps / div

(b)

Fig. 1. Example of a bit-error-rate test setup. In (a) are shown the essential elements: A pseudo-random bit sequence (PRBS) generator, a laser that can be modulated, an optical attenuator, the OEIC, and an error detector. In (b) an eye diagram is shown for the OEIC output.

The sensitivity is related to the input-referred noise current, i_n, by the relation,[19]

$$P_{\min}(dBm) = 10 \, \log((Qi_n/R_0)/1mW),$$ (1)

where Q is the signal-to-noise ratio required to achieve a given BER, e.g., Q = 6 for BER = 10^{-9}. The other parameter in the equation is the photodiode (PD) responsivity, $R_0(A/W)$, which measures the ability of the PD to respond to an optical stimulus. The input-referred noise current is given by,

$$i_n = \sqrt{\int_0^\infty \langle i_{in}^2 \rangle df},$$ (2)

where $\langle i_{in}^2 \rangle$ is the square of the input-referred noise current spectral density and the integration is over all frequencies, f. There are several contributions to the noise density and the sources of the more important ones are indicated in the optical front-end circuit shown in Fig. 2. They are: thermal noise current due to R_{fb} (feedback resistance used with transimpedance amplifier), shot noise current due to the DC input current to the first transistor of the preamplifier (in Fig. 2, base current, I_B, for a bipolar transistor, but would be gate-leakage current, I_{GL}, if a FET were used), shot noise current due to the DC output current of the first transistor (shown as collector current in Fig. 2, but would be drain current for a FET), thermal noise voltage due to the input resistance of the first transistor (shown as r_{bb} in Fig. 2 but would be r_{gg} in the case of a FET), and, shot noise current due to the PD dark current, I_{dark}.

Clearly the challenge in designing useful high-speed OEIC receivers (as well as hybrid receivers) is to minimize i_n, and maximize R_0, while still achieving the necessary bandwidth to operate at the required bit rate. Some recently reported OEIC receiver sensitivity data[3-5,7-11,13,15-17,20-32] are presented as a function of bit

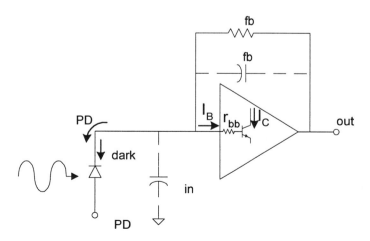

Fig. 2. An OEIC receiver front-end schematic showing noise sources (I_{dark}, I_B, I_C, R_{fb}, r_{bb}) and bandwidth limitations (C_{in}, C_{fb}). A transimpedance preamplifier serves as the gain stage.

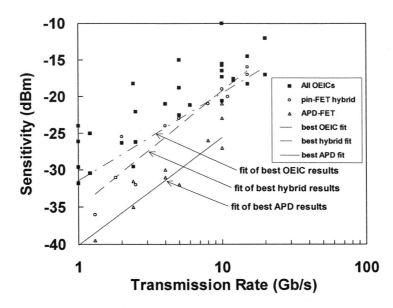

Fig. 3. A comparison of sensitivity data for OEIC receivers, hybrid receivers and receivers utilizing APDs.

rate in Fig. 3. Also shown are comparative data[36–39] for hybrid photoreceivers and for receivers using avalanche photodiodes (APDs). Least-squares fits for the best results in each category are also plotted in Fig. 3. It is evident that APD receivers exhibit the lowest sensitivities (by $\sim$ 5–8 dB) for B $\leq$ 10 Gb/s (as of this writing, there were no reported APD receivers operating above 10 Gb/s). As an alternative to APDs, erbium-doped fiber amplifiers (EDFAs) have been placed in front of optical receivers to achieve similar improvements in sensitivities.[5]

It is also apparent in Fig. 3 that OEICs and hybrid receivers are now comparable in performance (this hasn't always been the case, e.g., 3–5 years ago hybrids were 5–10 dB more sensitive than OEICs[36–38]). Further, the data in Fig. 3 indicate that the best OEIC front-ends exceed the performance of the best hybrids for data rates beyond 10 Gb/s.

3. Devices

It is informative to examine OEIC performances as a function of the device types used for the PDs and for the transistors, and such a breakdown is exhibited in Fig. 4 (the hybrid data is also included for comparison purposes). As can be seen a wide variety of devices have been employed: pin diodes (essentially a p-n diode with a low-doped, optically absorbing layer (i) interposed between the p and n layers), metal-semiconductor-metal (MSM) detectors (two back-to-back Schottky diodes), phototransistors (HPT), and, more recently, waveguided photodiodes (an optical waveguide feeds light to a pin diode)[40] as PDs, and heterojunction bipolar transistors (HBTs), high-electron-mobility transistors (HEMTs), junction-field-effect

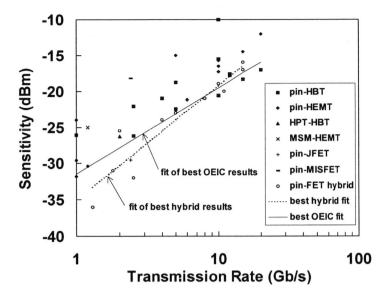

Fig. 4. Summary of recent sensitivity data for OEICs showing the performances for different device combinations. A comparison with hybrids is included.

transistors (JFETs), metal-insulator-semiconductor field-effect transistors (MIS-FETs) for the transistors. As can be seen in Fig. 4, the pin-HBT combination has achieved the best sensitivities for B > 10 Gb/s.[4,9,11]

The most important issue for the monolithic integration of a photoreceiver is to find a technique for combining PD and transistor on the same wafer with minimal performance degradation of either device type. Generally these techniques begin by growing sets of epitaxial layers for the devices, and, molecular-beam epitaxy (MBE), or metal-organic vapor-phase epitaxy (MOVPE), or sometimes combinations of the two are utilized. Of all the device combinations that have been used in the fabrication of OEICs, perhaps the simplest is an HBT-based approach in which the base-collector (B-C) junction is used as the absorption region of a pin photodiode. In this case only one set of device layers (HBT) are needed and the PD comes almost for free (two additional masks are needed to expose the optical absorption layer and to apply an antireflection coating).[2,4,10] A cross-section of an HBT-based OEIC is displayed in Fig. 5. The B-C PD plus HBT configuration has yielded quite good results, e.g., ≥ 18 GHz bandwidths.[2,7,10,14,18]

Other methods of integration include growing a stack of the PD and transistor layers, prepatterning the substrate then growing a layer stack, and selective area regrowth producing separate device layer stacks. It is also worth mentioning the flip-chip method. While not an OEIC technique per se, it has the potential of attaining low-parasitic photoreceivers. Good results have also been obtained with device sets developed by these methods, e.g., pin/HEMT,[6,8,12,17] MSM-HEMT,[26,33] a phototransistor (HPT)/HBT configuration,[29] pin-MISFET,[21] pin-JFET,[30] and waveguided pin-HFET.[40] This last OEIC attained an excellent 27 GHz bandwidth.

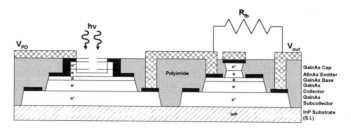

Fig. 5. InP-based OEIC cross-section of B-C PD and HBT.

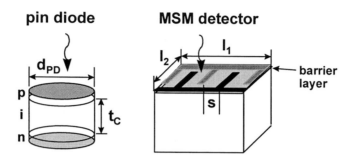

Fig. 6. The most common photodetector types found in InP-based OEICs.

The two basic PD types are depicted in simplified form in Fig. 6. The pin diode is shown on the left of the figure and the MSM detector is shown on the right. While the pin diode can be illuminated from either the top or the bottom areas, we will, for convenience, assume top illumination onto the region with area $\pi(d_{PD}/2)^2$ which produces a photocurrent that flows mostly vertically across the i region of thickness t_C. The MSM device is illuminated on the top surface defined by $I_1 \times I_2$ minus the areas occupied by the metal electrodes and produces a mostly lateral photocurrent flow. The favorable characteristics of a photodiode are low capacitance (high bandwidth), short transit time (high bandwidth), high responsivity (high sensitivity), and low dark current (low shot noise).

The capacitance of a pin diode is largely due to the parallel-plate structure defined by d_{PD} and t_C. Hence it would be desirable to make d_{PD} small and t_C large. But a short transit time favors small t_C. So the actual bandwidth achieved will be a function of a capacitance-transit time tradeoff. In addition, d_{PD} cannot be made infinitely small because one would like to capture all of the available radiation, i.e., $d_{PD} \geq$ beamwidth (hence, single-mode fiber is preferred in high-speed applications). A high responsivity favors, on the other hand, a large t_C and d_{PD}. Hence there is also a transit time-responsivity tradeoff to be made. It is this last tradeoff that is causing the exploration of waveguided PD structures.[40] An example of these tradeoffs is presented in Fig. 7 for a top-illuminated HBT base-collector pin diode.

The MSM is characterized by a fringe-field capacitance because the electrodes are in the same plane and are separated by relatively large distances (~ 3–5 μm,

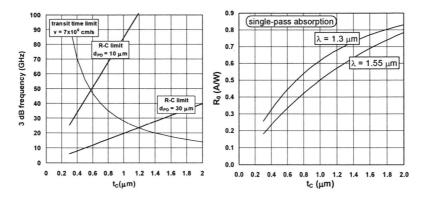

Fig. 7. Graphs showing the capacitance (R-C limit) — transit time tradeoff (on the left) and responsivity (on the right) as a function of collector thickness for a B-C PD.

whereas for a pin diode, $t_C <\sim 1 \mu m$). This means MSM detectors can have relatively small capacitance and large illumination areas, although the blockage of light by the electrodes acts to reduce the responsivity. A low dark current is a strong function of the details of fabrication and will not be discussed here other than to say epitaxial barrier layers are often used to establish adequate Schottky barrier heights (e.g., in MSMs).

The favorable characteristics of the transistors are low input current (low shot noise), i.e., low base current for the bipolars or low gate leakage current for the FETs, high unity current gain frequency, f_T, and high maximum frequency of oscillation, f_{max} (both for high bandwidth). High current-gain also helps to lower the input shot noise current. An example of one tradeoff involving a B-C PD and an HBT is the choice of t_C. Increasing t_C causes the PD responsivity to increase while the HBT's f_T decreases. This is shown in the graph in Fig. 8.

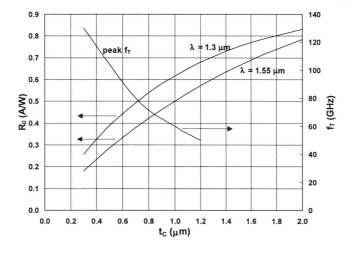

Fig. 8. Responsivity of a B-C PD and f_T of an HBT as functions of t_C.

A significant amount of progress has been made over the past few years with respect to the performance of transistors in OEIC technologies. Table 1 contains a list of some of the better OEIC technologies and it is evident that they are quite competitive with "electrical-only" technologies in terms of transistor speeds.

The decision as to which combination of devices will yield the ultimate high-speed receiver front-end is a complex one, and depends on the progress being made in the various technologies, as well as on the circuit design.

Table 1. A list of OEIC technology performances. The electrical device speeds are compara-ble to "electrical-only" technologies. Also shown are PD characteristics and the bandwidths of demonstrated OEIC receivers.

Authors	Ref	Year	Devices	f_T(GHz)	$d_{PD}(\mu m)$	R_0(A/W)	f_{3dB}(GHz)
Muramoto		1995	w-pin-HBT	30	4×25	0.60	8.3
Walden	[14]	1995	pin-HBT	60	15	0.60	19
Gutierrez-Aitken	[10]	1995	pin-HBT	67	9	0.40	19.5
Lunardi	[11]	1995	pin-HBT	70	15	1.06	10.4
Lai		1991	pin-HEMT	72	28		6.4
Klepser	[12]	1996	pin-HEMT	100		0.50	18
Sano	[7]	1994	pin-HBT	104	10	0.27	23
van Waasen	[40]	1997	w-pin-HFET	37	5×20	0.35	27

4. Circuit Architectures

The most commonly used circuit approach for OEIC front-ends is a photodiode connected to a transimpedance amplifier (see Fig. 2). The advantages of this circuit are high dynamic range and relatively high output voltage (high transimpedance gain). Expressions for low-frequency transimpedance, $Z_{TR}(0)$, input resistance, R_{in}, and 3 dB bandwidth, f_{3dB}, are as follows:

$$Z_{TR}(0) = -R_{fb}/(1 + 1/A) \tag{3}$$

$$R_{in} = R_{fb}/(1 + A) \tag{4}$$

$$f_{3dB} = (1 + A)/(2\pi R_{fb}(C_{in} + (1 + A)C_{fb})). \tag{5}$$

It is clear that R_{fb} determines $Z_{TR}(0)$ when A is large, and that high f_{3dB} is achieved when A is large and C_{in} and C_{fb} are small.

The design tradeoffs center on the value of the feedback resistance, R_{fb}: low noise (thermal noise current) and high transimpedance gain are attained with high values of R_{fb}, but, high bandwidths require low values of R_{fb}. A figure-of-merit that is often quoted for transimpedance amplifiers is the product of $Z_{TR}(0)$ and f_{3dB}. This idea can be extended to OEICs by including R_0 in the product:

$$F_{\text{OEIC}} = R_0 Z_{TR}(0) f_{3dB} \quad (V - Hz/W). \tag{6}$$

Table 2 presents a tabulation of some of the higher values of $F_{\rm OEIC}$ achieved to date. It is evident that most of the high $F_{\rm OEIC}$ values are associated with the pin-HBT combination, although pin-HEMT and pin-JFET receivers are also competitive.

A practical limitation of the transimpedance amplifier approach to OEIC receiver design is that as f_{3dB} increases toward 30 GHz (for 40 Gb/s receivers), the corresponding R_{fb} values tend toward $\leq 50\ \Omega$. This means that (1) in a 50 Ω environment the gain derived from the amplifier tends toward $\leq$ unity and (2) the thermal noise current due to R_{fb} becomes dominant. In addition, careful attention to the physical layout of such circuits is paramount because the electronic wavelengths are approaching the chip dimensions (mmwave design). Thus, it appears other approaches may be needed to implement receivers that can operate at $\geq$ 40 Gb/s. Two such possible alternatives are being examined: (1) a high-input-impedance amplifier,[12] and, (2) a traveling-wave amplifier (TWA).[34,40] Inductive peaking can also be used to extend bandwidth[35] (this reference discusses a GaAs-based OEIC, but the design approach can be applied to InP-based systems as well). Finally, as a precursor to future developments, preamplifiers (not OEICs) have been demonstrated with $f_{3dB} > 30$ GHz.[41,42]

Table 2. OEIC figure of merit data.

Authors	Ref	Year	Devices	R_0(A/W)	f_{3dB}(GHz)	$Z_{TR}(\Omega)$	$F_{\rm OEIC}$(V−Hz/W)
van Waasen	[40]	1997	w-pin-HFET	0.35	27	100	9.50E+11
Akahori	[8]	1994	pin-HEMT	1.06	9	144	1.37E+12
Gutierrez-Aitken	[10]	1995	pin-HBT	0.40	19.5	200	1.56E+12
Chandrasekhar	[15]	1996	pin-HBT	0.87	2	900	1.57E+12
Walden	[14]	1995	pin-HBT	0.60	19	180	2.05E+12
Sano	[7]	1994	pin-HBT	0.27	23	363	2.25E+12
Blaser	[30]	1992	pin-JFET	1.00	1.9	1500	2.84E+12
Walden	[16]	1996	pin-HBT	0.60	2.5	2800	4.20E+12

5. Arrays

Multiwavelength applications such as WDM networks have spurred interest in OEIC receiver arrays (as well as laser arrays for transmission). Table 3 contains a list of some OEIC array results. 1×8 OEIC arrays have been demonstrated for 2.5 Gb/s per channel operation[15,16] using B-C pin/HBT structures. Average per channel sensitivities in the range of -22 dBm to -26 dBm have been observed. A pin-HEMT 1×4 array has been reported with a 4 GHz bandwidth per channel[13] and has operated at 5 Gb/s with -15 dBm sensitivity. In addition, a B-C pin/HBT 1×4 array has been reported with a 19 GHz bandwidth per channel[14] which implies 100 Gb/s aggregate throughput.

A variation on the latter is a four-channel OEIC with a 3 dB frequency band extending from 1.2 GHz to 18 GHz wherein the lower frequency cutoff was realized

Table 3. A summary of demonstrated OEIC arrays.

Authors	Ref	Year	Devices	R_0(A/W)	f_{3dB}(GHz)	$Z_{TR}(\Omega)$	B(Gb/s)	P_{min}(dBm)	Channels
Berger		1993	pin-ModFET	0.41	0.8		1	−31.8	8
Walden	[16]	1996	pin-HBT	0.60	2.5	2800	2.5	−22.1	8
Chandra-sekhar	[15]	1996	pin-HBT	0.87	2	900	2.5	−26.2	8
Yano	[13]	1992	pin-HEMT	1.00	4	227	5	−15	4
Takahata	[17]	1996	w-pin-HEMT	0.61	8	155	10	−15.7	2
Walden	[14]	1995	pin-HBT	0.60	19	180	25		4

by inserting an integrated bias-tee between the photodiode and the input HBT device.[18] This circuit was built for an analog receiver application.

6. Conclusion

Excellent progress is being made with InP-based OEIC receiver front-end circuits, especially those employing the pin-HBT device combination. It is evident that they can be useful in WDM-type systems requiring arrays with data rates of ≥ 2.5 Gb/s, and, in single-channel digital systems at ≥ 10 Gb/s and in analog systems at ≥ 10 GHz.

References

1. R. F. Leheny et al., "Integrated In$_{0.53}$Ga$_{0.47}$As p-i-n F.E.T photoreceiver", Electron. Letts. **16** (1980) 353.
2. R. H. Walden et al., Optical Fiber Communication Conference Technical Digest, **4** (1994) 33.
3. Y. Akahori et al., "10-Gb/s high-speed monolithically integrated photoreceiver using InGaAs p-i-n PD and planar doped InAlAs/InGaAs HEMTs", IEEE Photon. Technol. Lett. **4** (1992) 754–756.
4. S. Chandrasekhar et al., "A OEIC photoreceiver using InP/InGaAs heterojunction bipolar transistors at 10 Gb/s", Electron. Lett. **28** (1992) 466–468.
5. Y. Akatsu et al., "A 10 Gb/s high sensitivity, monolithically integrated p-i-n-HEMT optical receiver", IEEE Photon. Technol. Lett. **5** (1993) 163–165.
6. A. L. Gutierrez-Aitken et al., "High-performance monolithic PIN-MODFET transimpedance photoreceiver", IEEE Photon. Technol. Lett. **5** (1993) 913–915.
7. E. Sano, M. Yoneyama, S. Yamahata, and Y. Matsuoka "23 GHz bandwidth monolithic photoreceiver compatible with InP/InGaAs double heterojunction bipolar transistor fabrication process", Electron. Lett. **30** (1994) 2064–2065.
8. Y. Akahori, M. Ikeda, A. Kohzen, and Y. Akatsu, "11 GHz ultrawide-bandwidth monolithic photoreceiver using InGaAs pin PD and InAlAs/InGaAs HEMTs", Electron. Lett. **30** (1994) 267.
9. L. Lunardi et al., "A 12 Gb/s high-performance, high-sensitivity monolithic p-i-n/HBT photoreceiver module for long-wavelength transmission systems", IEEE Photon. Technol. Lett. **7** (1995) 182–184.
10. A. L. Gutierrez-Aitken et al., "Wide bandwidth InAlAs/InGaAs monolithic pin-HBT photoreceiver", IEEE/LEOS Summer Topical Meetings: ICs for New Age Lightwave Communications, Keystone, CO, August 1995.

11. L. Lunardi *et al.*, "20 Gb/s monolithic p-i-n/HBT photoreceiver module for 1.55 mm applications", *IEEE Photon. Technol. Lett.* **7** (1995) 1201–1203.

12. B.-U. H. Klepser *et al.*, "Monolithically integrated InP-based pin-HEMT OEIC receiver with a bandwidth of 18 GHz", *Optical Fiber Communication Conference Technical Digest*, **6** (1996) 63–64.

13. H. Yano *et al.*, "5 Gbit/s four-channel receiver optoelectronic integrated circuit array for long-wavelength lightwave systems", *Electron. Lett.* **28** (1992) 503–504.

14. R. H. Walden *et al.*, "An InP-based HBT 1 × 8 OEIC array for a WDM network", *IEEE/LEOS Summer Topical Meetings: ICs for New Age Lightwave Communications*, Keystone, CO, August 1995, paper FC1.

15. S. Chandrasekhar *et al.*, "Investigation of crosstalk performance of eight-channel p-i-n/HBT OEIC photoreceiver array modules", *IEEE Photonics Tech. Letts.* **8** (1996) 682–684.

16. R. H. Walden, C. Dreze, K. Warbrick, and C. Chew, "An OEIC-based, 8-channel, optical receiver submodule for a 2.5 Gb/s WDM network access module", *Engineering Foundation High Speed Opto-Electronics for Communications II*, Snowbird, UT, August 1996.

17. K. Takahata *et al.*, "10-Gb/s two-channel monolithic photoreceiver array using waveguide p-i-n PD's and HEMT's", *IEEE Photon. Technol. Lett.* **8** (1996) 563–565.

18. D. Yap *et al.*, "RF-optoelectronic receiver arrays for wideband, large-dynamic-range links", *1996 Government Microcircuit Appl. Conf. Digest of Papers* **21**, Orlando, FL, (1996) 404.

19. R. G. Smith and S. D. Personick, "Receiver design for optical fiber communication systems", Chapter 4 in *Semiconductor Devices for Optical Communication,* Springer-Verlag, New York, 1980.

20. G. Sasaki *et al.*, *IEEE J. Lightwave Tech.* **7** (1989) 1510–1514.

21. V. D. Mattera *et al.*, "Monolithic InGaAs p-i-n InP metal-insulator-semiconductor field-effect transistor for long-wavelength optical communications", *Appl. Phys. Lett.* **57** (1990) 1343–1344.

22. H. Yano *et al.*, "Low-noise current optoelectronic integrated receiver with internal equalizer for gigabit-per-second long-wavelength communications", *IEEE J. Lightwave Tech.* **8** (1990) 1328–1333.

23. S. Chandrasekhar *et al.*, "4 Gbit/s pin/HBT monolithic photoreceiver", *Electron. Lett.* **26** (1990) 1880–1882.

24. S. Chandrasekhar *et al.*, "A monolithic long wavelength photoreceiver using heterojunction bipolar transistors", *IEEE J. Quantum Electron.* **27** (1991) 773–777.

25. H. Hayashi *et al.*, "Giga-bit rate receiver OEICs grown by OMVPE for long-wavelength optical communications", *IEE Proc.* **138** (1991) 164–170.

26. W.-P. Hong *et al.*, "Monolithically integrated waveguide-MSM detector-HEMT amplifier receiver for long-wavelength lightwave systems", *IEEE Photon. Technol. Lett.* **3** (1991) 156–158.

27. S. Chandrasekhar *et al.*, "A monolithic 5 Gb/s p-i-n/HBT integrated photoreceiver circuit realized from chemical beam epitaxial material", *IEEE Photon. Technol. Lett.* **3** (1991) 823–825.

28. P. R. Berger *et al.*, "1.0 GHz monolithic p-I-n MODFET photoreceiver using molecular beam epitaxial regrowth", *IEEE Photon. Technol. Lett.* **4** (1992) 891–894.

29. S. Chandrasekhar, A. H. Gnauck, R. A. Hamm, and G. J. Qua, "The phototransistor revisited: all bipolar monolithic photoreceiver at 2 Gb/s with high sensitivity", *IEEE Trans. Electron Devices* **39** (1992) 2677.

30. M. Blaser and H. Melchior, "High-performance monolithically integrated In$_{0.53}$-

$Ga_{0.47}As/InP$ p-i-n/JFET optical receiver front-end with adaptive feedback control", *IEEE Photon. Technol. Lett.* **4** (1992) 1244–1247.

31. W. S. Lee and S. A. Rosser, "Monolithically integrated long wavelength optical receiver OEICs using InAlAs/InGaAs heterojunction MESFETs (HFETs)", *Electron. Lett.* **38** (1992) 365–367.

32. S. Chandrasekhar *et al.*, "High-speed monolithic p-i-n/HBT photoreceivers implemented with simple phototransistor structure", *IEEE Photon. Technol. Lett.* **5** (1993) 1316–1318.

33. G.-K. Chang *et al.*, "A 3 GHz transimpedance OEIC receiver for 1.3–1.55 mm fiberoptic systems", *IEEE Photon. Technol. Lett.* **2** (1990) 197–199.

34. B. Kerzer *et al.*, "Ultra wide band optical receiver frontend", *EFOC & N Technical Digest* (1994) 199–202.

35. H. Kamitsuna, "Ultra-wideband monolithic photoreceivers using HBT-compatible HPT's with novel base circuits, and simultaneously integrated with an HBT amplifier", *J. Lightwave Technol.* **13** (1995) 2301–2307.

36. L. E. Larson and J. J. Brown, *GaAs IC Symposium Technical Digest* (1992).

37. T. V. Muoi, "Fiber optic transmitter and receiver design", *Optical Fiber Commun. Conf.* Short Course 128, 1994.

38. K. Pedrotti, *Optical Fiber Commun. Conf.*, Feb. 1994.

39. S. Chandrasekhar, *Optical Fiber Commun. Conf. Technical Digest* **2** (1996) 63.

40. S. van Waasen *et al.*, "27-GHz bandwidth high-speed monolithic integrated optoelectronic photoreceiver consisting of a waveguide fed photodiode and an InAlAs/InGaAs-HFET traveling wave amplifier", *IEEE J. Solid-State Circuits* **32** 1394–1401.

41. Y. Imai *et al.*, "40-Gb/s-class InP HEMT ICs for very-high-speed optical communications", in *7th Int. Conf. InP and Related Materials Proc.* 1995, pp. 89–92.

42. Y. Suzuki *et al.*, "An HBT preamplifier for 40-Gb/s optical transmission systems", *GaAs IC Symp. Technical Digest* **18** (1996) 203–206.

International Journal of High Speed Electronics and Systems, Vol. 9, No. 2 (1998) 643–670

ULTRAHIGH f_{max} AlInAs/GaInAs TRANSFERRED-SUBSTRATE HETEROJUNCTION BIPOLAR TRANSISTORS FOR INTEGRATED CIRCUITS APPLICATIONS

BIPUL AGARWAL, RAJASEKHAR PULLELA, UDDALAK BHATTACHARYA*
DINO MENSA, QING-HUNG LEE, LORENE SAMOSKA
JAMES GUTHRIE and MARK RODWELL
*Department of Electrical and Computer Engineering, University of California,
Santa Barbara, CA 93106, USA*

Transferred-substrate heterojunction bipolar transistors (HBTs) have demonstrated very high bandwidths and are potential candidates for very high speed integrated circuit (IC) applications. The transferred-substrate process permits fabrication of narrow and aligned emitter-base and collector-base junctions, reducing the collector-base capacitance and increasing the device f_{max}. Unlike conventional double-mesa HBTs, transferred-substrate HBTs can be scaled to submicron dimensions with a consequent increase in bandwidth. This paper introduces the concept of transferred-substrate HBTs. Fabrication process in the AlInAs/GaInAs material system is presented, followed by DC and RF performance. A demonstration IC is shown along with some integrated circuits in development.

1. Introduction

Heterojunction bipolar transistors (HBTs)[1] have high transconductance, extremely reproducible DC parameters and wide bandwidth. These attributes make HBTs the device of choice for many precision high speed circuits. Important HBT applications include analog-to-digital conversion,[2] chip-sets for fiber-optic transmission,[3,4] and direct digital frequency synthesis. These are all medium-scale integrated circuits operating at frequencies of several tens of GHz and incorporating both precision broadband analog and very high clock-rate digital subcircuits. In these applications, both the transistor current gain cut-off frequency f_τ, and the power gain cut-off frequency f_{max} must be considerably higher than the signal frequencies involved. 100 Gbps optical-fiber transmission ICs will require HBTs having f_τ and f_{max} greater than 200–250 GHz.[5] A second-order $\Sigma - \Delta$ analog-digital converter having a 50 GHz sample rate would require HBTs with f_τ and $f_{max} \sim$ 200–300 GHz, but might provide 12 bits resolution at 1 GHz bandwidth. Broadband amplifiers for 40 and 100 Gbps communication would require HBTs with similar performance. HBTs with bandwidths of several hundred GHz will benefit many similar applications.

Progressive improvements in device bandwidths are needed to keep pace with the demand for integrated circuits operating at higher frequencies. High electron

*Now with Intel Corporation, Hillsboro, OR 97125.

mobility transistors (HEMTs) with short gate lengths ($\sim 0.1\ \mu$m) are presently the largest bandwidth three-terminal devices. HEMTs with a power gain cutoff frequency ($f_{\max}$) of 450 GHz have been reported.[6] The superior bandwidths of HEMTs is a result of the rapid improvement in HEMT bandwidth with deep submicron scaling.

Reducing the lithographic dimensions (lateral) and/or the semiconductor layer thicknesses (vertical) of a device is termed device scaling. Scaling pertinent device dimensions is central to high frequency semiconductor device design. With several important semiconductor devices, the device bandwidth increases as critical lithographic dimensions and layer thicknesses are reduced. Examples of highly scaled devices with large bandwidths are 0.1 μm gate length high electron mobility transistors (HEMTs),[6] 0.25 μm gate length complementary metal-oxide-semiconductor (CMOS) transistors, 0.1 μm Schottky-collector resonant tunnel diodes (SRTDs)[7] and submicron Schottky diodes used as sub-millimeter wave mixers.[8]

Figure 1 shows $f_{\max}$ versus emitter width for some HBTs reported in the literature. It can be seen that there is no strong correlation between $f_{\max}$ and emitter width, at least for narrow emitters. Consequently, HBTs are not fabricated with deep submicron dimensions except where bias currents have to be limited for low power operation.

We will now explore the relationship between $f_{\max}$ and emitter and collector widths. A simple hybrid-π model of an HBT is shown in Fig. 2 with the model components related to the device parameters and biasing conditions. The expression for short circuit current gain corresponding to this device model is

$$A_I = \frac{-\beta}{1 + j\omega\beta\left[(1/g_m)\left(C_{be,\text{tot}} + C_{cb}\right) + (r_{ex} + r_c)C_{cb}\right]}, \tag{1}$$

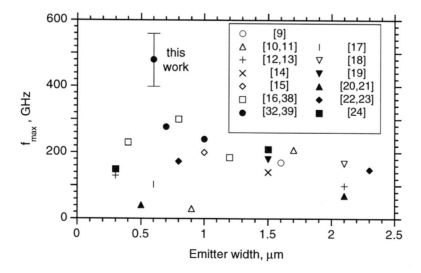

Fig. 1. Comparison of $f_{\max}$ and emitter width for some HBTs in the literature.

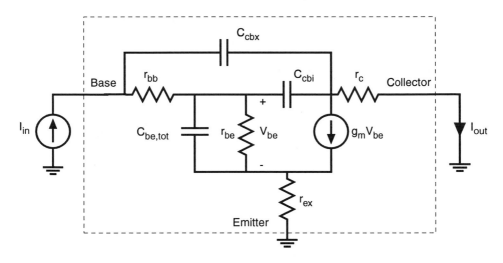

r_{bb} base resistance	$C_{be,tot} = g_m(\tau_b + \tau_c) + C_{be,depl}$
r_c collector contact resistance	C_{cbi} intrinsic collector-base capacitance
r_{ex} emitter contact resistance	C_{cbxi} extrinsic collector-base capacitance
$r_{be} = \beta/g_m$	$g_m = qI_c / kT$

Fig. 2. Hybrid-π model of an HBT, connected externally for calculation of the short-circuit current gain.

where $C_{cb} = C_{cbi} + C_{cbx}$ and the following assumptions are made: DC short circuit current gain $\beta \gg 1$, $\omega C_{cb} \ll g_m/(1 + g_m r_{ex})$, only first order terms in angular frequency ω considered and the extrinsic C_{cb} charging time is small.

The short circuit current gain cutoff frequency f_τ corresponds to the frequency at which the magnitude of A_I is unity. Near this frequency, the imaginary part of the denominator in Eq. (1) is much larger than unity. The expression for f_τ as a function of device parameters and biasing conditions is therefore approximately[25]

$$\frac{1}{2\pi f_\tau} = \tau_b + \tau_c + \frac{kT}{qI_c}(C_{be,depl} + C_{cb}) + (r_{ex} + r_c)C_{cb}. \qquad (2)$$

Here τ_b is the base transit time, τ_c the collector transit time, kT/q the thermal voltage, I_c the collector current, $C_{be,depl}$ the base-emitter depletion capacitance, C_{cb} the total collector-base capacitance, r_{ex} the emitter contact resistance, and r_c the collector contact resistance. At a given current density, all terms except r_c in the above equation are independent of lateral scaling. The base and collector transit times can be reduced by reducing the appropriate semiconductor layer thicknesses. Hence, f_τ can be improved by vertical scaling of the device, but is independent of lateral scaling.

The power gain cutoff frequency $f_{\max}$ is another important figure-of-merit of high-frequency HBT performance. $F_{\max}$ defines the maximum frequency at which a device can provide power gain. F_{max} not only depends on f_τ, but also on the base-resistance-collector-base-capacitance time constant as[25]

$$f_{max} = \frac{1}{2}\sqrt{f_\tau f_{cb}}, \tag{3}$$

where

$$f_{cb} = \frac{1}{2\pi r_{bb} C_{cbi}}, \tag{4}$$

where, r_{bb} is the base resistance and C_{cbi} is the fraction of the collector-base capacitance that is charged through the base resistance. No matter how large the value of f_τ is, power gain is available only at frequencies below f_{max}. Hence, to improve device bandwidth, is it important to improve f_τ and the $r_{bb} C_{cb}$ time constant. Note that while the extrinsic collector-base capacitance has no impact on f_{max}, it does impact the performance of many circuits, and should be minimized.

The cross-section of a double-mesa HBT is shown in Fig. 3. The base resistance r_{bb} has three components. These are: the contact resistance from the base Ohmic contact, the sheet resistance from the gap between the emitter mesa and the base Ohmic contact, and the spreading resistance of the base layer underneath the emitter mesa. The contact resistance is given by[26]

$$R_{contact} = \frac{\sqrt{\rho_{bc}\rho_{bs}}}{2l}, \tag{5}$$

where ρ_{bc} is the specific contact resistance per unit area of the metal-semiconductor interface (units of Ω-cm^2), and ρ_{bs} is the base sheet resistivity (units of $\Omega/\square$). The gap resistance is given by[26]

$$R_{gap} = \rho_{bs} \frac{W_{gap}}{2l}, \tag{6}$$

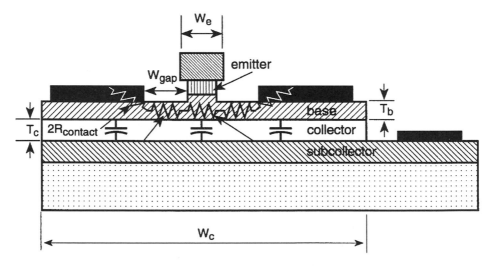

Fig. 3. Schematic cross-section of a double-mesa HBT.

where W_{gap} is the separation between the emitter mesa and the base Ohmic contact. The spreading resistance is given by[26]

$$R_{\text{spread}} = \rho_{bs} \frac{W_e}{12l}. \tag{7}$$

For HBTs fabricated with a self aligned base-emitter process, $W_{\text{gap}} \simeq 0.1 \ \mu$m and R_{gap} is negligible. We note also that R_{spread} is proportional to the emitter width W_e, but that R_{contact} is independent of W_e. Hence, for narrow-emitter devices, the base resistance is dominated by R_{contact} and is only weakly dependent upon W_e. The traditional "base spreading resistance" is not an appropriate description for the r_{bb} of a typical narrow-emitter HBT. The total collector base capacitance is given by

$$C_{cb} = \frac{\epsilon l W_c}{T_c}, \tag{8}$$

where, W_c is the width of the collector ($=$ width of the base mesa), l is the length of the emitter stripe and T_c is the thickness of the collector depletion region. Hence, C_{cb} is proportional to the width of the base mesa which in turn is much wider than, and, independent of W_e. Note that only a fraction C_{cbi} of C_{cb} is charged through the base resistance r_{bb}. Determination of the C_{cbi}/C_{cb} ratio from the HBT physical dimensions has not been fully explored in the literature. We estimate that

$$C_{cbi} \simeq \frac{\epsilon l (W_e + 2W_{\text{gap}} + 2l_{\text{contact}})}{T_c}, \tag{9}$$

where

$$l_{\text{contact}} = \sqrt{\rho_{bc}/\rho_{bs}} \tag{10}$$

is the transfer length of the base Ohmic contact. This approximation is based upon taking the fraction of C_{cb} whose charging current shares a common path through the base with the currents associated with charging C_{be}. We note that the collector-base junction area of double-mesa HBTs is further increased by the presence of a base contact pad area necessary to bring interconnect metallization onto the base Ohmic metal. Lateral scaling of the emitter width below $\sim 1 \ \mu$m does not substantially improve the bandwidth of double-mesa HBTs.

We have developed a transferred-substrate HBT which can be scaled laterally to improve bandwidth.[27] The process involves transferring the transistor epitaxial layers from the original substrate to a new carrier substrate, hence the name "transferred-substrate". The cross-section of a transferred-substrate HBT is shown in Fig. 4. Transferred-substrate HBTs have lithographically defined narrow emitter and collector stripes aligned to each other on opposite sides of the base epitaxial layer.[28] This is the special feature of the transferred-substrate HBT in contrast to the double-mesa HBT. The fabrication of transferred-substrate HBTs requires access to the emitter and the collector sides of the epitaxial film. The process of substrate transfer allows this access and is an essential step in the fabrication process.

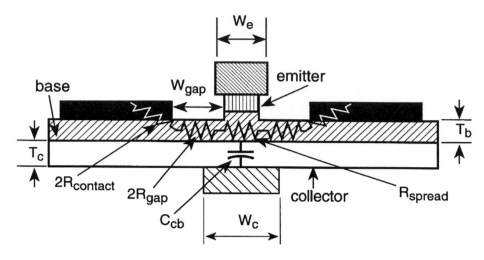

Fig. 4. Schematic cross-section of a transferred-substrate HBT.

The base resistance is dominated by R_{contact} and is independent of W_e, as for the double-mesa HBTs. The collector-base capacitance C_{cb} is proportional to the width of the collector stripe W_c, if fringing capacitance is negligible (Fig. 4); $C_{cb} = \epsilon l W_c / T_c$, where l is the length of the collector/emitter stripe and T_c is the thickness of the collector depletion region. If a constant ratio is maintained between the emitter and the collector widths, C_{cb} is proportional to the emitter width W_e. Noting that $r_{bb} \simeq R_{\text{contact}} = \sqrt{\rho_{bc}\rho_{bs}}/2l$, and $C_{cb} = \epsilon l W_c / T_c$ (where $W_c \propto W_e$), the $r_{bb} C_{cb}$ time constant is proportional to W_e. This, when used in Eq. (4) yields the relationship

$$f_{cb} \propto \frac{1}{W_e}. \tag{11}$$

Using Eq. (11) in Eq. (3), it is observed that the maximum frequency of oscillation $f_{\max}$ depends on W_e as

$$f_{\max} \propto \frac{1}{\sqrt{W_e}}. \tag{12}$$

Hence, $f_{\max}$ rapidly improves with submicron lateral scaling of the device.

The cross-section of the transferred-substrate HBT in Fig. 4 shows a direct Schottky contact to the collector depletion layer. An Ohmic collector contact having the same width as the Schottky contact will also show the same variation of $f_{\max}$ with scaling. The two devices are almost identical except for an extra potential drop across the collector-base junction of the Schottky-collector device. This potential drop, due to the Schottky contact, is the difference between the work function of the metal Schottky contact and the electron affinity of the collector semiconductor. A Schottky collector contact provides a slight improvement in f_τ relative to the Ohmic collector contact because of the absence of a collector contact resistance (Eq. (2)). It is also easier to fabricate deep submicron Schottky collector contacts (e.g., 0.1 μm T-gate) than deep submicron Ohmic contacts.

The scaling law of Eq. (12) suggests that the operating bandwidth of HBTs can be increased without bounds by lithographic scaling alone. In fact, to obtain usable devices, vertical scaling of the epitaxial layer thicknesses must accompany the lithographic scaling. Except in the case of reactively-matched amplifiers and distributed circuits where f_{max} is the sole determinant of circuit bandwidth, both f_τ and f_{max} are generally important for optimum circuit performance in a given technology. Devices with $f_{max} \ll f_\tau$, obtained by thinning the epitaxial layers *without* lateral lithographic scaling will show circuit bandwidth determined by r_{bb} and C_{cb}. Devices with $f_{max} \gg f_\tau$ (e.g., devices with relatively thick epitaxial layers and significant lateral lithographic scaling) will show circuit bandwidth dominated by $(\tau_b + \tau_c)$. Reduction of $(\tau_b + \tau_c)$ is obtained by thinning the epitaxial layers, which unfortunately increases $r_{bb}C_{cb}$. The transferred-substrate HBT allows for the subsequent reduction of $r_{bb}C_{cb}$. A device having high values for *both* f_τ and f_{max} is thus possible.

Several approaches have been reported for reducing the collector-base capacitance of HBTs, and thereby improving f_{max}. One approach is the reduction of the width of the base mesa.[29] This relies on improvements in base contact technology because a narrow base mesa results in a smaller base Ohmic contact area. The base contact width must be at least one transfer length if the contact resistance is to be kept small. In contrast, the transferred-substrate technique provides independent control of the base and collector contact widths. There have been other approaches to make the collector contact width independent of the size of the base mesa, such as selective etching to undercut the collector,[30] collector isolation implant,[31] contacting the base with a very narrow L-shaped contact,[32] and selective lateral oxidation of the emitter[33] for a collector-up growth. However, for deep-submicron scaling, powerful fine-line lithography is likely to be the technique of choice for defining the collector and emitter contacts of HBTs.

In addition to high device bandwidths, integrated circuit processes must incorporate a low capacitance wiring environment and low ground return inductance. Wire lengths, and hence, transistor spacings must be small. Given that fast HBTs operate at a current density $\sim 10^5$ A/cm^2, efficient heat sinking is then vital. The transferred-substrate HBT IC process[34] uses benzocyclobutene (BCB), a low-loss, low dielectric-constant ($\epsilon_r = 2.7$), spin-on dielectric as the substrate for microstrip interconnects, thus providing low capacitance. The thin ($\simeq 10$ μm) BCB substrate with a gold ground plane underneath also provides low inductance ground vias. Transistor heat-sinking is through electroplated gold thermal vias.

2. Fabrication

The AlInAs/InGaAs material system was chosen for the fabrication of transferred-substrate HBTs. The HBT epitaxial layer structure (Fig. 5(a)) is grown by molecular beam epitaxy on a Fe-doped semi-insulating (100) InP substrate, starting with a 2500 Å AlInAs buffer layer. The GaInAs collector is 2700 Å thick, is Si-doped at

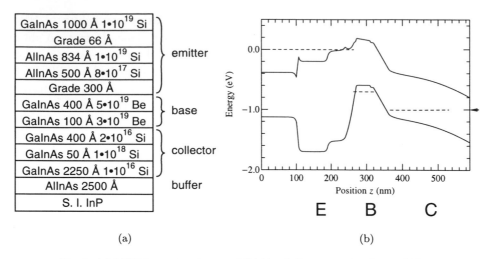

<table>
<tr><td>GaInAs 1000 Å 1•10^{19} Si</td></tr>
<tr><td>Grade 66 Å</td></tr>
<tr><td>AlInAs 834 Å 1•10^{19} Si</td></tr>
<tr><td>AlInAs 500 Å 8•10^{17} Si</td></tr>
<tr><td>Grade 300 Å</td></tr>
</table>
emitter

GaInAs 400 Å 5•10^{19} Be
GaInAs 100 Å 3•10^{19} Be
base

GaInAs 400 Å 2•10^{16} Si
GaInAs 50 Å 1•10^{18} Si
GaInAs 2250 Å 1•10^{16} Si
collector

AlInAs 2500 Å
buffer

S. I. InP

(a)   (b)

Fig. 5. (a) MBE layer structure and (b) Band diagram under forward bias.

$1 \times 10^{16}/\text{cm}^3$ and contains a $5 \times 10^{11}/\text{cm}^2$ Si pulse-doped layer 400 Å from the base. This pulse-doped layer inhibits the onset of base push-out.[35] The 500 Å GaInAs base is graded in both doping and bandgap. The 100 Å of the $\text{Ga}_{0.47}\text{In}_{0.53}\text{As}$ base immediately adjacent to the collector is Be-doped at $3 \times 10^{19}/\text{cm}^3$. The remaining 400 Å of the base is Be-doped at $5 \times 10^{19}/\text{cm}^3$. By increasing the Ga cell temperature progressively during growth of the 400 Å layer, the Ga:In ratio is gradually increased, introducing a ~ 0.03 eV bandgap gradient across the 400 Å layer. The base is then graded in 300 Å to the AlInAs emitter. The first 66 Å of the grade is Be-doped at $2 \times 10^{18}/\text{cm}^3$ and the remainder is Si-doped at $8 \times 10^{17}/\text{cm}^3$. The AlInAs emitter is about 1350 Å thick. The first 500 Å are Si-doped at $8 \times 10^{17}/\text{cm}^3$ and the remainder is Si-doped at $1 \times 10^{19}/\text{cm}^3$. This is graded in 66 Å to the InGaAs emitter cap. The emitter cap is n^+ doped with Si at $1 \times 10^{19}/\text{cm}^3$.

The band diagram corresponding to the layer structure of Fig. 5(a) under forward bias is shown in Fig. 5(b). The biasing conditions are as follows: base-emitter voltage $V_{BE} = 0.7$ V, collector-emitter voltage $V_{CE} = 1.0$ V, and a emitter current density of 1×10^5 A/cm^2. The collector current density is assumed to be the same as the emitter current density, as is the case of a narrow-collector HBT. The effect of the electrons in the collector space charge layer due to the collector current is included while calculating the electric field and the electrostatic potential in the collector space charge layer. An electron velocity of 3×10^5 m/s is assumed in these calculations.

Figure 6 shows sequentially, the steps involved in fabrication of transferred-substrate HBTs. The fabrication process starts with the evaporation of Ti/Pt/Au emitter contacts. A combination of a dry etch, a selective wet etch, and a non-selective wet etch is then used to etch down to the base epitaxial layer. Figure 7 shows a SEM photomicrograph of the device cross-section after this step. The

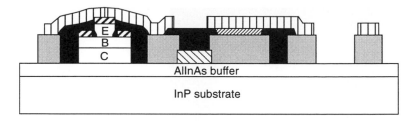

(a)

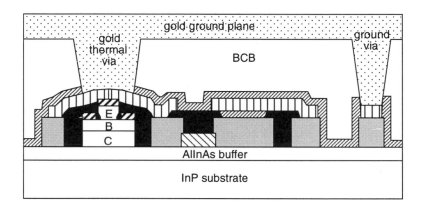

(b)

(c)

Fig. 6. Fabrication process for transferred-substrate HBTs.

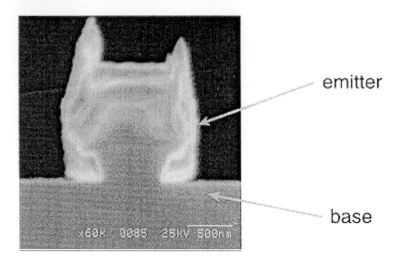

Fig. 7. SEM cross-section of device after emitter-base etch.

non-selective wet etch causes a lateral undercut and reduces the emitter-base junction area to below the lithographically defined area as seen in this cross-section.

Self-aligned Ti/Pt/Au base metal is evaporated and sintered at 300°C for one minute. Transistors are then isolated by forming mesas using a dry etch, stopping on the AlInAs buffer layer. Thin film NiCr is evaporated on the wafer to form resistors with 50 $\Omega/\Box$ sheet resistivity. The first level of metallization (metal1) is done at this point. This metal forms most of the transmission lines, interconnect wiring, probe pads, capacitor bottom plate and resistor contacts. The devices are passivated and planarized with polyimide. Polyimide also passivates the NiCr resistors and serves as interconnect crossovers for multiple wiring levels. The capacitor dielectric (1000 Å SiN) is then deposited and etched away in unwanted regions. Emitters and bases are contacted by electroplated Au. The electroplated Au also forms the second level of metallization (metal2) and the top plate for MIM capacitors. The process at this point is shown in Fig. 6(a). Figure 8(a) shows a photomicrograph of a device after this step.

The substrate transfer process starts with the deposition of a 4000 Å SiN (insulator) layer by PECVD which serves to insulate the wafer. The wafer is then coated with a thick layer of Benzocyclobutene (BCB, $\epsilon_r = 2.7$). Thermal vias are formed on the emitters by dry etching openings in the BCB. This also etches the SiN insulator layer in the vias. The vias are then filled with thick Au by electroplating. The electroplated Au also forms an electrically and thermally conducting ground plane, thus grounding all emitters. Ground vias are also formed in this step. Microstrip interconnect lines on BCB provide a low-capacitance environment wiring for dense circuits. Ground vias to the thick electroplated ground plane provide low ground return inductance. Thermal vias provide efficient heat-sinking for devices operating at very high current densities. Figure 6(b) shows the process at this point.

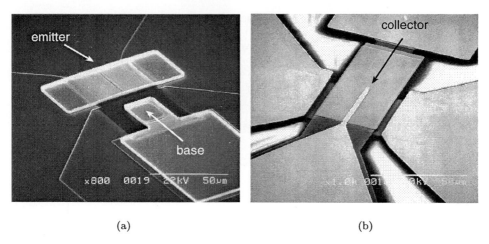

(a)   (b)

Fig. 8. SEM photomicrograph of device after (a) partial fabrication and (b) complete fabrication.

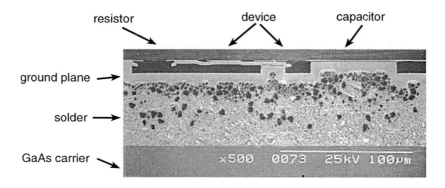

Fig. 9. SEM photomicrograph of cross-section through an IC.

The wafer is then inverted and bonded to a GaAs transfer substrate. The InP substrate is removed in an aqueous solution of HCl. The AlInAs buffer layer is also removed by the same etch. This etch stops selectively on the InGaAs collector layer. Ti/Pt/Au Schottky collector contacts are then deposited. Outside the active collector area, 1500 Å of the collector drift region are then removed by a self-aligned wet etch to reduce fringing capacitance. The process at this step is shown in Fig. 6(c). Figure 8(b) shows a photomicrograph of a completed device. Figure 9 shows a SEM photomicrograph of the cross-section of an integrated circuit with active and passive components.

Transistors with non-grounded emitters (required in ICs) are fabricated by protecting the SiN insulator with a metal layer (Fig. 10) before applying BCB. Without this layer, the SiN insulator is removed during the via formation etch. This metal layer (metal3) also forms power supply bypass capacitors and forms a third level of metallization. Figure 11 shows a SEM photomicrograph of the cross-section of a fully fabricated device with a non-grounded emitter.

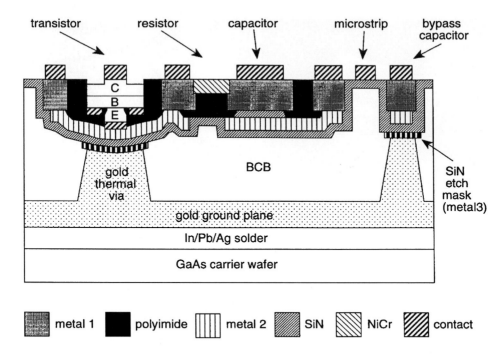

Fig. 10. Fabrication process for devices with non-grounded emitters.

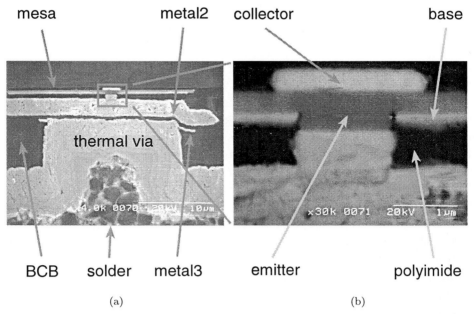

Fig. 11. SEM photomicrograph of (a) cross-section through a device and (b) close-up of intrinsic device.

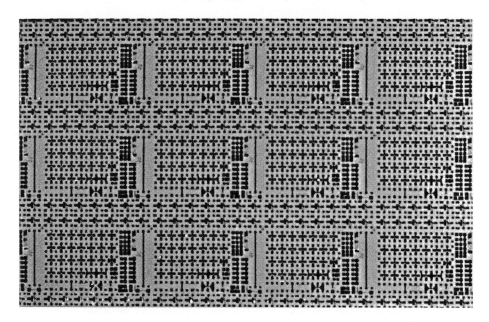

Fig. 12. Photomicrograph of part of a wafer with ICs and discrete devices.

Devices with non-grounded emitters should have similar performance and heat-sinking as the devices with grounded emitters. SiN has $\epsilon_r = 6$ and ~ 10–30 W/m-K thermal conductivity. For an HBT with a 0.6×25 μm^2 emitter, the thermal via is 10×24 μm^2. For non-grounded emitter devices, the calculated capacitance from the emitter airbridge to the grounded substrate is 25 fF, which is much smaller than 730 fF base-emitter capacitance. The calculated 67–200 K/W thermal resistance of the SiN layer should result in less than 3.5°C additional temperature rise for a device biased at 10^5 A/cm^2 and 1.0 V.

For integrated circuit applications, high device yield is required. The transferred-substrate HBT IC technology is being improved towards this goal. The technology is currently capable of yielding circuits with few transistors. Figure 12 shows a photograph of part of a wafer with discrete devices and small integrated circuits.

The device and circuit results reported here are from different wafers with small variations in the MBE layer structure and the fabrication process, and do not significantly affect the central theme of this work.

3. DC and RF Measurements

A variety of devices have been fabricated with different emitter and collector widths. It is important in characterization to determine the emitter-base and collector-base junction areas fairly accurately. Wet chemical etches with lateral undercuts are used to reduce both the emitter-base and collector-base junction areas to below their lithographically defined dimensions. Junction dimensions are determined by measuring (with a microwave network analyzer) the junction capacitances versus the lithographically defined junction widths. Devices were fabricated with 0.6 μm $\times$ 25 μm emitters and with collector dimensions of 0.8 μm $\times$ 29 μm (narrow-collector) and 1.8 μm $\times$ 29 μm (wide-collector). Devices with very wide collectors ($\sim$ 5 μm), as is normally the case in double-mesa HBTs, were also fabricated. All devices have their emitters grounded by thermal vias, except where explicitly stated. While most of the data presented here is for the above devices, some data will also be presented from devices with different geometries and configurations.

3.1. *DC measurements*

Figure 13 shows the Gummel plots for transferred-substrate HBTs with the three different collector widths. The independence of these plots to collector width can be observed. The collector current ideality factor n_c is close to unity. The base current ideality factor n_b is 1.2.

DC common-emitter characteristics of the HBTs are shown in Fig. 14. The small signal current gain at DC, β, is 55. At high current densities, narrow-collector devices show significantly larger collector-emitter saturation voltages ($V_{CE,\text{sat}}$), arising from screening of the collector electrostatic field by the electron space charge. Screening occurs at a *collector* current density J_C satisfying the relationship (V_{CB} + ϕ) $= T_C^2(J_C/v_{\text{sat}} - qN_d)/2\epsilon$, where T_C is the collector depletion layer thickness, N_d the collector doping, ϕ the junction built-in potential, and v_{sat} the electron velocity. In wide-collector HBTs, there is significant lateral spreading of the electron flux

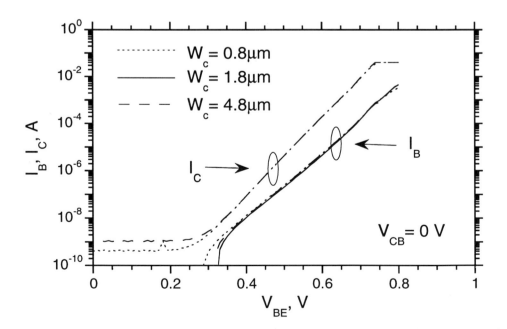

Fig. 13. Gummel plots of transferred-substrate HBTs with 0.6 μm $\times$ 25 μm emitters and different collector widths.

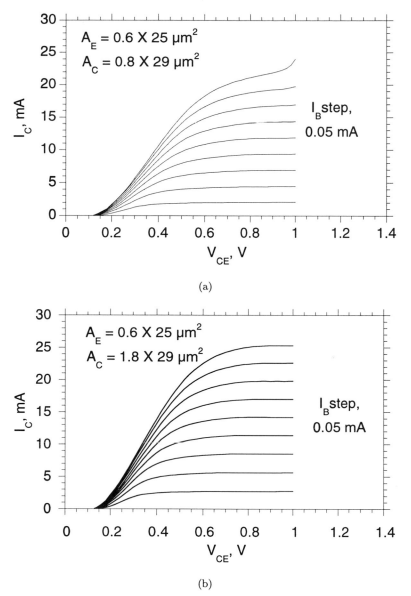

$A_E = 0.6 \times 25 \ \mu m^2$

$A_C = 0.8 \times 29 \ \mu m^2$

$I_B$step, 0.05 mA

(a)

$A_E = 0.6 \times 25 \ \mu m^2$

$A_C = 1.8 \times 29 \ \mu m^2$

$I_B$step, 0.05 mA

(b)

Fig. 14. DC common-emitter characteristics of devices with (a) narrow collector and (b) wide collector.

at high current densities,[36] reducing the collector space-charge density. Lateral current confinement in narrow-collector HBTs results in both increased $V_{CE,\text{sat}}$ and decreased emitter current density at the onset of f_τ collapse (Kirk effect), resulting in increased emitter charging times and reduced f_τ.

The common-emitter breakdown voltage BV_{CEO} is ~ 3 V at ~ 0 current density, decreasing to 1.5 V at 10^5 A/cm². The low breakdown voltage is due to the narrow-bandgap InGaAs collector material. InP collectors with a InGaAs/InAlAs linear grade would provide superior breakdown.[37] Our current facilities do not permit InP growth. Given that vertical scaling must accompany lateral scaling to obtain commensurate improvement in both f_τ and f_{max}, scaled 0.1 μm devices will demand very thin (1500–2000 Å) collectors, and InP must be employed.

DC characteristics of the devices with 0.6 μm × 25 μm emitters and 1.8 μm × 29 μm collectors were measured in the common-base configuration also. Figure 15

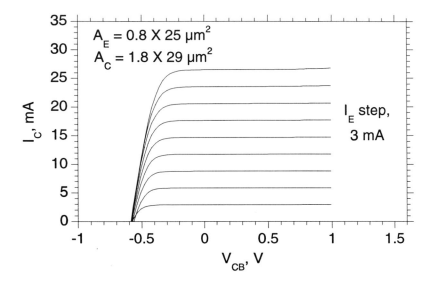

Fig. 15. DC common-base characteristics of devices.

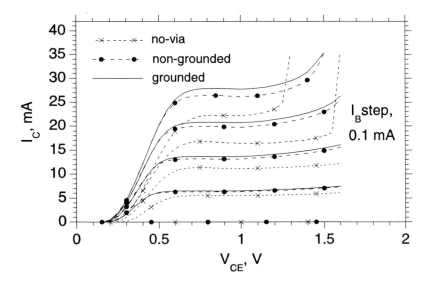

Fig. 16. DC common-emitter characteristics of device with different heatsink structures; $A_E = 1$ μm × 25 μm, $A_C = 4.8$ μm × 29 μm.

shows these characteristics. The common base current gain α is 0.98. The common-base breakdown voltage BV_{CBO} is ~ 6 V at ~ 0 current density, decreasing to 1.5 V at 10^5 A/cm^2.

We now compare the DC characteristics of devices with non-grounded emitters (with thermal vias) and devices without thermal vias. Three devices are compared. All have 1 μm $\times$ 25 μm emitters and 4.8 μm $\times$ 29 μm collectors. The first device has its emitter grounded with a thermal via (grounded), the second device has metal 3 to protect the insulator layer (non-grounded), and, the third device does not have a thermal via (no-via). Figure 16 shows their common-emitter characteristics. There is very little difference between the grounded and non-grounded devices, as was expected from the temperature calculations in the previous section. The no-via device has lower breakdown voltage and lower current gain due to self-heating effects. Hence, a thermal via is essential for high power operation.

3.2. *RF measurements*

The devices were characterized by on-wafer network analysis to 50 GHz. Figure 17 shows the short-circuit current gain h_{21}, maximum stable gain (MSG), and Mason's[38] invariant (unilateral) power gain U. Pad parasitics have not been stripped. Bias conditions are as shown. Extrapolating at -20 dB/decade, $f_{\max} = 400$ GHz and $f_\tau = 164$ GHz for the wide-collector devices (Fig. 17(b)). The high $f_{\max}$ is due to the reduced $r_{bb}C_{cb}$ time constant which is the result of low resistance base Ohmic contacts and the low intrinsic collector-base capacitance. We have used Mason's gain for extrapolating $f_{\max}$ because of its characteristic -20 dB/decade slope, its independence of the transistor configuration (common-base versus common-emitter), and its independence of pad inductive and capacitive parasitics.

RF measurements (Fig. 17(a)) of the narrow-collector devices yield 134 GHz f_τ and 520 GHz extrapolated $f_{\max}$. Such measurements are however at the limits of reliability for a 50 GHz instrument. We now estimate $f_{\max}$ by calculation from $f_{\max} \simeq \sqrt{f_\tau/8\pi r_{bb}C_{cbi}}$. Here C_{cbi} is the intrinsic base-collector capacitance, the fraction of C_{cb} charged through the base resistance r_{bb}. R_{bb} consists of sheet resistance and contact resistance (neglecting gap resistance). The measured base sheet resistance is 600 $\Omega/\square$. Because of poor test structure design, the specific base Ohmic contact resistivity falls below levels which can be reliably measured by the (TLM) test structures employed on the present wafer, and we therefore take reported values of specific contact resistivity for test structures ($23~\Omega - \mu$m^2)[39] having similar levels of Be-doping. With these parameters, we calculate a 2.4 Ω base spreading resistance and a 2.4 Ω base contact resistance. C_{cb} is extracted from s-parameter measurements by plotting the imaginary part of the reverse admittance parameter y_{12} versus frequency. The C_{cbi}/C_{cb} ratio is not readily predicted from device geometry,[40] but is often fitted by comparing measured and modeled s-parameters. Here, we roughly estimate the C_{cbi}/C_{cb} ratio as being equal to the ratio of the emitter-base and

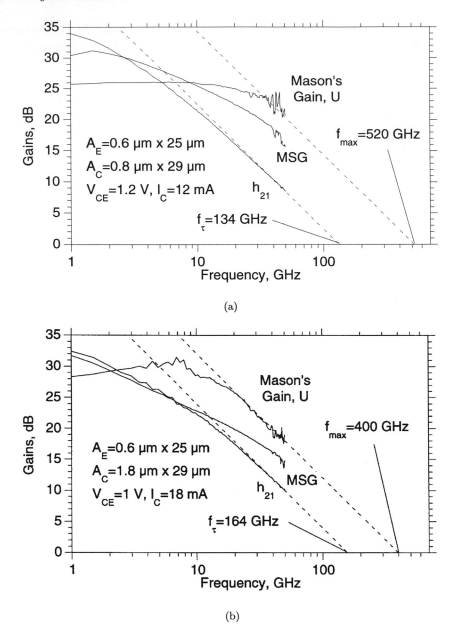

Fig. 17. RF characteristics of devices with (a) narrow collector and (b) wide collector.

collector-base junction areas. For wide-collector devices, measured $f_\tau = 164$ GHz, $C_{cb} = 22.7$ fF, $C_{cbi} = 7.5$ fF and $f_{\max} = 426$ GHz. For narrow-collector devices, $f_\tau = 134$ GHz, $C_{cb} = 8.2$ fF, $C_{cbi} = 6.2$ fF, and $f_{\max} = 423$ GHz. Both the above calculations and the microwave measurements indicate that $f_{\max}$ is at least 400 GHz. The MSG is given by

$$MSG = \left\| \frac{s_{21}}{s_{12}} \right\| \simeq (\omega C_{cb})^{-1} \left(r_{ex} + \left(\frac{kT}{qI_E} \right) \right)^{-1} \tag{13}$$

and is higher for the narrow-collector devices because of the reduced C_{cb}.

Figure 18 shows the variation of f_τ and $f_{\max}$ with bias. F_τ is similar for both devices. The Kirk effect threshold is high due to the presence of the pulse-doped

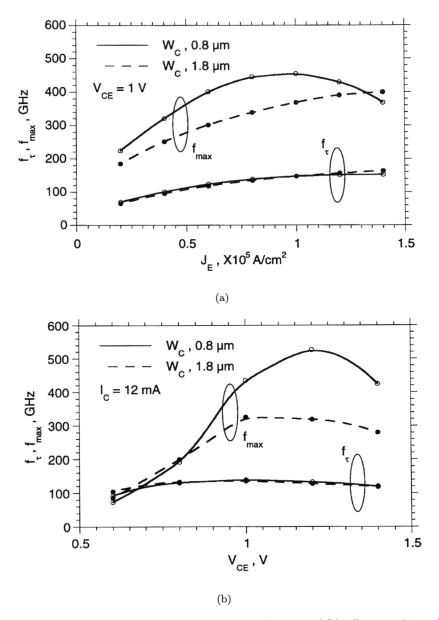

(a)

(b)

Fig. 18. Variation of f_τ and $f_{\max}$ with (a) emitter current density and (b) collector-emitter voltage.

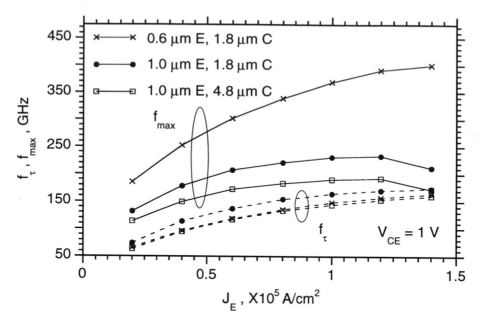

Fig. 19. Variation of f_τ and f_{max} with J_E for devices with different geometry.

layer in the collector close to the base.[35] The narrow-collector devices exhibit the Kirk effect at lower current density than wide-collector devices. Figure 18(b) shows a plot of f_τ and f_{max} versus collector-emitter voltage, V_{CE}. At low V_{CE}, the collector is partially depleted leading to increased C_{cb} and reduced f_{max}. At high V_{CE}, f_τ and f_{max} decrease, suggesting a decrease in the collector electron velocity at high electric fields. By plotting $1/2\pi f_\tau$ versus $1/J_E$, it is determined that the sum of the base and collector transit times $(\tau_b + \tau_c)$ is 0.75 ps.

Figure 19 shows how the performance is affected by lateral scaling in transferred-substrate HBTs. Shown in the figure is the variation of f_τ and f_{max} with emitter current density for three different devices. The first one has a 1 μm $\times$ 25 μm emitter and 4.8 μm $\times$ 29 μm collector. The second one has the same emitter but a narrower 1.8 μm $\times$ 29 μm collector. The third one has the same collector as the second one but a narrower 0.6 μm $\times$ 25 μm emitter. From Fig. 19 we can see that f_τ is similar for all three devices. Some difference is present due to the $r_{ex}C_{cb}$ term in the expression for f_τ (Eq. (2)). Reducing the collector width alone does not significantly improve f_{max}, as the intrinsic part of the collector-base capacitance remains the same. Scaling both the emitter and the collector improves f_{max} by a large amount. These devices are from different wafers, and thus have slightly different parameters, but the general scaling trend can clearly be observed here. It is expected that with further submicron scaling of the emitter and collector dimensions, a f_{max} much higher than 400 GHz can be obtained.

A small-signal hybrid-π model of the transferred-substrate HBT was developed and is shown in Fig. 20. The device with 0.6 μm $\times$ 25 μm emitter and 1.8 μm $\times$ 29 μm

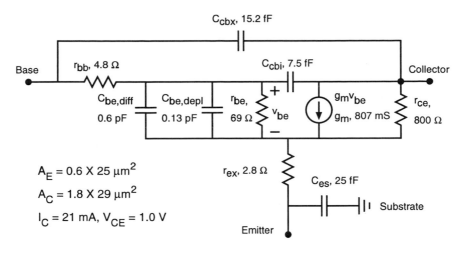

Fig. 20. Small-signal hybrid-π model of a transferred-substrate HBT

collector was used as a representative device because of good RF and DC characteristics suitable for circuit design. All the parameters except r_{ce} were extracted from bias dependence of s-parameters rather than being fitted on a computer. The C_{cbi}/C_{cb} ratio is determined as described above. Pad parasitics are negligible. This model is simple enough to facilitate quick circuit design and analysis, and also provides a reasonable fit to the measured s-parameters. We note that it is difficult to fit measured HBT s-parameter data to a hybrid-π model for any HBT operating close to the Kirk threshold. Charge-control analysis of an HBT operating in the region of collector field screening indicates that the base stored charge is modulated by the collector potential, an effect not modeled in the hybrid-π circuit.

4. Integrated Circuits

Transferred-substrate HBTs have been shown to have excellent high-frequency performance[41] and have great potential for future high frequency analog, digital, and mixed-signal integrated circuits. Analog circuits of interest include preamplifiers, variable-gain amplifiers, broadband traveling-wave amplifiers, voltage controlled oscillators and mixers. Digital circuits like D-flip-flops, frequency dividers, multiplexers/demultiplexers, selectors, digital phase-locked loops are being developed. In the mixed-signal category, analog-digital data converters are very important applications and are under development.

For integrated circuit applications, the process must have very high device yield and uniformity of DC and RF parameters. In addition to a high-performance device, there are other critical ingredients of an IC technology. Some of these are: passive components like resistors and capacitors, low-loss transmission lines, a low-capacitance multiple-level wiring environment, low ground return inductance vias and good heat-sinking. The transferred-substrate HBT technology has these key features as explained in the section on fabrication process. Device yield and uniformity is currently at levels suitable for fabricating circuits with few devices and is constantly being improved.

4.1. *Analog circuits*

The first demonstration integrated circuit to be fabricated in this technology was a feedback amplifier.[42] Feedback amplifiers are suitable for technology demonstration because of their simple design, low integration levels, few passive components and easy testability. A Darlington amplifier with series and shunt resistive feedback and 50 Ω input/output impedance[43] was designed and fabricated.

Figure 21 shows a schematic circuit diagram of the amplifier. The dotted line shows the chip boundary. Q1-Q2 form the Darlington pair. The emitter stripe lengths of Q1 and Q2 are selected to maximize bandwidth. If the emitter stripe length of Q1 is large, its input capacitance is large, degrading bandwidth; if it is too small, its base and emitter resistances are large, increasing the driving impedance for Q2 and degrading bandwidth. Large Q2 emitter stripe length increases its Miller-multiplied base-collector capacitance whereas a small emitter stripe length increases the base resistance, through which the (degenerated) device input capacitance must be charged. Hence, there are optimum emitter stripe lengths for Q1 and Q2. R_f, the shunt feedback resistor, is chosen to provide an input impedance of 50 Ω. R_1 is the series feedback resistor in the emitter of Q1 and sets the Q1 emitter current

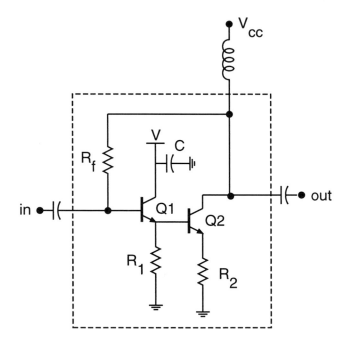

Fig. 21. Circuit diagram of Darlington feedback amplifier.

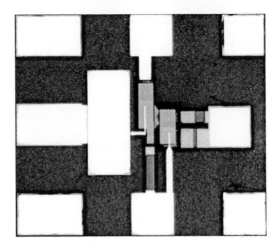

Fig. 22. Photograph of the amplifier IC (0.4 mm × 0.37 mm).

density close to peak f_τ bias. R_2 sets the degenerate transconductance of Q2 (and hence of the circuit) to provide the desired gain and 50 Ω output impedance. The circuit is biased with V_{CC} and an off-chip resistor connected through a bias-tee at the output. The collector of Q1 is not connected to the output, but is biased with an independent supply to eliminate Miller multiplication of its base-collector capacitance. The IC consumes 46 mW DC power. Figure 22 shows a photograph of the amplifier. The chip dimensions are 0.4 mm × 0.37 mm.

The amplifier characteristics were measured with a network analyzer to 50 GHz. Figure 23(a) shows the forward gain s_{21} of the amplifier. The low-frequency gain is about 13 dB and the 3-dB bandwidth (relative to the low-frequency gain) is 50 GHz. The gain peaking at high frequencies is due to the second (internal) pole in the feedback loop created by the emitter resistance of Q1 and the base resistance and input capacitance of Q2.

The bandwidth is below the potential of the HBT technology for several reasons. First, $f_{\max}$ is low on the current ICs due to high base Ohmic contact resistance. Secondly, the resistor current carrying capability was lower than expected, forcing us to bias the transistors at less than the intended design conditions. At the bias conditions at which the amplifier measurements were taken, the transistor f_τ and $f_{\max}$ are 120 and 175 GHz respectively. Figure 23(b) shows the input and output return losses and the reverse isolation of the amplifier. Technology, device, and circuit design improvements are currently underway, and are expected to yield amplifiers with higher bandwidths.

Other circuits currently in development are broadband traveling-wave amplifiers, variable gain amplifiers, optical receiver preamplifiers and monolithic VCOs. Target applications include fiber-optic transmission systems for 40 and 100 Gbits/s and radar links.

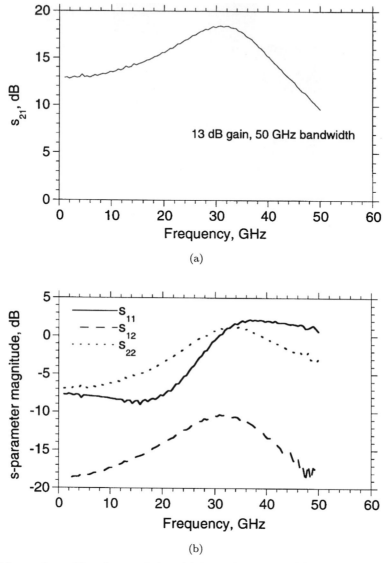

Fig. 23. Measured amplifier characteristics: (a) forward gain, and (b) return losses and reverse isolation.

4.2. *Digital circuits*

Digital circuits typically require higher levels of integration than analog circuits. Figure 24 shows a SEM photomicrograph of a transferred-substrate HBT digital technology in development. Master-slave D-flip-flops, both in ECL and CML configurations, with projected clock rates of 80 GHz have been designed. D-flip-flops are building blocks for a variety of digital circuits. Target applications include direct digital frequency synthesizers, MUX/DEMUX and clock-data recovery circuits for wireless LANs and radar links.

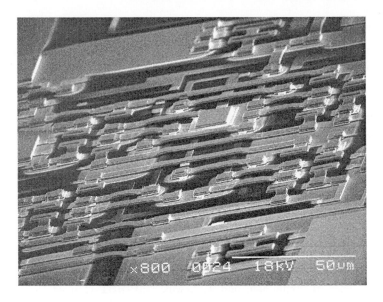

Fig. 24. Transferred-substrate HBT digital IC in development.

5. Future Work

We have demonstrated *scalable* transferred-substrate HBTs with very high $f_{\max}$ in an IC technology. A demonstration IC in this technology has been shown. A variety of circuits have been designed and are currently in development. Current efforts seek to improve the process technology and device performance. Scaling of emitter and collector dimensions to deep submicron dimensions is being pursued to achieve $f_{\max}$ in the region of 700 GHz. Vertical scaling including reducing the base and collector layer thicknesses to improve f_τ is in progress. To improve the breakdown voltage, wide bandgap InP must be incorporated into the collector. Devices with Ohmic collectors instead of Schottky collectors are being developed to facilitate low V_{CE} operation for CML digital circuits. Improved heat spreading in resistors with high thermal conductivity encapsulation materials is being investigated. The substrate-transfer process involving In/Pb/Ag solder is a significant factor limiting yield, and the GaAs transfer substrate impairs heat-sinking. Substrate-transfer with a few hundred microns of electroplated high thermal conductivity copper, rather than solder, is being investigated. The thick electroplated Cu itself acts as the carrier wafer, eliminating the GaAs carrier wafer. Another major limitation to the yield is the critical alignment required between the emitter and collector on either side of the base epitaxial layer. The work presented here used a contact aligner with crude lithographic capabilities. Projection lithography systems should enable fine-line lithography and deep submicron alignment precision.

With the demonstrated high bandwidth of transferred-substrate HBTs, and with the process features delineated above, MSI-scale circuits at 100 GHz clock rates should be feasible.

Acknowledgments

The authors would like to thank Prof. H. Kroemer, Dr. W. E. Stanchina and Dr. M. G. Case for invaluable discussions during the course of this work. We would like to acknowledge the contribution of Dr. M. J. Mondry in early material growth during the genesis of this project. This work was supported by the AFOSR under grant F4962096-1-0019, ONR under grant N00014-95-1-0688, and DARPA under the Thunder & Lightning, MOST, and OTC programs.

References

1. H. Kroemer, "Heterostructure bipolar transistors and integrated circuits", *Proc. IEEE* **70** (1982) 13–25.
2. J. F. Jensen, G. Raghavan, A. E. Cosand, and R. H. Walden, "A 3.2-GHz second-order delta-sigma modulator implemented in InP HBT technology", *IEEE J. Solid-State Circuits* **30** (1995) 1119–1127.
3. M. Mokhtari, T. Shawn, R. H. Walden, W. E. Stanchina, M. Kardos, T. Juloha, G. Schuppener, H. Tenhunen, and T. Lewin, "InP-HBT chip-set for 40-Gb/s fiber optical communication systems operational at 3 V", *IEEE J. Solid-State Circuits* **32** (1997) 1371–1378.
4. K. C. Wang, S. Beccue, C. Chang, K. Pedrotti, A. Price, K. Runge, D. Wu, R. Yu, P. M. Asbeck, and A. Metzger, "HBT technologies and circuits for TDM and WDM optical networks", *LEOS Annual Meeting Proc.*, 1996, pp. 209–210.
5. S. Yamahata, K. Kurishima, H. Ito, and Y. Matsuoka, "Over-220-GHz-f_τ-and-f_{max} InP/InGaAs double-heterojunction bipolar transistors with a new hexagonal-shaped emitter", *GaAs IC Symp. Tech. Dig.*, 1995, pp. 163–166.
6. H. Heiß, D. Xu, S. Kraus, M. Sexl, G. Böhm, G. Tränkle, and G. Weimann, "Reduction of the output conductance in InAlAs/InGaAs HEMTs with 0.15 μm gates", *IPRM Tech. Dig.*, 1996, pp. 470–473.
7. M. Reddy, M. J. Mondry, M. J. W. Rodwell, S. C. Martin, R. E. Muller, R. P. Smith, D. H. Chow, and J. N. Schulman, "Fabrication and dc, microwave characteristics of submicron Schottky-collector AlAs/In$_{0.53}$Ga$_{0.47}$As/InP resonant tunneling diodes", *J. Appl. Phys.* **77** (1995) 4819–4821.
8. I. Mehdi, T. H. Lee, D. A. Humphrey, S. C. Martin, R. J. Dengler, J. E. Oswald, A. Pease, R. P. Smith, and P. H. Siegel, "600 GHz planar-Schottky-diode subharmonic waveguide mixers", *IEEE MTT-S Intl. Microwave Symp. Tech. Dig.*, 1995, pp. 377–380.
9. T. Sugiyama, Y. Kuriyama, N. Iizuka, K. Tsuda, K. Morizuka, and M. Obara, "High f_{max} AlGaAs/GaAs HBTs with Pt/Ti/Pt/Au base contacts for DC to 40 GHz broadband amplifiers", *IEICE Trans. Electron.*, **E78-C** (1995) 944–948.
10. Y. Amamiya, H. Shimakawi, N. Furuhata, M. Mamada, N. Goto, and K. Honjo, "Lateral p^+/p regrown base contacts for AlGaAs/InGaAs HBTs with extremely thin base layers", *Device Research Conf. Tech. Dig.*, 1995, pp. 38–39.
11. Y. Ueda, N. Hayama, and K. Honjo, "Submicron-square emitter AlGaAs/GaAs HBTs with AlGaAs hetero-guardring", *IEEE Electron Device Lett.* **15** (1994) 66–68.
12. W. E. Stanchina, J. F. Jensen, R. H. Walden, M. Hafizi, H.-C. Sun, T. Liu, G. Raghavan, K. E. Elliott, M. Kardos, A. E. Schmitz, Y. K. Brown, M. E. Montes, and M. Young, "An InP-based HBT fab for high-speed digital, analog, mixed-signal and optoelectronic ICs", *GaAs IC Symp. Tech. Dig.*, 1995, pp. 31–34.

13. M. Hafizi, "Submicron, fully self-aligned HBT with an emitter geometry of 0.3 μm", *IEEE Electron Device Lett.* **18** (1997) 358–360.

14. R. Yu, S. Beccue, P. J. Zampardi, R. L. Pierson, A. Petersen, K. C. Wang, and J. E. Bowers, "A packaged, broad-band monolithic variable gain amplifier implemented in AlGaAs/GaAs HBT technology", *IEEE J. Solid-State Circuits* **31** (1996) 1380–1387.

15. K. W. Kobayashi, J. Cowles, L. T. Tran, A. Gutierrez-Aitken, T. R. Block, A. K. Oki, and D. C. Streit, "A 50 MHz–50 GHz multidecade InP-based HBT distributed amplifier", *IEEE Microwave Guided Wave Lett.* **7** (1997) 353–355.

16. K. Kurishima, H. Nakajima, S. Yamahata, T. Kobayashi, and Y. Matsuoka, "Growth, design and performance of InP-based heterostructure bipolar transistors", *IEICE Trans. Electron.* **E78-C** (1995) 1171–1181.

17. L. W. Yang, P. D. Wright, H. Sen, Y. Lu, P. R. Brusenback, S. K. Ko, L. Calderon, D. Laude, W. D. Hartzler, and M. Dutt, "Design and fabrication of submicrometer Al-GaAs/GaAs HBTs grown by LPOMVPE", *Cornell Conf. High-Speed Devices Circuits*, 1991, pp. 295–304.

18. H. Shigematsu, T. Iwai, Y. Matsumiya, H. Onishi, O. Ueda, and T. Fujii, "Ultrahigh f$_\tau$ and f$_{\max}$ new self-alignment InP/InGaAs HBT's with a highly Be-doped base layer grown by ALE/MOCVD", *IEEE Electron Device Lett.* **16** (1995) 55–57.

19. I. Aoki, K. Tezuka, H. Matsuura, S. Kobayashi, T. Fujita, and A. Miura, "80 GHz AlGaAs HBT oscillator", *GaAs IC Symp. Tech. Dig.*, 1996, pp. 281–284.

20. B. Jalali, R. N. Nottenburg, Y.-K. Chen, D. Sivco, D. A. Humphrey, and A. Y. Cho, "High-frequency submicrometer Al$_{0.48}$In$_{0.52}$As/In$_{0.53}$Ga$_{0.47}$As heterostructure bipolar transistors", *IEEE Electron Device Lett.* **8** (1989) 391–393.

21. J. Lin, Y. K. Chen, A. Humphrey, R. A. Hamm, R. J. Malik, A. Tate, R. F. Kopf, and R. W. Ryan, "Ka-band monolithic InGaAs/InP HBT VCOs in CPW structure", *IEEE Microwave Guided Wave Lett.* **5** (1995) 379–381.

22. B. Willen, M. Mokhtari, and U. Westergren, "New planarization process for low current, high-speed InP/InGaAs heterojunction bipolar transistors", *Electron. Lett.* **32** (1996) 266–267.

23. R. Bauchnecht and H. Melchior, "InP/InGaAs double HBTs with high CW power density at 10 GHz", *IPRM Tech. Dig.*, 1997, pp. 28–31.

24. T. Oka, K. Ouchi, and T. Nakamura, "Small InGaP/GaAs heterojunction bipolar transistors with high-speed operation", *Electron. Lett.* **33** (1997) 339–340.

25. S. M. Sze (ed.), *High-speed Semiconductor Devices* (John Wiley and Sons, 1990), pp. 341–342.

26. R. Williams, "Modern GaAs Processing Methods" (Artech House, 1990), pp. 234–235.

27. U. Bhattacharya, M. J. Mondry, G. Hurtz, I.-H. Tan, R. Pullela, M. Reddy, J. Guthrie, M. J. W. Rodwell, and J. E. Bowers, "Transferred-substrate Schottky-collector heterojunction bipolar transistors: first results and scaling laws for high f$_{\max}$", *IEEE Electron Device Lett.* **16** (1995) 357–359.

28. P. M. Enquist and D. B. Slater Jr., "Symmetric Self-aligned Processing", *U.S. Patent* **5,318,916** (1994).

29. S. Yamahata, K. Kurishima, H. Nakajima, T. Kobayashi, and Y. Matsuoka, "Ultra-High f$_{\max}$ and f$_\tau$ InP/InGaAs double-heterojunction bipolar transistors with step-graded InGaAsP collector", *GaAs IC Symp. Tech. Dig.*, 1994, pp. 345–348.

30. W. Liu, D. Hill, H. F. Chau, J. Sweder, T. Nagle, and J. Delany, "Laterally etched undercut (LEU) technique to reduce base-collector capacitance in heterojunction bipolar transistors", *GaAs IC Symp. Tech. Dig.*, 1995, pp. 167–170.

31. M.-C. Ho, R. A. Johnson, W. J. Ho, M. F. Chang, and P. M. Asbeck, "High performance low-base-collector capacitance AlGaAs/GaAs heterojunction bipolar

transistors fabricated by deep ion implantation", *IEEE Electron Device Lett.* **16** (1995) 512–514.

32. M. Yanagihara, H. Sakai, Y. Ota, M. Tanabe, and A. Tamura, "L-shaped base electrode with 0.5 μm contact width for high f_{max} HBTs", *Wide Bandgap Semicon. and Device Symp. Tech. Dig.*, 1995, pp. 408–415.

33. A. R. Massengale, M. C. Larson, C. Dai, and J. S. Harris, "Collector-up AlGaAs/ GaAs HBTs using oxidized AlAs", *Device Research Conf. Tech. Dig.*, 1996, pp. 36–37.

34. B. Agarwal, D. Mensa, R. Pullela, Q. Lee, U. Bhattacharya, L. Samoska, J. Guthrie, and M. J. W. Rodwell, "A 277 GHz f_{max} transferred-substrate heterojunction bipolar transistor", *IEEE Electron Device Lett.* **18** (1997) 228–231.

35. T. Ishibashi, H. Nakajima, H. Ito, S. Yamahata, and Y. Matsuoka, "Suppressed base-widening in AlGaAs/GaAs ballistic collection transistors", *Device Research Conf. Tech. Dig.*, 1990, pp. VIIB-3.

36. P. J. Zampardi and D.-S. Pan, "Delay of Kirk effect due to collector current spreading in heterojunction bipolar transistors", *IEEE Electron Device Lett.* **17** (1996) 470–472.

37. C. Nguyen, T. Liu, R. Virk, and M. Chen, "Bandgap engineered InP-based power double heterojunction bipolar transistors", *IPRM Tech. Dig.*, 1997, pp. 15–19.

38. S. J. Mason, "Power gain in feedback amplifiers", *IRE Trans. Circuit Theory* **CT-1** (1954) 20–25.

39. G. Stareev and H. Künzel, "Tunneling behavior of extremely low resistance nonalloyed Ti/Pt/Au contacts to n(p)-InGaAs and n-InAs/InGaAs", *J. Appl. Phys.* **74** (1993) 7592–7595.

40. Y. Matsuoka, S. Yamahata, K. Kurishima, and H. Ito, "Ultrahigh-speed InP/InGaAs double-heterostructure bipolar transistors and analyses of their operation", in *J. Appl. Phys.* **35** (1996) 5646–5654.

41. Q. Lee, B. Agarwal, D. Mensa, R. Pullela, J. Guthrie, L. Samoska, and M. J. W. Rodwell, "A > 400 GHz f_{max} transferred-substrate heterojunction bipolar transistor IC technology", *IEEE Electron Device Lett.* **19** (1998) 77–79.

42. B. Agarwal, D. Mensa, Q. Lee, R. Pullela, J. Guthrie, L. Samoska, and M. J. W. Rodwell, "A 50 GHz feedback amplifier with AlInAs/GaInAs transferred-substrate HBT", *IEDM Tech. Dig.*, 1997, pp. 743–746.

43. M. Rodwell, J. F. Jensen, W. E. Stanchina, R. A. Metzger, D. B. Rensch, M. W. Pierce, T. V. Kargodorian, and Y. K. Allen, "33-GHz monolithic cascode AlInAs/GaInAs heterojunction bipolar transistor feedback amplifier", *IEEE J. Solid-State Circuits* **26** (1993) 1378–1382.